Global change is a byword of our times – the earth is c[...]
under our eyes, and confused by this perspective we wo..,
consequences of the danger cries that we hear. Yet change is not new,
the earth has been continuously changing for billions of years, often
swiftly and dramatically. The study of the earth is also changing, through
the acceptance of continental drift 25 years ago to the latest studies of
our environment. In this new edition of his award-winning book on the
earth's history, Professor van Andel updates and expands his earlier text,
drawing on a wealth of new knowledge that has become available in the
last decade. The major events in the earth's history are examined – the
evolution of the solid earth, the changing oceans and atmospheres, and
the progression of life – to render a historical account of the earth's
evolution. These major strands of change taking place on the earth are
interwoven in prose that is both concise and compelling. This book of
commentary and ideas provides an understanding of how the environment changed naturally in the past, and will help the reader to place
current environmental changes and concerns in perspective.

Much new knowledge has been gained in the last decade; accordingly
while little material has been deleted, this new edition has grown to cover
the key topics, including a chapter on how we can improve our grasp
on geological time. Also, mindful of the current interest in global change,
new sections describe the green-house effect and address its possible
future ramifications.

In *New Views on an Old Planet: A History of Global Change*, Professor
van Andel makes earth history comprehensible and appealing to the
general reader. Careful to avoid mathematics and technical language, he
uses illustrations, thought experiments, and analogies to convey his account of the earth's history. This will also serve as an excellent text for
introductory courses in the earth and environmental sciences.

New views on an old planet

NEW VIEWS ON AN OLD PLANET

A history of global change

Second edition

TJEERD H. VAN ANDEL
University of Cambridge

CAMBRIDGE
UNIVERSITY PRESS

PUBLISHED BY THE PRESS SYNDICATE OF THE UNIVERSITY OF CAMBRIDGE
The Pitt Building, Trumpington Street, Cambridge, United Kingdom

CAMBRIDGE UNIVERSITY PRESS
The Edinburgh Building, Cambridge CB2 2RU, UK   http://www.cup.cam.ac.uk
40 West 20th Street, New York, NY 10011-4211, USA   http://www.cup.org
10 Stamford Road, Oakleigh, Melbourne 3166, Australia
Ruiz de Alarcón 13, 28014 Madrid, Spain

First published 1985
Second edition 1994
Reprinted 1995, 1997, 1999, 2000

Printed in the United States of America

Typeset in Galliard

A catalog record for this book is available from the British Library

Library of Congress Cataloging in Publication Data is available

ISBN 0 521 44243 5 hardback
ISBN 0 521 44755 0 paprback

Contents

Contents

Contents

Preface

The study of geology was a science which suited
idle minds as well as though it were history.
  Henry Adams, *The Education of Henry Adams*

The earth is changing. It is changing before our eyes, and life is changing with it at an ever-faster rate. Awareness of this fact has been slow in coming, because we have always thought of nature as dependable and constant, of the earth as rigid, and of the sea as eternal. We are confused by this new perspective and worried by the many voices that cry danger but seldom offer true solutions.

Yet change is not new in the long history of the earth; it began when the earth began, and has been continuous. It has also often been swift and dramatic, contrary to what geologists have traditionally believed.

These realizations derive from a science of the earth that is itself changing greatly after the adoption, a quarter century ago, of a ruling theory that includes continental drift. This revolution has brought together scholars of many backgrounds, skills, and interests who are learning to understand each other, knowing that otherwise there will be no more progress. Our view of the history of the earth's surface and of all that lived and lives on it is being altered forever by this process.

It is of this eternally changing earth that I write here, in the belief that there are many who, though not professional earth scientists, wish to understand our planet as it is now and are willing to make an effort to learn how it came to its present state. I shall try to discuss this fascinating but complicated subject simply, without jargon and excessive technicalities, but the reading cannot always be easy. Still, anyone endowed with some perseverance should find their way through it and be rewarded for the effort.

Earth science is history, because all that happens today is the product of what happened yesterday. Every feature of the earth is part of a chain of events that reaches deep into the past, each dependent on what went before, each influencing what came after. Therefore I will treat geology as a historical science, but shall begin at the end rather than at the beginning, in the conviction that to go from our familiar world into the unknown one step at a time is easier than to be dropped abruptly on an alien embryo planet four billion (4,000,000,000) years ago. Also, instead of following a straight course, I have attempted to weave the major strands that force change into a tapestry: the drift of continents, the fluctuations of climate, the evolution of the oceans, and the march of life.

Time on many scales is a major preoccupation of geologists. The weather changes by the hour and the day, the climate does so by decades, centuries, and millennia. Life suffers sharp setbacks overnight and takes millions of years to recover, but then rises to greater glory than ever before. Long times and wide spaces, rather than brief events and local details, dominate this book.

The discoveries of science cannot be properly understood unless we know how science, any science, works, and I have tried to illuminate that wherever it was appropriate. Questions have been raised more often than answered, because questions rather than answers drive knowledge forward, and I have not shied away from controversy, because controversy rejuvenates us.

The first edition of this book was written a decade ago. It touched what were then the frontiers of science, but the passing of time left its mark on the contents. Issues then subject to heated debate have been resolved, but other, greater ones have taken their places. Expanding like the universe, the earth sciences have added much new knowledge while discarding little, and so the book has grown. A small chapter on the future of plate tectonics has been deleted, but a new one on how to improve our grasp on geological time has been added. There are many new sections, and few are the paragraphs that have not been substantially revised. Mindful of the current interest in global change, I have, in Chapter 3, even taken one short step into the future. In many cases I hope to have found better ways to explain or to illustrate, but as before jargon has been held to a minimum, and mathematics kept out.

Several colleagues whose opinions I value have urged me to cite the sources of my information and inspiration, but I have resisted the temptation. This book was and is primarily intended for those who are not or not yet geologists, and they would not benefit from innumerable citations in the text. If anyone should want to see the list, several thousand titles long, I should be pleased to provide a copy.

I am conscious of the fact that all recommended reading is in English, but that is not to be regarded as evidence that I believe that the only worthwhile geology is published in that language. I have drawn widely from French, German, and some Spanish and Russian sources, but the dominantly English-speaking readership would have little use for reading recommendations in those languages.

This book was born in the early 1980s at Stanford University from an undergraduate course for students who did not intend to become geologists, and 16 years of teaching such an audience are built into it. Although not originally intended as a course text, it has been so used and, wishing to be helpful, I have enriched the lists of further reading with this use also in mind.

In some ways the preparation of this second edition has been distressing. I have come to realize ever more that most major issues concerning the earth require an interdisciplinary skill so wide-ranging that few are now able to create a synthesis of substance. Given our ever-increasing specialization, this causes me to fear that we may gradually lose the benefit of the vast pool of information we have so painstakingly gathered.

A truly unhappy experience was the amount of bad, unripe, or irresponsible science I encountered, especially on my way to the revision of Chapter 19. Camps taking opposite views refuse to pay attention to each other, dramatic hypotheses that should have been abandoned after the most minimal search for the evidence are published without comment, and referees who are either totally negligent or fear to speak up are too common for comfort. And a fascination with the sensational and an unseemly hankering for the limelight seem to have overwhelmed the scientific conscience of many authors. I do know that generating new ideas is more pleasurable than the slow process of testing them against the evidence, but respect for that evidence is the essence of scholarship. The science of the earth will surely wither if fewer and fewer people accept that as their ethic. Informing the public is a worthy goal, but coveting popularity through the media at any price is not.

Collectively, I thank those who wrote or said what over the years my mind has selectively absorbed. Students in my Stanford courses helped me to find out whether or not a discourse would work, and several non-geologists tested successive versions of the text. Alfred Kröner, Michael O. McWilliams, the late Thomas J. M. Schopf, James D. Valentine, and, for the second edition, Simon Conway Morris, Peter Friend, David Norman, Nick Shackleton, Alan G. Smith, and Rachel Wood have rescued me from many errors and infelicitous expressions, and I have been inspired by conversations with Dan McKenzie. Catherine Flack, my editor at Cambridge University Press, has been a steady source of moral support. Louis Lerman and indirectly Sherwood Chang helped to improve Chapter 15, the one most distant from my own expertise. Marjorie van Andel generously prepared those drawings in Chapters 17 and 18 that were beyond my competence, and Sharon Capon raised the level of the text figures well beyond what I achieved in the first edition.

It is customary to acknowledge that mistakes and aberrant views are entirely one's own. I believe that the reader will find this self-evident, and understand that many an interesting topic was left out by necessity, not by choice.

T. H. van Andel
Cambridge, United Kingdom
August 1993

Acknowledgments

The following publishers have given permission to reprint material from copyrighted works: From "The Waste Land" in *Collected Poems* 1909–1962 by T. S. Eliot, copyright 1936 by Harcourt Brace Jovanovich, Inc., copyright 1963, 1964 by T. S. Eliot (reprinted by permission of Harcourt Brace Jovanovich, Inc., and Faber & Faber); from "Burnt Norton" in *Four Quartets* by T. S. Eliot, copyright 1943 by T. S. Eliot, renewed 1971 by Esme Valerie Eliot (reprinted by permission of Harcourt Brace Jovanovich, Inc., and Faber & Faber); from *The Abyss of Time* by Claude Albritton, copyright 1980 by Freeman, Cooper & Company; from *The Pulse of the Earth* by J. H. F. Umbgrove, copyright 1947 by Martinus Nijhoff, Publishers; from *The Creative Mind* by Henri Bergson, copyright 1946 by the Wisdom Library, a division of the Philosophical Library; from *Ode to the Sea and Other Poems* by Howard Baker, copyright 1966 by Swallow Press (reprinted by permission of the Ohio University Press, Athens, for the United States, its dependencies and territories, and Canada, and by permission of Howard Baker); from "Bubbles Upon the River of Time" by M. M. Waldrop in *Science*, 215, 1082–83 (February 26), copyright 1982 by the American Association for the Advancement of Science.

At the start

What are the roots that clutch, what branches grow
Out of this stony rubbish? Son of man,
You cannot say, or guess, for you know only
A heap of broken images

T. S. Eliot, "The Waste Land"

Human beings have used and valued rocks for a million years and more, but not all kinds serve us equally well. To find the best flint or obsidian for tools or the finest soapstone for carving requires skill and experience, so quarries and quarrymen go back at least 20,000 years. Much later the search for metals called for a sophisticated expertise to which medieval books on mining bear eloquent witness.

This does not, however, imply a medieval birth for geology. Until the Renaissance brought a fresh spirit of inquiry, scholarly issues tended to be considered in a rigid philosophical or religious context, and logic rather than experiment or observation was thought to hold the key to the truth. Religion kept its hold on our perceptions of the earth longer than in any other discipline and has not fully lost it even today, as the debate on evolution versus creation shows. Only in the late 18th century did geologists learn to base their hypotheses solidly on observations, and test them by experiment or further observation. It was then that the science of the earth was born.

In medieval times, the acknowledgment of an ancient life different from that of the present would have raised troubling questions of a theological nature. Fossils were therefore viewed as quirks of nature formed from inanimate matter as minerals are, although Theophrastus knew better in the 6th century BC. Later pioneers, such as Leonardo da Vinci, asked modern questions from the rocks and obtained modern answers. Recognizing that shells and bones were indeed the remains of animals, da Vinci concluded that the sea had been where the land is now: "Above the plains of Italy, where flocks of birds are flying today, fishes once moved in large shoals," and saw that the Great Flood of the Bible did not suffice to explain this. Its traditional forty days are much too short for sedentary clams and corals to migrate and set down so far from the original shore. Also, the alternation of barren and fossil-bearing beds seemed to exclude a single flood.

Over the next two centuries the common view, fed by a desire to reconcile as best possible the geological evidence with the Scriptures, was that of a series of floods, each putting an end to life, each followed by a fresh creation. And indeed, the rocks record many great extinctions (although never a complete one), often accompanied by mountain building and invasions of the sea, and followed by the spread of new life forms. As much as two hundred years ago, pragmatic geologists in Britain, France and Germany used those breaks to erect a framework of earth history that still stands today, refined but essentially unchanged.

Gradually stripped of its supernatural elements and calling on less metaphysical forms of upheaval than acts of God, the concept of many floods became the theory of catastrophism held to be true by most paleontologists since the early 19th century. It was left to Charles Darwin to explain the progressive changes in life forms with time, but most scholars had long disassociated those from the notion of repeated creations.

Late in the 18th century, a new view of the history of the earth had emerged, propounded among others by James Hutton (1726–1797), a Scottish physician and gentleman farmer. After much experience with the geology of his native land and a great deal of travel abroad, Hutton became convinced that no more was required to explain the past than the processes we can observe every day. The dictum "the present is the key to the past" sums up what is now called uniformitarianism, and still influences geological thinking. In Hutton's view, past states of the earth were not so different from the present one, catastrophes were not required, and the history of our planet saw only gradual change. For the earth to have accomplished all that the rocks tell us about obviously did require much time, but just how much would have surprised even Hutton himself.

It fell to a Huttonian follower, Charles Lyell (1797–1875), to convert geologists en masse to uniformitarianism. In his book *The Principles of Geology*, first published in 1830, Lyell put his stamp on geological thought for much more than a century. His uniformitarianism was more rigid than Hutton's, impossibly so by today's standards. In its simplest form, Hutton's dictum says no more than that the laws of nature, independent as they are of space and time, operated in the past as they do today. It is a wise directive that keeps us from invoking supernatural or at least unprovable events as explanations. Lyell went much further with his belief that not only were the processes the same, but that they operated forever in the same combinations, at the same rates and with the same intensities. His was a static world, but we know now that the earth has had a far more lively time of it than he envisaged. We have also learned that some processes have shaped the earth in the past that are not evident in the brief instant of eternity that is ours to observe.

We shall return to uniformitarianism in the Epilogue, but to achieve a proper vantage point we shall trace the history of the earth in terms of three themes. These are (1) the evolution of the solid earth, an engine driven by its internal heat and slowed by gradual cooling, (2) the history of oceans and atmosphere, another engine, this one fueled by the heat of the sun, and (3) the evolution of life which, while dependent on the first two, greatly modifies the external earth and so reshapes itself as well. Eventually, earth and sun must die when their fuel becomes exhausted, and life will die with them.

We begin with the latest installment in an unfinished story, the Great Ice Age. The ice age world is easily imagined and can still be verified in Antarctica or Greenland. Its record is fresh and fairly complete and we can hone our interpretive skills on it. The early earth on the other hand, as alien as another planet, is not so accessible.

What causes an ice age? We have but parts of the answer, one of which is the changing behavior of the ocean. What might cause the ocean to change? Here we are on firmer ground; rearrange the continents, close or open seaways, and the ocean will be diverted from its present habits. This takes us to continental drift and the scientific revolution of a quarter century ago that made it the present ruling theory. Driven by the earth's internal heat, continental drift is behind many events in the history of the earth, the rise and fall of the sea, geographic change and the building and collapse of mountains.

Innumerable advances and retreats of the sea across the edge of the continents mark the course of history. At times, ours has been a planet almost covered with water; at other times the continents rise high above the sea as they do now. The record is to be found in the deposits of the continental margins, a record so rich that we are shocked to find that it fails to tell us the causes of the endless inundations and withdrawals.

Far stranger is the earth's childhood, the Precambrian when everything began: ocean basins and continents, rivers and seas, the atmosphere and life, the earth itself. But the early rocks are so old and so much has been lost, altered or deformed that so far logical reasoning rather than the rock record must guide us much of the time. Childhood's end came rather abruptly, when some 600 million years ago life suddenly proliferated and the world took on a familiar aspect.

Last we look at life. Its history can be discussed from many points of view, but our perspective will be the interaction between life in its ever greater complexity and the changing external environment of the earth. Life and environment are mutually dependent, and mutually modify each other, and it is the environment that drives evolution through Darwin's process of natural selection. Tracing this forward in

5

time from misty beginnings, we shall return to the present and contemplate again whether what we have seen best suits a uniformitarian or a catastrophic view.

During this voyage we shall be able to look not only at the findings of science but also at its methods, its philosophy, and at the attitude and behavior of its practitioners. Most often one is fascinated by the ingenuity, skill and courage brought into play, but sometimes there is reason to feel discouraged by what appears an almost willful carelessness. This discomfort has been expressed in the words of H. L. Mencken "For every complex problem there is a solution that is simple, neat and wrong."

Foundations

It has not been easy for man to face time. Some, in recoiling from the fearsome prospect of time's abyss, have toppled backward into the abyss of ignorance.

                      Claude Albritton, *The Abyss of Time*

ROCKS, EVENTS, AND TIME

To reconstruct the history of the earth, one must know what happened and when it happened. Because the only remains of the earth's past are rocks, we hope that time and events have been recorded there and that we can read what they have to say.

To decipher the archives of the earth geologists draw on the uniformitarian principle that "the present is the key to the past." With knowledge of modern environments and their deposits I can reconstruct those of the past, provided the record has not been too much ravaged by time. From time to time I should remind myself that the present is brief. Events may have occurred for which I know of no modern counterpart, but it would open the door to sheer fantasy if I were to assume that too readily. The *deus ex machina* of ancient tragedy has no place in science.

Since rocks contain the record, we must bring order to their infinite variety. Some crystallize from molten magma;* they are called igneous. If they form in the depths of the earth, they are intrusive, and because they cool slowly there, their crystals grow large, making them coarse-grained. Granite is an intrusive igneous rock. If the magma pours out at the surface as lava, fast cooling yields a fine-grained, sometimes glassy, extrusive igneous rock, such as basalt or obsidian, the latter also known as volcanic glass.

At the surface of the earth, all rocks are attacked by wind and water, by frost and heat, and by organisms, including humans. They weather, break down into ever smaller fragments, or dissolve altogether. Carried away by water or wind, the gravel, sand, silt and clay are deposited elsewhere to form sediments. These harden in time into sedimentary rocks, sand becoming sandstone and clay converting to mudstone or shale. Dissolved substances precipitate and form evaporites such as rock salt. Organisms build biogenic rocks, forming reefs or shell limestones, bone beds, and coals.

Rocks of any kind may be returned to the depths of the earth by burial under a thick layer of younger deposits or by mountain building. High temperatures and pressures then transform them into metamorphic rocks. Granite is metamorphosed to gneiss, clay turns into slate, limestone into marble. Eventually, those are raised once more to the surface, are eroded, and the fragments are carried away.

* Definitions of terms can be found in the Glossary at the end of the book.

Foundations

This is the geological cycle: uplift to erosion to transport to deposition to burial to re-emergence through uplift and mountain building, and once more to erosion, without end. At any given spot or time we observe only a small slice of the everlasting change; this forces us to piece the history of the world together from many, many fragments, like a jigsaw puzzle.

In this manner the events that brought those fragments into being can be deciphered and assigned to their proper positions in the history of the earth. Did this limestone form before or after that granite? Did fish precede dinosaurs? Arranging the recorded events in their correct sequence is the most substantial of the many tasks of the field of stratigraphy.

Having established the order of events at a given spot, we must relate them to the whole of earth history so that we will know what, for example, was happening in England when the sea flooded California. Relating distant fragments of the rock record to each other is called correlation, and to accomplish it we use mainly fossils. But mere sequencing, place A before B which predates C, is not enough; there is no history without time. We must measure time in units: in seconds, hours, and especially years, millions and billions of years.

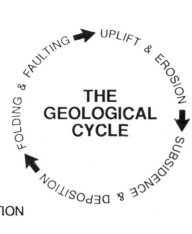

THE GEOLOGICAL CYCLE

FOLDING & FAULTING → UPLIFT & EROSION → SUBSIDENCE & DEPOSITION →

① DEPOSITION

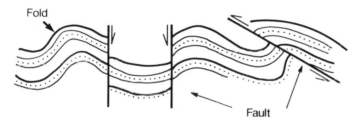

HIGH

Erosion → Transport → → → Deposition

LOW

Beds or strata

② DEFORMATION

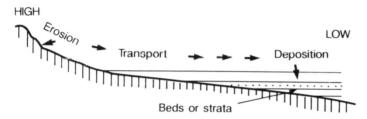

Fold

Fault

③ UPLIFT & EROSION

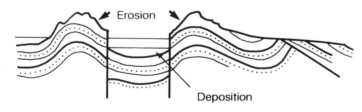

Erosion

Deposition

I
Reading the record of the rocks

Unpromising as a barren mountainside seems at first, it contains rich information about an astonishing array of subjects. Pick up some fossil mollusks or corals. Clams from the intertidal zone add a thin calcareous layer to their shells when the tide is high, then stop as the water falls. Each layer represents one tidal cycle, two tides per day. High spring tides produce thicker layers that mark off the lunar month. Where the winter is cold, growth slows or even stops, allowing us to identify the seasons and count the days of the year. In warmer waters, corals respond to the sun: a growth layer for each day, but little at night. The length of the year can be inferred for the remote past, for as long as corals and mollusks inhabited the earth (Figure 1.1).

Why should one wish to know the number of days or of lunar months for years so long ago? As we go farther into the past, we find more days in the year and fewer hours in the day. This is because the rotation of the earth has gradually slowed as a result of tidal friction. Also, there is reason to believe that the moon has not maintained a constant distance from the earth; if that is so, it ought to be reflected in the length of the lunar month. And so 400 million years (my) ago the earth year had about 400 days, the lunar month was 30 rather than 28 days long, and there were 22 hours in each day.

But what if there are no fossils or if they have no story to tell? In that case the rock itself must be consulted and will, at times, respond most amazingly. In southern Australia the Elatina Formation is about 650 my old, much older than the first clams and corals. To the uninitiated it is a boring sequence of clayey, silty and sandy layers, each only millimeters thick, but the careful observer notes that similar sets of thin laminae keep repeating, interrupted after every 13 or 26 repetitions by a dark clay band. If we assume that each set represents a year, those numbers are familiar: the cycles in the activity of the sun today also last 13 and 26 years.

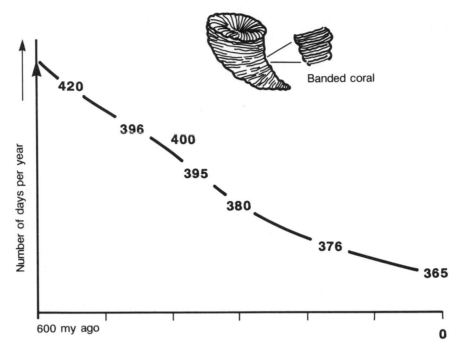

Figure 1.1. *Fine ribs mark many corals and mollusks as a result of the daily deposition of a layer of calcium carbonate. The thickness of this layer varies with the seasons and the lunar months. If we count the number of days per year in fossil specimens, we find more days per year as we go farther back in time.*

Do these thin layers of silt and sand in a distant sea then record the behavior of the sun 650 my ago? But can solar cycles really affect sedimentation? Another, more thorough inspection shows that the Elatina layers were laid down daily by the tide, thicker layers marking the spring tide every lunar month. This is not unusual; tidal deposits are common in the rock record. Now the banding of shells and corals has shown that one hundred million years later the earth still had 13 lunar months per year instead of 12. Therefore we conclude that the 13-year cycles reflect the tides of those early days, and not the less probable sunspots. The scientific procedure is elegantly illustrated here: observation, an ingenious working hypothesis, further observation, and a new working hypothesis resting on what is known as the principle of minimum astonishment. A fitting introduction to what this book is about.

1.1   ENVIRONMENT AND FACIES

The earth expresses itself in myriad ways. Running water, helped by sand and gravel, has cut river channels and the Grand Canyon, glaciers turn jagged mountains into rounded hills. Amazed Mexican farmers saw smoke rise from a potato field where now the volcano Paricutín stands. Two centuries of diligent observation have yielded a huge literature relating land-forms and rocks to the earth processes that produced them. This knowledge enables us to infer from a sedimentary rock the environment in which it was formed. The reflection of the original environment in a rock is called its facies; a sandstone may appear in a beach or a river facies, and the sediments of the seafloor exhibit facies related to depth or distance from shore. And so forth.

A detailed discussion of all facies and their meanings would fill the rest of this book, but we can achieve some understanding by looking at a few common sedimentary rocks. Limestones composed of coral and shell debris indicate a clear shallow sea. Coals are the products of thick layers of dead plants preserved under water and therefore indicating a swamp facies. Salt and other minerals precipitate from seawater. When we see these evaporites, we think of sheltered lagoons in a dry, hot climate. Redbeds, strikingly red, orange and brown gravel and sandstone, usually form in deserts or semi-arid environments.

To clarify this, let us try our hand at an interpretation of Figure 1.2. The non-fossiliferous limestone at the base merely says "marine," but the lack of land-derived material suggests either that land was distant or that no rivers entered this sea. A shore without streams implies a dry climate. The next limestone bears oyster shells and toward the top a little clay and sand; the land was near. Oysters do not live in deep water or far from shore; therefore the sea must have become shallower, but there is still no evidence for rivers bringing mud and sand in abundance. Thin-bedded sandstones followed by coal point to a beach, perhaps with a row of dunes here and there, then to a swampy coastal plain. Evidently the land was not so dry after all or the climate had changed. Dunes cover the coal, but their tops have been eroded; the record is broken here. Quite a bit of history must have been removed, because the next deposit is unmistakably marine. Did the land sink or the sea rise?

It is common for parts of a rock sequence to be missing, because many events destroy rather than create deposits, while others simply leave no trace (Figure 1.3). Sometimes, nothing happens for a while; no rock record is being formed today outside my window, nor is any being removed. Most often, however, erosion by wind, ice, rivers or waves, is the culprit. At times the work of erosion is difficult to spot – limestones or evaporites may simply dissolve – but more often it leaves tracks: a residue of cobbles, a hard ragged surface. A hiatus is any interval of time for which a record no longer exists, whether due to erosion or not. The

15

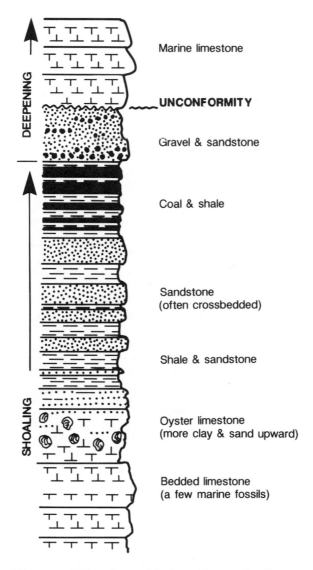

Figure 1.2. This rock record begins with a marine limestone deposited far from land but in shallow water (which its fossils prefer). It changes upward to nearshore sand, then a coal swamp. A time of erosion is marked by an unconformity, upon which rests another marine limestone. What we see is the record of a fall and subsequent rise of the sea.

gap includes the duration of the erosion itself as well as the length of time once recorded in the layers that have been removed.

Erosion may cut across folded beds, showing that much history has been lost forever: first the deposition of sediments, then their folding into mountains, and finally the removal of those mountains by erosion. Later, new sediments will bury the eroded surface which becomes an

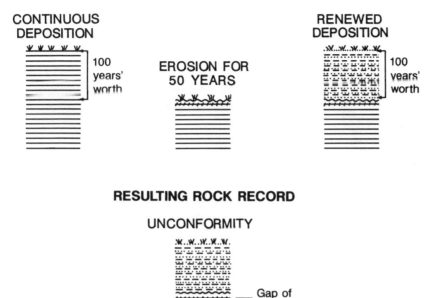

SEQUENCE OF EVENTS

CONTINUOUS
DEPOSITION
100 years' worth

EROSION FOR
50 YEARS

RENEWED
DEPOSITION
100 years' worth

RESULTING ROCK RECORD

UNCONFORMITY

Gap of
150 years

Figure 1.3. Deposition is often discontinuous, interrupting the rock record, and erosion may have removed part of what was there already. The hiatus, the gap in time represented by the break, is therefore almost always longer than the interval of non-deposition.

unconformity. On the deepsea floor sedimentation is slow and a single pulse of erosion may remove the record of tens of millions of years.

1.2 EVENTS IN TIME AND SPACE

Having deciphered the events recorded in Figure 1.2, how shall we determine whether the change of sea level was only local or more widespread, and find out what happened before or came after this little bit of history? To achieve this we have to fit our local fragment into a broader context of time and space by connecting it with sequences of rocks found elsewhere. Four simple principles or "laws" help us here (Figure 1.4). The first three are of venerable antiquity, having been formulated in 1669 by Nicolaus Steno, a Danish physician at the court of Florence who in this way laid the foundation of modern stratigraphy.

The first principle, known as the law of superposition, says that the age of a sequence of sedimentary rocks or lava flows decreases upward, each bed being younger than the one underneath and older than the one

17

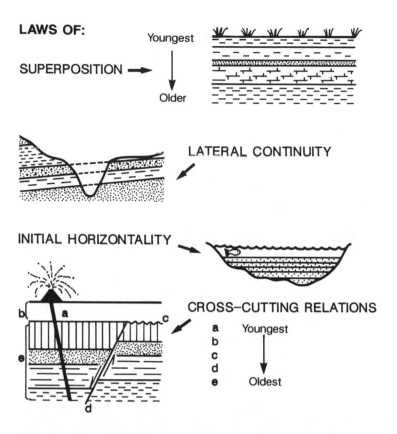

LAWS OF:

SUPERPOSITION ➡

LATERAL CONTINUITY

INITIAL HORIZONTALITY

CROSS–CUTTING RELATIONS

Figure 1.4. To establish the sequence of events recorded locally by rocks, and connect it with others observed elsewhere, we use four basic principles or "laws" that, simple as they are, provide a powerful tool for reconstruction of the geological record, the stratigraphy, of a region.

above. We have already applied this law in our discussion of Figure 1.2. It is simple, self-evident, and has wide application, but there are exceptions, such as the intrusion of a lava flow between two layers of sediment.

The second principle is that of initial horizontality. Most sediments, having been deposited in seas or lakes or on river plains, were originally essentially horizontal. If they are now tilted, we conclude that they were deformed after deposition.

The principle of lateral continuity states that sediments initially form continuous layers, changing their character only when the environment changes. If a bed abruptly ends, we must suspect that something intervened after deposition: dislocation by a fault, erosion on a shore, or interference by human beings, the greatest diggers of all time. This principle justifies the reconstruction across present gaps of strata that were once continuous.

The final principle says that if a bed is traversed by, for example, an

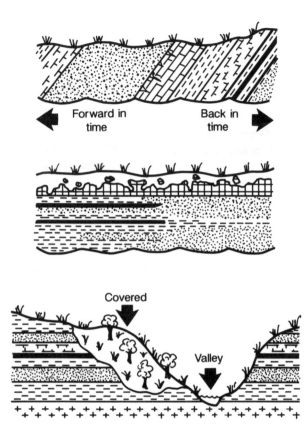

Figure 1.5. Walking along an exposed sequence of tilted beds is like a walk in time. Walking along a set of horizontal ones may take us from one ancient environment to the next. Vegetation and the work of erosion and human beings, however, may make life difficult for the geologist.

intrusive lava dike or a fault, the bed must be older than the cross-cutting event. Thus, the volcano and the fault in Figure 1.4 are younger than the rocks they cut across.

Often we can extend the scale of our observations by walking along an outcrop (Figure 1.5). If the beds are tilted, walking one way will carry us forward in time, the other back. If the beds are horizontal, time does not change but we range farther in space, perhaps noting facies changes. Nevertheless, every outcrop ends eventually, in a landslide, a road, a clump of trees or a stream channel, and we lose track of our sequence.

Having described our set of strata precisely, we are able to recognize similar sequences elsewhere. Making use of the law of lateral continuity, we postulate that all were once connected. Such a distinct set of strata, different from those above and below, and traceable across the countryside ("mappable" says the geologist), is called a formation. If we assign various outcrops of limestone to the same formation, we mean that they are, or at least once were, part of the same stratum.

Sooner or later, however, every formation must end, usually when it is replaced with rocks that are different because they were formed in different environments. In the early 19th century a famous case of such a facies change was debated at length in Britain. There, two major sediment series can be seen, a set of redbeds in South Wales, and a suite of fossil-bearing sandstones, shales and limestones in Devon. Today the two sequences, which are both about 350 my old, are separated by the Bristol Channel which conceals their contact. Is one older than the other (and if so, which one), or do the clearly continental redbeds gradually change into the more marine limy sequence? The argument raged for years until careful fieldwork turned up a few outcrops where one facies was interbedded with the other, so settling the matter. Different as they look, the two formations represent separate facies laid down at the same time.

Suppose we have established that the beds in two outcrops belong to the same formation. Can we also assume that they were deposited at the same time? Alas, no law requires that all parts of a formation must be of the same age, and quite often they are not. Consider the shore while the sea rises (Figure 1.6). Soon, a beach will form where dry land existed before, and mud will be laid down offshore on top of the sand of the former surf zone. As the sea rises, the shore moves inland and also forward in time. A series of beds forms, each continuous and throughout of the same distinctive nature, but progressively younger in the landward direction. Clearly, if we wish to know whether or not events in separate places are contemporaneous, we need tools other than the physical properties of formations.

1.3   FOSSILS AND CORRELATIONS

The time-honored solution to the problem of correlation over long distances is a stratigraphy based on fossils (Figure 1.7). Long before Darwin explained why life changes continuously, geologists knew that the fossil assemblages of older beds differed from those in younger ones. Species become extinct, new ones appear and die off in their turn. With fossils we can erect zones to subdivide the historical record of the earth, provided that the species appear or become extinct everywhere simultaneously, "geologically speaking," a phrase that allows us considerable leeway. Fossils can be used to assign a formation or an outcrop to its proper place in the history of the earth, the geological time scale, assum-

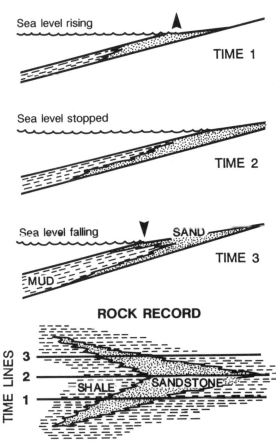

SEQUENCE OF EVENTS

Sea level rising — TIME 1

Sea level stopped — TIME 2

Sea level falling — SAND — TIME 3

MUD

ROCK RECORD

TIME LINES 3 2 1

SHALE — SANDSTONE

Figure 1.6. A set of distinctive sedimentary beds that can be traced over a large area is called a formation. It is a useful concept in geological mapping, but does not imply that all parts of a formation are of the same age. Here, as the sea rose and the shore migrated inland, beds of sand and mud migrated landward and forward in time with it. As a result, the landward parts are not of the same age as the seaward ones; the boundary between sandstone and mudstone crosses time lines 1, 2 and 3.

ing that adequate fossils can be found. That is possible only for the last 600 my of earth history.

There are, as always, difficulties and impediments. Fossils, too, are subject to facies changes. One only need think of the difference between a tide-pool fauna and the dwellers on the open sandy beach next to it to realize the problem. Some species need a limited, narrowly defined environment; others are tolerant and therefore widely distributed, or occupy a vast environment such as the open ocean. Obviously, widely distributed species make better stratigraphic markers than narrowly spe-

PRINCIPLES OF BIOSTRATIGRAPHY

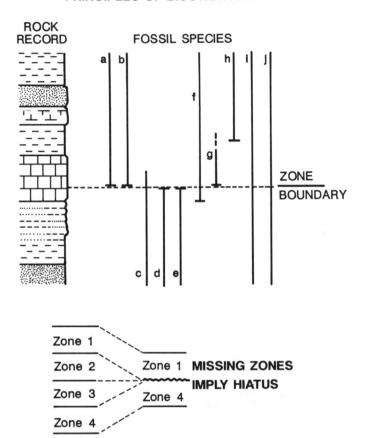

Figure 1.7. To determine whether a rock outcrop is younger, older, or of the same age as another, we can use fossils. Fossil zones, based on the first appearance of some species and the extinction of others, often extend over large areas and are a key tool in stratigraphy. Missing fossil zones confirm the existence of an unconformity and provide an estimate of its duration.

cialized ones, but there are other prerequisites for a good fossil zone. Its fossils should be abundant, they must not have been washed out of an older bed and redeposited, and they must be in good condition so that one can identify them with confidence. To have many species is better than to have a few: one coral tells us less than ten different kinds of mollusks. Microscopic fossils (microfossils), found by the hundreds or thousands in a single sample, are better than mammoth bones. It is also helpful if the assemblage evolved rapidly, because there will then be many extinctions and new arrivals to define our zones.

How well do fossils serve the tasks of correlation and of subdividing earth history? Quite well indeed; numerous fossil zones mark the last 600 my, many of them are valid for large areas and not unduly influenced by facies differences. Just two groups of calcareous oceanic plankton, microscopic organisms with calcareous shells living near the surface of the sea, allow us to subdivide the last 65 my of history into no fewer than 50 to 60 zones. With so fine a time scale we can pinpoint events with fair precision, within a few hundred thousand years as we shall see.

It is more difficult to deal with rocks without fossils. Fossils are usually rare in sediments deposited on land, because the chance that anything will be buried at all is slim, and the opportunity for permanent preservation even slimmer. There are spectacular terrestrial fossil beds, the frozen mammoths of the Siberian tundra, for example, the dinosaur bone beds of Wyoming, or the limestone that yielded *Archaeopteryx*, the first bird, but they are exceptions. And what is one to do with events that do not even make rocks but only deform them, such as folding or faulting? In such cases stratigraphy rests on the principle of superposition and the cross-cutting rule, and requires an ingenuity worthy of a Sherlock Holmes. The best way to understand the tricks of the trade is to study a few examples of the kind geologists encounter all the time (Figure 1.8).

The framework of earth history, the geological time scale, rests mainly on fossils. This device was developed early in the 19th century, and although much fine detail has been added since, it has withstood almost two centuries of use. The largest advance has been the development of techniques that permit a calibration in years. Before we turn to that subject in the next chapter, however, let us first examine the speed with which geological events proceed.

1.4 RATES OF GEOLOGICAL PROCESSES

As a natural consequence of Hutton's emphasis on an infinite amount of time and Lyell's insistence that there had been no catastrophes, geological processes have been and often still are considered to be slow, capable of producing impressive results only if much time is available. But many years devoted to the study of the present world have taught us that the actions of the earth are not always slow like the tortoise, that in fact some of them are fast like the hare (Table 1.1). A steady drip may hollow the stone, but a single landslide can remove a million stones in an instant.

Tectonic events, the folding and uplifting of mountains, the rise and fall of a continent, are momentous occurrences involving huge masses of rock or water. They imply very large forces and we cannot conceive of them as rapid. Yet continents drift across the surface of the earth at

Foundations

PLACING STRUCTURES AND IGNEOUS
ROCKS IN THE TIME SCALE

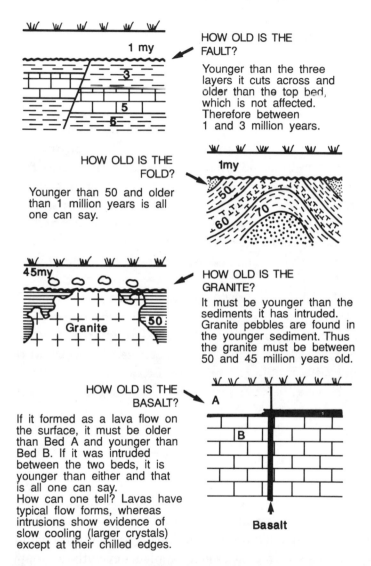

HOW OLD IS THE FAULT?

Younger than the three layers it cuts across and older than the top bed, which is not affected. Therefore between 1 and 3 million years.

HOW OLD IS THE FOLD?

Younger than 50 and older than 1 million years is all one can say.

HOW OLD IS THE GRANITE?

It must be younger than the sediments it has intruded. Granite pebbles are found in the younger sediment. Thus the granite must be between 50 and 45 million years old.

HOW OLD IS THE BASALT?

If it formed as a lava flow on the surface, it must be older than Bed A and younger than Bed B. If it was intruded between the two beds, it is younger than either and that is all one can say.
How can one tell? Lavas have typical flow forms, whereas intrusions show evidence of slow cooling (larger crystals) except at their chilled edges.

Figure 1.8. Often there are no fossils, and most of our stratigraphic laws do not apply to igneous and metamorphic rocks, nor to the traces left by such events as folding or faulting. The four cases shown here depict how geologists deal with these situations.

measurable rates; each century western California, for example, moves another 4–8 m northward relative to the rest of North America. And since Columbus crossed it, the Atlantic Ocean has widened by more than 20 m. Vertical movements can also be impressive. The island of Espiritu Santo in the South Pacific rises from the sea at a rate of 5.5 m/1,000

24

Table 1.1. *Most geological processes are faster than we thought*

Process	Rate
Glacio-eustatic sea level change	up to 10 m/1,000 yrs (fast)
	2 m/1,000 yrs (slow)
Regional erosion	meters/1,000 yrs
Shelf deposition	centimeters/yr
Deep-sea deposition	millimeters/1,000 yrs
Uplift and subsidence	
in subduction zones	up to 10 m/1,000 yrs
common rate	10 cm/1,000 yrs
Continental drift	centimeters/yr

years. The coast of Oregon has been rising at 100–200 m/my for millions of years. A submarine ridge off the coast of Peru pierces the sea floor at a rate of 6 m/100 years, and the Netherlands sink one more meter into the North Sea with each passing century.

How fast do sediments accumulate? What length of time is represented by a limestone or shale? Naturally, the rates depend on the environment, ranging from remarkably fast to exceedingly slow. Near New Orleans, the Mississippi delta is about 100 m thick, but around 1500 AD, when the Spaniards first saw the site, it was still on the coast. Since then, this coast has moved 25 km south, adding 50 m of land per year to the grateful State of Louisiana. At sea, sedimentation is not as fast, centimeters per year, and in the deep-sea it is even slower, centimeters per millennium. This seems little enough, but in "oceanic deserts" far from the places where plankton is abundant and its tiny shells sink to the seafloor by the billions, a little windblown dust, a few fish teeth, and an occasional tiny meteorite speck accumulate at a mere millimeter per 1,000 years.

On land it is more difficult to give typical sedimentation rates, because the environments are so unstable and so localized. Rivers meander back and forth, eroding their own deposits; dunes bury a road, but the wind sweeps them away again, and in the end erosion planes everything down to near sea level, the final stage in a geological cycle that began with mountains. Unless and until another uplift intervenes, of course. Sedimentation is but a trivial interlude in this scenario. In Australia, erosion has worked undisturbed for a hundred million years, and much of the continent is now close to its final, low, planar stage. Erosion is most rapid in mountain country, where even over large areas an average of 1–3 m of rock may be removed every 1,000 years. In very high, young mountains such as the Himalayas this may be as much as 15 m, and unless they keep rising, the range will be gone in a few million years, and so will the Alps and the Rocky Mountains. The whole of North America is being stripped down some 5–10 cm per millennium. Ultimately, the debris, all of it, finds its final resting place in the sea.

Hutton was thus wrong, things do not by any means always happen in a very sedate way. This becomes a critical issue when we remember that our ability to tell time precisely diminishes as we go farther into the past. Going back a hundred million years, we are quite content if we are able to say that two events were simultaneous give or take a million years. But that is as long as it took for icecaps to spread more than twenty times across the northern hemisphere, only to melt away again, each time followed by conditions like those of the present. The farther back we go, the more we are forced to study only long-lived events.

It is therefore important to know how fast things do happen. Whether an ancient delta was built in a few centuries by a river as large as the Mississippi or by a small one in a few million years is no trivial matter in the oil business. Some faunal extinctions seem dramatically abrupt, but that may not be true, because from the vantage point of a hundred million years, a single million may seem like an instant.

2

Perspective on time

It is a habit peculiar to geologists to speak of millions of years as casually as politicians dispose of billions of dollars. For the latter, it does not seem to be real money, and I suspect that to us it is not really time. But to be comfortable with very long time is a habit which it took a century to develop.

There is, indeed, a need for much time. James Hutton thought that the earth was very old, with "no vestige of a beginning, no prospect of an end," but the record of the rocks is ambiguous about time. The building of a large delta may take only a few centuries, but raising the Alps or the Sierra Nevada obviously demands more time. Early in the 19th century, a clear outline of the history of the earth had been established by the methods we have just discussed, but the time dimension remained elusive. Darwin's theory of evolution heightened the sense that much time had been involved in making humans out of single-celled algae, but no one could say just how much. It remained for the discovery of radioactivity in the last years of the 19th century to provide a firm time perspective of the geological past.

2.1   HOW OLD IS THE EARTH?

In 1654, James Ussher, Archbishop of Armagh in Ireland, computed on the basis of the Scriptures, and with much guidance from his imagination, that the Creation took place in the year 4,004 BC, an estimate later refined by a vice-chancellor of the University of Cambridge to the 26th of October at 9 a.m., a sensible hour.

Thus began a slow increase in our conception of the age of the earth (Figure 2.1). For a while, Ussher's age was regarded as a reasonable value, but gradually it became obvious that 6,000 years would not suffice for all that had been recorded in the rocks. In 1749, the French naturalist

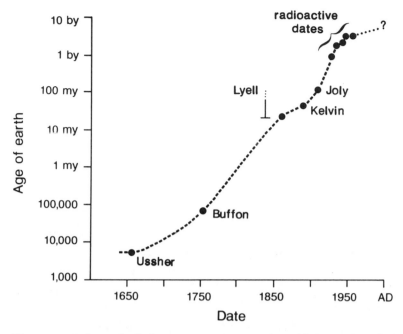

Figure 2.1. Only gradually have we come to accept how old the earth really is. In the 17th century, 6,000 years was thought to be sufficient, but gradually this rose to 100,000 years. In the 19th century, scientific methods raised the age to tens of millions, but only the discovery of radioactivity in 1896 gave us the means to measure it. As older and older rocks were found, the known age of the earth rose to 4.5 billion years, in agreement with astronomical considerations.

G. L. de Buffon estimated that at least 75,000 years were needed to produce all the fossil- bearing strata he knew of. Late in the 19th century, the Irish chemist J. Joly reasoned that rivers continuously deliver salt to the sea, while little goes out, and calculated that 90 million years were required to reach the present salinity. Today, with better data we would make this 260 million if we use common sea salt, but 45 million with magnesium, another element abundant in the sea. Silica, on the other hand, hangs around for a mere 8,000 years. The method is obviously flawed, the flaw being the assumption that nothing leaves the ocean. We shall return to this in Chapter 14.

The 90 million years of Joly were deemed far too many by the great physicist William Thomson, Lord Kelvin, who was provoked by the strict uniformitarianism that claimed that the processes of the earth have neither slowed down nor speeded up with time. To him this smacked of a perpetual-motion machine, and such machines violate basic laws of physics. The earth is an engine driven by heat and when the earth cools, as it must, the engine slows and eventually stops. Kelvin could think of only one fuel, the initial heat of the planet when it formed as a molten ball. The cooling rate could be estimated, and after several attempts he settled on an age of 20 or 30 million years.

Geologists deal with a complex subject in which certainty is difficult to achieve. They tend to be easily overawed by their peers in physics and chemistry and loath to argue against them on geological grounds. Thus, when Lord Kelvin spoke, geologists did not like his conclusions, but neither did they wish to break the laws of physics. One of the few challengers was an American by the name of T. C. Chamberlin, who in 1899 calmly announced that if so brief a history was prescribed by physics, then physics must be wrong. Some fuel or process other than the original heat would, he felt, be found to explain why the engine of the earth had already run much longer than 30 million years.

Chamberlin was fortunate, because the discovery of radioactivity in 1896 almost immediately supplied not only the missing source of energy for the earth's engine, but also the means of calibrating geological history in years. When the fuss died down, Lord Kelvin was shown to have been right when he derided the age estimates of the geologists, but wrong when he thought them too long. It is worth noting, however, that the heat released by radioactivity alone does not suffice for an age of billions of years. The interior of the earth itself, turning over like a boiling pot of soup, supplies the surface with a great deal of heat drawn from the whole interior rather than just from the outer skin.

A chemist at Yale, Bertram Boltwood, a participant in the lively discussion that followed the discovery of radioactivity and the development of early atom models, was the first to see the opportunity it presented to the earth sciences. In 1907, only a few years after the Kelvin–Chamberlin argument, he used a simple technique of radioactive dating to show that the earth was more than 400 and perhaps as much as 2,000 million (or two billion) years old. In subsequent years the method was refined and, as the number of analyses grew, so did the estimated age of the earth. Just after World War II, I was taught that it was two billion; over the next 15 years, that was raised to 3.5 billion, and today we accept 4.5 billion years. Compared to the universe, thought to have an age about 15 billion years, that still leaves the earth a mere child.

2.2 THE PAST MEASURED IN YEARS

In principle, the technique of dating rocks with radioactive elements is simple. The nucleus of an atom consists of protons and neutrons. The

29

number of protons, the atomic number, controls the chemical behavior of the element. The number of neutrons varies, giving rise to isotopes with virtually identical chemical behavior but slightly different mass. The sum of protons and neutrons is the mass number. Oxygen has 8 protons and hence an atomic number of 8. It has three isotopes, of which two (with 8 and 10 neutrons) have considerable interest for geology. These isotopes have mass numbers 16 and 18 and are usually written as ^{16}O and ^{18}O (or oxygen-16 and oxygen-18). Carbon has three isotopes: ^{12}C, ^{13}C and ^{14}C.

Most isotopes are stable, but not all. The isotopes of oxygen and two carbon isotopes are stable, but ^{14}C is not; it is said to be radioactive. A radioactive isotope decays; that is, it changes to some other element or isotope. Uranium isotopes decay to lead, ^{14}C to nitrogen. The rate of decay is characteristic for each isotope and cannot be changed by any known force. It is customary to express the rate of decay as the half-life of the isotope, the time it takes for half the number of atoms initially present to decay to daughter atoms plus one or more kinds of radiation (Figure 2.2). Half-lives vary enormously: 26.8 minutes for an unstable isotope of lead, 1,600 years for the radium in a watch, and 4.5 billion years for uranium-238 (^{238}U).

To calculate the age of a rock using one or more of several convenient isotopes is simple, but each of them has advantages and disadvantages. Some are widely available, but others, although less common, are attractive because their age ranges are convenient or because their chemical behavior in minerals and rocks is straightforward (Table 2.1). We are well equipped to date older rocks, especially beyond 100 my, but for younger ones ^{14}C belongs to a very small set of possibilities, and we could definitely use more methods to cover the last half million years.

The unstable isotope of carbon is different from most others because it has not, as they have, been around from the beginning. Instead, it is being created continuously in the upper atmosphere by collisions between cosmic rays and the nuclei of nitrogen. The small amount of radiocarbon so produced diffuses throughout the atmosphere as carbon dioxide (CO_2), is dissolved in the oceans, converted by plants into organic matter, and ingested by animals. Radiocarbon begins to decay to nitrogen-14 as soon as it appears, but living organisms maintain an equilibrium with the reservoir in the atmosphere or the ocean until they die. Then their content of ^{14}C decreases by radioactive decay, so that the age of dead tissue, wood or shell can be determined from whatever activity remains. Ordinarily, the limit of usefulness of radiocarbon dating lies at

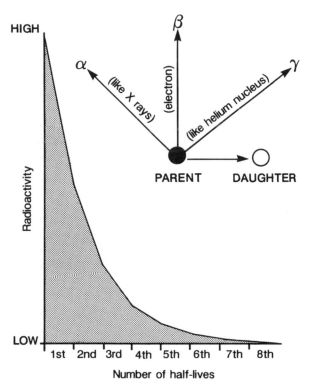

Figure 2.2. Radioactive isotopes decay to stable daughters, while emitting radiation or particles or both. The rate of decay is expressed as the half-life, the time over which half of the initially present amount decays to the daughter isotope. The result is an exponential decay curve which drops rapidly at first, then declines more slowly.

about 40,000 years, or a little more than six half-lives, but newer techniques allow a maximum of up to 60,000–70,000 years. Beyond that we still have problems.

Dating a rock does require more than the application of a simple equation. The rock or mineral must be impermeable; neither parent nor daughter must have been lost or gained except by radioactive decay since the mineral was formed. Leak the daughter away and the date will be too young; leak the parent and it will be too old. Lose or gain both and nothing useful can be said.

If a rock is metamorphosed, the radioactive clock is reset. Consider granite. The magma cools; minerals that contain uranium, but not its daughter isotope lead, crystallize; the clock ticks. Before we come to read it, however, the granite is metamorphosed to gneiss by heat and pressure, its minerals are converted into different ones, and the lead and uranium are redistributed. If we analyze one of the new minerals, the date we find is that of the metamorphic event. Still, if the rock as a whole did

Table 2.1. *Some radioactive isotopes widely used to determine the ages of minerals and rocks*

Parent	Daughter	Half-life	Usable for	Comments
Rubidium-87	Strontium-87	49 by[a]	>100 my	Good in granite
Thorium-232	Lead-208	14 by	>200 my	
Uranium-238	Lead-206	4.5 by	>100 my	Widely used
Potassium 40	Argon-40	1.3 by	>0.1 my	Good in basalt
Uranium-235	Lead-207	0.7 by	>100 my	Widely used
Samarium-147	Neodymium-147	110 by	>1 by	
Carbon-14	Nitrogen-14	5,370 yr	<40,000 yr	In archaeology

[a] by = billion years; my = million years; yr = years.

not leak parents or daughters, a bulk analysis may give us the time when the original granite was formed.

Isotopic age data are subject to errors that cannot easily be spotted. Radioactive decay is a statistical phenomenon and repeat measurements give slightly different answers. If we see a date of 100 ± 6 my, we know that, were we to repeat the measurement many times, two-thirds of the answers would lie between 94 and 106 my. Also, rocks may leak or be contaminated, but the calculation still yields a date which thus needs to be viewed with caution.

Returning to the table of radioactive isotopes, one notes an interesting feature. In Chapter 1, using formations and fossil zones, we found it easy to place sedimentary rocks in the context of earth history, but encountered difficulties with igneous and metamorphic rocks which bear no fossils. Radioactive dating, on the other hand, applies very largely to metamorphic and igneous rocks. Those have great interest of their own, of course, but leave us with the problem of how to dovetail them with the ages of sediments. In most cases we are reduced to the inverse of the procedure discussed in the previous chapter, viz. inferring ages "older than" and "younger than" from igneous rocks associated by good fortune with the sediments that need to be placed in the time scale. This procedure is best understood by studying a few examples of the kind geologists encounter all the time (Figure 2.3). It is a slow, painstaking job for which some people seem to have a special aptitude and tolerance. To them the rest of us owe a great deal of gratitude.

2.3   THE GEOLOGICAL TIME SCALE

Today we know that the earth is 4.5 billion years old. It is not easy to come to grips with this astonishing number, but we must not accept it

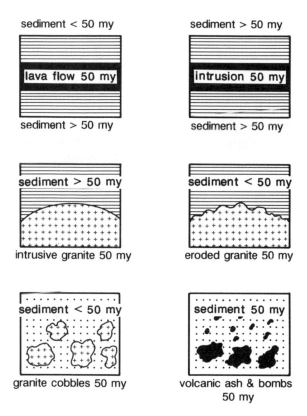

sediment < 50 my

lava flow 50 my

sediment > 50 my

sediment > 50 my

intrusion 50 my

sediment > 50 my

sediment > 50 my

intrusive granite 50 my

sediment < 50 my

eroded granite 50 my

sediment < 50 my

granite cobbles 50 my

sediment 50 my

volcanic ash & bombs
50 my

Figure 2.3. Radioactive ages are obtained mainly on igneous and metamorphic rocks. To tie them to the stratigraphy based on fossils and sediments is not easy, because the two rock types are not associated with each other as often as one would wish. Here are six common ways in which it is done.

without trying to grasp what it means, or everything that follows will make little sense. How shall we understand evolution, for example, if we have no feeling for the million-year life-span allotted to many species? And numerous are the geologists who have constructed histories for a 10 my period that would have fitted comfortably into a mere few thousand!

Let us transform the 4.5 billion years into something we can cope with on a human scale. Suppose we set the entire 4.5 billion equal to a single year (Table 2.2). A second will then represent 143 years. At midnight, when the new year starts, the earth is born. At midnight on December 31 we reach the here and now. The experience is sobering. The length of recorded human history, if we stretch it, is 30 seconds. The diverse life of higher plants and animals that we take for granted did not develop until late in November. The part of earth history we know fairly well, the Phanerozoic, is less than two months long.

33

Foundations

Table 2.2. *How long is geological time?*

Event	Years ago	Calendar date
Origin of the earth	4,500 my	1 Jan., 0000 hrs
Oldest dated rocks	3,940 my	10 Feb.
Earliest life forms	3,200 my	16 April
Ocean and atmosphere "normal"	1,000 my	1 Oct.
Primitive "higher" animals	800 my	1 Nov.
Beginning of the Phanerozoic	570 my	12 Nov.
First land plants	400 my	28 Nov.
Single supercontinent	250 my	10 Dec.
Death of the dinosaurs	65 my	26 Dec.
Human beings appear	4 my	31 Dec., 1800 hrs
End of last ice age	13,000 yr	31 Dec., 2356 hrs 20 seconds
Birth of Christ	2,000 yr	31 Dec., 14 seconds before midnight

For more than a century the geological time scale evolved without benefit of calibration and its subdivisions are therefore of unequal lengths (Figure 2.4). Even today, geologists usually talk about time not in years but in names, names designating intervals of greatly varying lengths, and say "during the Eocene" rather than "between 35 and 56 my ago." Consequently, the task of familiarizing oneself with the history of the earth involves a not inconsiderable vocabulary. The system is hierarchical; it begins with a few units so broad that they are rarely used, and ends with such fine subdivisions that only a few specialists have heard of them. The general categories, however, bearing scholarly Greco-latinized names such as Paleozoic (time of old life) and Proterozoic (time of early life), or a whiff of the Welsh countryside in Cambrian, become quickly familiar.

The units and major boundaries of the time scale are well established and the ages of the main divisions are no longer much debated either. As regards the finer subdivisions, however, the dating is still in flux and the uncertainties are larger than we would like. For the latest Cenozoic, we trust our boundary dates to within 100,000 years, but the uncertainty increases to half a million in the Paleocene and a few million in the middle of the Mesozoic. Inevitably also, with increasing age the statistical uncertainty of the isotopic dates increases too, limiting even more the precision we can achieve.

Time is continuous and has no gaps. The geological time scale, therefore, has no gaps either. The information on which it is based, however,

EON	ERA	PERIOD	EPOCH	AGE(my)
PHANEROZOIC	CENOZOIC	QUATERNARY	Holocene	10,000 yrs
		QUATERNARY	Pleistocene	
				1.6
		NEOGENE	Pliocene	
				5.2
		NEOGENE	Miocene	
				23.3
		PALEOGENE	Oligocene	
				35.4
		PALEOGENE	Eocene	
				56.5
		PALEOGENE	Paleocene	
				65
	MESOZOIC	CRETACEOUS		
				146
		JURASSIC		
				208
		TRIASSIC		
				245
	PALEOZOIC	PERMIAN		
				290
		PENNSYLVANIAN	CARBONIFEROUS	
				323
		MISSISSIPPIAN	CARBONIFEROUS	
				362
		DEVONIAN		
				408
		SILURIAN		
				439
		ORDOVICIAN		
				510
		CAMBRIAN		
				570
PRE-CAMBRIAN		PROTEROZOIC		
				2,500
		ARCHEAN		
				4,500

Figure 2.4. This geological time scale is very abbreviated, but gives the names of the units and the ages of their boundaries that are necessary to follow the discourse in this book. Note that the Carboniferous is the European equivalent of the Mississippian and Pennsylvanian combined. The diagram is based on the standard scale compiled by W. B. Harland and collaborators (A Geologic Time Scale 1989, Cambridge, Cambridge University Press, 1990), except where we discuss the transition from Precambrian to Cambrian (Chapters 17 and 18). There, significant changes are required. As refinement continues, many minor boundary ages will also be revised again and again.

does not come from a clock that always runs, but from a discontinuous rock record. Many major boundaries, such as between the Devonian and the Carboniferous or the Permian and the Triassic, have been placed at unconformities that are almost global, at least in shallow marine formations. At such points, the continuity of time and time scale is not matched by the continuity of the rocks. This distinction between time and the rock record needs to be kept firmly in mind.

2.4 DATING SEDIMENTARY ROCKS

Although the time scale based on radioactive decay is quite satisfactory and has been used wherever possible, it is almost wholly restricted to igneous and occasionally metamorphic rock samples. This is inconvenient, especially when we wish to date very long sequences of sediments that have no igneous intrusions or lava flows. Therefore, the search for other methods continues, with sometimes surprising success. We may consider two here that have come into wide use in recent years, the magnetic polarity reversal time scale and a time scale based on temporal changes in the ratio of the two stable isotopes of strontium.

Many igneous rocks, when they cool and crystallize, become magnetized parallel to the earth's magnetic field. Basalt above all preserves the field well, and it is not difficult to measure the ancient north and south directions in rocks. Many sediments also exhibit magnetization because, as they settle peacefully through the water, small magnetic grains have time to line up with the magnetic field.

Since the beginning of this century it has been known that if one measures magnetic orientations in thick stacks of almost continuous lava flows, the observed magnetic field periodically appears to reverse its polarity; that is, magnetic north becomes magnetic south and vice versa. This means that the magnetic field of the earth itself reverses its polarity from time to time. Reversals are common, occurring irregularly but on the average about every 700,000 years (Figure 2.5), and each reversal takes only a few thousand years. Careful measurements of lava sequences on land dated by the potassium–argon method (Table 2.1) by the late Allan Cox of Stanford University, among others, have led to a polarity reversal time scale now extending back some five million years. Beyond that, continuous lava sequences are scarce and the uncertainties of isotopic dating uncomfortably large, but even short sections floating free in time can help in detailed dating of old rock sequences.

Magnetic polarity reversal dating can be applied to many igneous and some sedimentary rocks. The second method is usable only in sediments and rests on the ratio of two stable isotopes ^{86}Sr and ^{87}Sr of the element strontium. The isotopes derive from the weathering of continental and oceanic rocks and are found everywhere in ocean water although in small amounts. Solutions weathered out of continental rocks have a high $^{87}Sr/$ ^{86}Sr ratio, whereas oceanic islands and volcanoes like those of the Andes have a low one. The contributions of the two sources vary over time and the strontium isotope ratio in seawater varies with them, but the oceans are well mixed and at any time the ratio is the same everywhere.

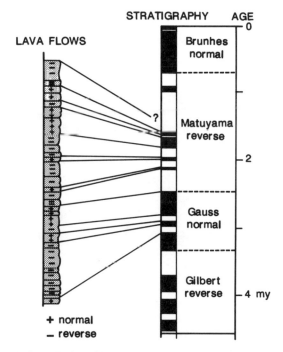

STRATIGRAPHY   AGE

LAVA FLOWS

Brunhes
normal

?   Matuyama
reverse

2

Gauss
normal

Gilbert
reverse   4 my

+ normal
− reverse

MAGNETIC POLARITY STRATIGRAPHY

Figure 2.5. Measurements of the magnetic field of the earth, frozen into long sequences of cooling lava flows, show that its direction, its polarity, has episodically reversed, magnetic north becoming magnetic south and vice versa. Dated by means of radioactive isotopes, the measurements furnish a magnetic polarity reversal time scale, part of which is shown at the right. It has become an important tool in stratigraphy.

Strontium is similar to calcium in its chemical behavior and enters the calcareous skeletons and shells of corals, mollusks and other marine fossils with the ratio it has in the seawater of that time. Analyses of fossils from sedimentary rocks dated by other means provide a standard curve (Figure 2.6) of the changing $^{87}Sr/^{86}Sr$ ratio over the last 65 my (the Cenozoic). Because the method does not care what kind of fossil it deals with and can be applied to very small samples, it has many practical uses.

The end is not in sight for new dating methods; recently a rarely used pair of radioactive isotopes, uranium-238 and lead-206, have been used to date corals throughout the Phanerozoic, a potentially very useful tool. On and on it goes, an important enterprise with much potential.

2.5 TOO MUCH TIME, TOO FEW EVENTS?

The history of the earth is longer, more than forty times longer, than anyone thought before the discovery of radioactivity. Do we now have

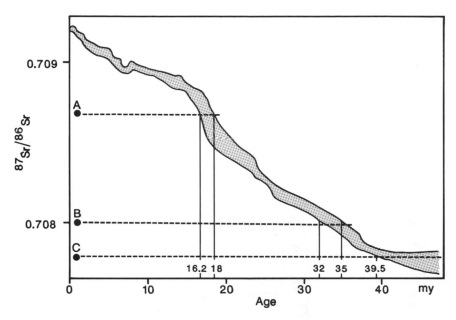

Figure 2.6. Sediments can be dated directly with the aid of their strontium isotope ratios (⁸⁷Sr/⁸⁶Sr), and their ages obtained by comparison with a standard curve based on dated rock samples. Each point on the standard curve has a statistical uncertainty, and so the curve is a band (shaded) rather than a line. Formation A is between 16.2 and 18 my old, while Formation B, being in a flatter part of the curve, is less precisely determined. Formation C is poorly placed; all we can say is that it is older than 39.5 my.

enough time? Does this enlarged perspective accommodate all that has happened in the history of the earth? It does, and it does more: it seems to leave us with time unaccounted for.

If we know the thickness of an ancient series of deposits and the environments in which they formed (their facies), and we use appropriate deposition rates derived from modern equivalents, we can estimate the time needed to deposit them. We have to take account of some uncertainties and errors, but the exercise is simple and the answer ought to be reasonably accurate. We can then compare the result with the time that was available, taken from the geological time scale. In Wyoming a series of Cretaceous sandstones and shales closely resembles the coastal deposits of the modern Gulf of Mexico. Doing the exercise, we find that a mere 100,000 years would suffice to lay down the entire sequence, but it occupies a stratigraphic interval six million years long. We can repeat the experiment elsewhere; in the majority of cases, the rock record needs only 1–10 percent of the available time. Evidently sedimentation, unlike work according to Parkinson's Law, does not expand to fill the time available.

What happens the rest of the time? Unconformities can be seen some-times, or at least invoked to explain part of the deficit. Are they common enough to explain it all? In the abyss, an environment so quiet that we would expect hardly any erosion, the campaigns of the drill ship *Glomar Challenger* have shown that often as much as half the record is missing. In shallow seas it is even worse, but the sum of all detectable breaks is usually less than 75 percent. Perhaps the geological record consists of rare major events separating long intervals when nothing happens at all. That is not easy to accept, because one would expect to see at least some evidence for long times without deposition, a soil profile, burrowing by organisms, even only a hardening of the surface. But more often than not we see nothing at all. Alternatively, innumerable minor events may have left imprints so faint that each was obscured instantly by the next one.

Whatever the explanation, the notion that the archives of the earth mainly record brief major events and omit the long quiet intervals is not by itself too disquieting. An atom also consists mainly of empty space, yet a chair, although made of atoms, usually has adequate solidity. The elimination of daily trivia from history need not be harmful. Even if earth history proceeds mainly by leaps and bounds, our interpretation, based on two centuries of study, seems consistent and makes a lot of sense.

More disturbing is that this fragmentary record, if real, cannot be relied upon to retain adequate traces of brief yet important events, a hurricane or the impact of an asteroid. And if we must rely on a highly intermittent account to record a phenomenon that itself varies in rate, such as the evolution of life with its sudden changes alternating with long periods of no change, our chance of finding the information we seek is small.

For these reasons, the completeness of the geological record attracts much attention. How can we know how much and what is missing? Are we failing to see mainly brief unimportant events or do we miss brief important ones too? These questions occupy many a statistically minded geologist, but the products of their labors, important as they are, tend to be among the less readable pronunciations of the profession. For the intrepid I have listed in the Special Topics section of the reading list a few samples.

PERSPECTIVE

Rocks have their stories to tell. Faults and folds describe the building of mountains, and fossils tell us about environments of deposition. Fossils can also be used to arrange rocks in their proper sequence, and simple ground rules permit us to connect a sequence of strata in one place with others elsewhere. Generally, though not always, the present is the key to the past, and younger strata rest on top of older ones. To place the geological history so deduced in the context of its time, we use dates based on the decay of radioactive isotopes. That enables us to estimate the duration of events and the rates of processes, important clues in understanding history.

This suffices here as far as methods are concerned; others will be introduced at their appropriate points. Let us turn now to the first case of change: the climate of the Great Ice Age. Its last cold phase is recent and vivid and still affects our present world, and another one might be waiting in the wings. To understand the Great Ice Age, we need first to grasp what makes weather and climate – a little tedious, perhaps, but useful, and welcome later when we examine how the ocean works.

FOR FURTHER READING

Among the many popular books on geology I cite only one, because its emphasis on the history of life complements mine so well: Nisbet, E. G. (1991). *Living Earth: A Short History of Life and its Home* (London: Harper Collins Academic).

For basic geology the great classic is *Holmes' Principles of Physical Geology* (revised and republished in 1992 by P. McL. D. Duff; London: Chapman & Hall). A resource much consulted by my own first-year students is Smith, D. G. (ed.) (1981). *The Cambridge Encyclopedia of Earth Sciences* (Cambridge: Cambridge University Press).

A good start on the history of geology is Albritton, C. (1980). *The Abyss of Time* (San Francisco: Freeman Cooper & Company). On uniformitarianism, catastrophism and evolution in 19th century society see Gillespie, C. C. (1959). *Genesis and Geology* (New York: Harper); and on geological controversies see Hallam, A. (1992). *Great Geological Controversies* (New York: Oxford University Press).

For earth history and its methods a fine book is Stanley, S. M. (1986). *Earth and Life through Time* (New York: W. H. Freeman), while the following three books enlarge on Chapters 15 through 19 at a beginner's level but without loss of quality: Eicher, D. L. (1976). *Geologic Time*; Laporte, L. F. (1979). *Ancient Environments*; and McAlester, A. L., Eicher, D. L. & Rottman, M. L. (1982). *The History of the Earth's Crust* (all published by Prentice-Hall, Englewood Cliffs, NJ).

At the college level, readable texts are: Dott Jr., R. H. & Batten, R. L. (1981). *Evolution of the Earth* (New York: McGraw-Hill); Matthews, R. K. (1984). *Dynamic*

Stratigraphy: An Introduction to Sedimentation and Stratigraphy (Englewood Cliffs, NJ: Prentice-Hall); Schoch, R. M. (1989). *Stratigraphy: Principles and Methods* (New York: Academic Press).

For dating see: Harland, W. B., Armstrong, R. L., Cox, A. V., Craig, L. E., Smith, A. G. & Smith, D. G. (1990). *A Geologic Time Scale 1989* (Cambridge: Cambridge University Press); and Dalrymple, C. B. (1991). *The Age of the Earth* (Stanford, CA: Stanford University Press).

Geology relies much on field observation and the rich collection of color photos in the following book adds needed reality: Moores, E. M. & Wahl, F. M. (eds.) (1989). *The Art of Geology* (Boulder, CO: Geological Society of America).

SPECIAL TOPICS

Anders, M. H., Kruger, S. W. & Saddler, P. M. (1987). A new look at sedimentation rates and the completeness of the stratigraphic record, *Journal of Geology*, **95**, 1–14.

De Paolo, D. J. & Ingram, B. L. (1985). High resolution stratigraphy with strontium isotopes, *Science*, **217**, 938–41.

Hawkesworth, C. J. (1992). Geological time, in *Understanding the Earth*, G. C. Brown, C. J. Hawkesworth & R. C. L. Wilson (eds.), pp. 132–44 (Cambridge: Cambridge University Press).

Hess, J., Bender, M. L. & Schilling, J.-G. (1986). Evolution of the ratio of strontium-87 to strontium-86 in seawater from the Cretaceous to the present, *Science*, **231**, 979–84.

Kukal, Z. (1990). Rates of geological processes, *Earth Science Reviews*, **28**, 7–284.

McShea, D. W. & Raup D. W. (1986). Completeness of the geological record, *Journal of Geology*, **94**, 569–74.

Rosenberg, G. D. & Runcorn, S. K. (eds.) (1975). *Growth Rhythms and the History of the Earth's Rotation* (London: Wiley).

Sonett, C. P., Finney, S. A. & Williams, C. R. (1988). The lunar orbit in the late Precambrian and the Elatina sandstone laminae, *Nature*, **335**, 806–09.

Williams, G. E. (1981). The solar cycle in Precambrian time, *Scientific American*, **255**, 88–96.

Climate past and present: the Ice Age

If the glacial period were uniformity, what was
catastrophe?

    Henry Adams, *The Education of Henry Adams*

THE SNOWS OF YESTERYEAR

Twenty thousand years ago, an observer viewing the earth from space would have seen an unfamiliar world. The outlines of the continents were different from those of today, the Gulf of Mexico was much smaller and the North Sea not present at all. Indonesia, where warm seas now wash tropical islands, was a large and rather arid extension of southeast Asia. The rain-forests of the Amazon and the Congo were mere oases in a vast savanna, and an enormous shining cap of ice concealed much of Europe and North America. Trade-winds still whipped whitecaps on the blue ocean around coral islands, but the Atlantic was covered with pack ice down to the latitude of Spain. Glaciers crowned Hawaii and many another mountain in the tropics, and huge lakes shimmered in the winter sun between the rain clouds over California, Nevada and Utah.

On the northern hemisphere, glaciation has followed on glaciation for more than two million years; the most recent one left us a mere 10,000 years ago. This recent ice age is a classic example of a major climate change. Antarctica and Greenland have never left the glacial condition since they entered it, and help us visualize the world as it was then. Yet the climate change that brought this cold world about was not very large, nor was it the only example of its kind in the history of our planet.

Thirty glacial intervals have passed already, but nothing suggests that this is the end. Over the eons of its existence, the world has seen several ice ages, and ours is not so far the longest. Another advance of icecaps on the northern continents is beyond doubt. Could the droughts of central Africa, the severe winters in New England, the excessive rains followed by years of drought in California, the dry winters and bright summers that are lying in a diagonal band across northwestern Europe, be the harbingers of another one? The question cannot be answered, because other climatic changes, some induced by ourselves, are in the offing. Besides, since it is unlikely that another glaciation will come within the next few hundred years, it hardly matters.

No ice age is ever an unrelieved state of polar weather. Times of severe cold and vast icecaps, the glacials, alternate with interglacials when milder climates, like our own, prevail. These swings of climate do not only affect the high and middle latitudes; they leave their mark, albeit in a less dramatic manner, on tropical and subtropical lands and seas as well.

3

Climate and climate change

Our short memory causes us to think of our climate as dependable. It has been winter and it shall be spring; from year to year the differences are small and transient. This is the seasonal cycle of climate. We shall meet many climate cycles, but only this one is fully understood. Other climate changes are not so obviously cyclic. Not long ago the climatic backdrop to human history was not like today at all. I do not refer here to the glacial world of prehistoric cave dwellers, but to Viking days, to the golden 17th century, to Manifest Destiny and the Oregon Trail only a century and a half ago.

Climate varies on many scales, and the smaller variations close to us help us understand the large swings of the geological past. What do they do, what might they mean and, as far as we so far understand it, how do they come about?

3.1 THE INCONSTANT CLIMATE

We derive our confidence that the climate will not change much from its constancy and generally favorable state in the first half of this century. Before 1920 drought and famine could be counted on to devastate India on the average once every 8.5 years. That risk dropped to half between 1920 and 1960, and the population, previously controlled largely by famine, rose accordingly. Early in this century Californians expected a truly dry winter about once in seven years, but for the next 50 years droughts arrived less than half as often. The 1960s changed all that and the world's climate has returned to an earlier, less benign state. At the same time rapid population growth and an increasing need for advance planning in an ever more complex world keep reducing our capacity for a swift and flexible response.

All this is perfectly normal. During the past two millennia our climate

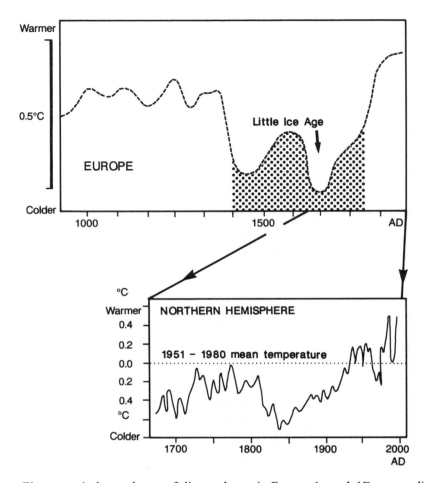

Figure 3.1. A thousand years of climate change in Europe. Around AD 1000 conditions were favorable, but gradually deteriorated to become the Little Ice Age, followed by a still continuing warming trend. The left scale, however, shows that the changes involve less than 1 °C. A detailed curve for the northern hemisphere since AD 1700 (bottom right) shows the short-term variation that makes it so difficult to say at any given time whether the average global temperature is rising or falling.

has changed often, rapidly, and often drastically. Across Europe, the Little Ice Age, lasting from about 1450 to 1850 AD, was much colder than it is now (Figure 3.1). Severe winters were common, the summers cool and damp, and glaciers crept down Alpine valleys. In northern England, Scotland and Scandinavia, growing grain above 200 m (600 ft) became impossible (Figure 3.2), and lost harvests caused widespread famines which stirred war and rebellion. In Iceland, where human subsistence is always marginal, a cold spell in the 14th century and the Little Ice Age converted the country permanently from a wheat-growing to a sheep-farming economy, although the temperature fell only 1.5–2.0 °C.

48

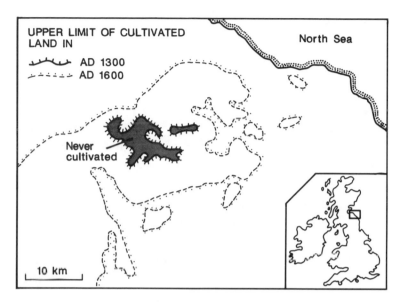

Figure 3.2. In northwestern Europe the transition from the warm medieval period to the Little Ice Age greatly reduced the available farmland. As the temperature dropped, the elevation above which grain could not be grown fell steadily, as this map of usable land in northeastern England shows. Between AD 1300 and AD 1600 the area where reasonably reliable harvests were possible was reduced by nearly half, causing great economic and political problems.

The cold 14th century also contributed to the demise of the first European colony west of the Atlantic. Vikings from Iceland had settled Greenland in the warm centuries between 800 and 1200 AD, but the change to a colder climate sharply reduced their yield from farming. Travel across an ocean sometimes ice-bound even in summer cut the supply from Norway of essentials such as timber and iron, and new blood for the small population failed to arrive. Economic and political pressures in northern Europe that shifted the focus away from the far west of the Viking world administered the *coup de grâce*.

Not everywhere was the Little Ice Age a bad time. Early in the 19th century, the prairies of North America were nourished by rain far more than now, and the tales about the cornucopia out west that seduced so many settlers in the 1840s were founded on reality. Certainly, some of the tale-tellers were unconscionable real estate crooks, but it was not their fault that the end of the Little Ice Age arrived together with the new immigrants, who found land and climate to be far less suitable than promised. The marked decline in rainfall that followed would have decimated the buffalo herds, had not the white man taken his toll a little earlier and somewhat more thoroughly.

The reconstruction of the climates of the recent past rests on a rich variety of sources: logs maintained by deck officers of the British navy and held for centuries in Admiralty archives, records of date and quality of wine harvests kept (where else?) in French monasteries, the chemistry of ocean sediments and glacier ice, and even the width of tree-rings. Most trees produce annually a layer of new wood which, by its thickness, reflects the weather conditions of the preceding and current growing seasons. If we calibrate tree-ring data for the last century with the weather records of that period, we can use earlier tree-ring sequences to discover the climate of the past, for some 8,000 years in the western United States, and almost as long in Europe.

Advances and retreats of mountain glaciers also tell us of temperature and precipitation changes. Decrease the snowfall or increase the summer melt and glaciers retreat rapidly up-valley, leaving bare rock behind. Lichens are the first to colonize these surfaces; they grow outward from their starting point at a rate that can be measured, and the size of their colonies measures the time elapsed since the glacier withdrew.

It has only recently dawned on historians that the climate of the past has affected human history and must be taken into account if we want to understand the course of human events. The current worry about the green-house effect extrapolates that concern into the future and has alerted public and politicians to the role played by climate in long-term social and economic planning, not to mention in war and peace.

3.2 LITTLE ICE AGES

How widespread was the Little Ice Age? To answer that question we cannot rely on historical records because they are available only for Europe, but must look at geological evidence from all over the world. The information is confusing at times and the dating imprecise, but we find that the Little Ice Age was indeed a global event, although its several peaks of severity did not occur everywhere at the same time.

Still, it might have been a fluke, a mere blip on a climate pattern that was otherwise steady since the demise of the last icecap. The matter is important, because the Little Ice Age arrived and departed so suddenly and had such a large effect. Can we expect another one soon, perhaps in time to relieve the green-house warming?

This question takes us beyond the limits of written history to the record of advances and retreats of mountain glaciers as the Holocene climate oscillated between wet and cool, warmer and drier. Since the end of the last glacial more than 10,000 years ago, the glaciers of Alaska and the Rocky Mountains, of the Alps and Scandinavia, New Zealand

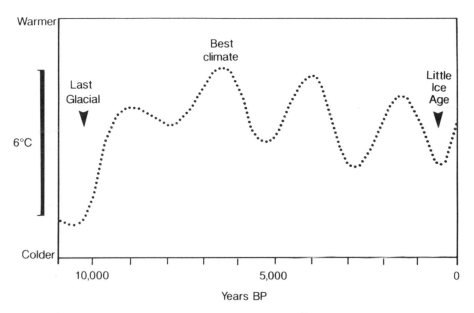

Figure 3.3. The optimal climate since the last glacial period came about 8,000 years ago and we are clearly not as well off now. Even the mild years of the early 20th century were one degree colder than the period between 6,000 and 7,000 years ago. The graph also shows that the Little Ice Age was only the most recent of four cold spells in the last 11,000 years. It seems that we ought to be approaching another cold phase, but the human-induced green-house warmth will compensate, and perhaps over-compensate, for that.

and South America have advanced and withdrawn many times within their valleys, stripping away soils and felling trees during advances, and depositing glacial debris during retreats. The record, dated and correlated across the world, shows glacier advances about 10,500, 7,000–8,000, 5,000, and 3,000 years ago (Figure 3.3), in addition to the Little Ice Age. Measuring their spacing with a slightly elastic yardstick, we find that little ice ages occur every 2,000–3,000 years, each lasting only a few centuries.

It should be said that, notwithstanding the evocative name, little ice ages are not very cold (about 1.5–2.0 °C below the annual average of today), and that the post-glacial climate has otherwise been fairly uniform. In Sweden, for example, the mean annual temperature has not deviated more than 3.5 °C from its present value.

What about the cause of this remarkable recurrence of cooler intervals? There are many candidates but the choice is restricted by the short duration of the events. The forces that caused ice to advance and recede across the northern hemisphere for the past few millions of years operate on time scales ten to one hundred times too long (Section 4.2). Volcanic eruptions, known to be able to cool the earth by injecting

51

dust and gases high in the stratosphere, run to years or at most decades. So what is left?

Sunspots, large disturbances of the solar surface that recur every 13 and 26 years, have long been suspected of bringing drought, although how they do it is not clear. Besides, sunspot cycles are too short, but a long interval when there were no sunspots at all, the Maunder Minimum, coincides roughly with the Little Ice Age. Might sunspot minima be behind little ice ages?

Radiocarbon dating of tree-rings has shown that ^{14}C is not formed at a constant rate by a never-varying flux of cosmic rays coming from the sun, as we used to think. Instead, the flux does vary and was, for example, low during the Little Ice Age. This may mean that earlier intervals with a low cosmic ray flux also had low sunspot levels. It also means that radiocarbon years vary in length and need calibration before they are equivalent to calendar years. Sampling and radiocarbon dating each tree-ring in an 8,000-year-long sequence has calibrated the radiocarbon calendar, and has shown that little ice ages indeed coincide with times of low cosmic ray flux.

This would settle it, were it not that the correlation is a bit sloppy, and that we still do not see how sunspots or their absence can influence climate.

3.3   A BRIEF DISCOURSE ON THE WORKINGS OF CLIMATE

Climate is the average weather over the long term, from human decades to geological eons. Today's rain is weather, this winter's excess of it merely a wet season, but many wet seasons in a row may spell a climate change.

A change in climate can take two forms, a long-term change for better or for worse ("better" or "worse" in human terms) or a change to greater year-by-year variability. It is often easier to adapt to a warmer or drier, cooler or wetter climate that remains reasonably stable than to a highly unpredictable one. When severe winters turn up frequently but at random, stocking adequate fuel is a major gamble that raises the cost in warm years or closes down industry and kills people in cold ones. If great droughts lasting several years come and go without pattern, the supply of water to densely populated regions or farming areas requires advance planning and much money, and that rarest of all things, political will. Anything that enhances our ability to foresee what the climate is likely to do in the next decades will thus be useful, even essential. To understand what this implies we need to know how the atmosphere works.

As watchers of the daily forecast know, our understanding of the weather is still deficient, and randomness in the atmosphere precludes, perhaps forever, reliable forecasts beyond days or weeks. Climate averages the short-term variation and, influenced by processes such as ocean currents that have greater constancy than airflow, it has a longer memory. Therefore we may hope to understand the climate from a long series of observations even though the behavior of the atmosphere is imperfectly known. If that series should contain persistent or periodic elements, we might be able to forecast the climate sooner than we can predict next season's weather. This perception has, quite suddenly, made climatology into a lively and even crowded field.

Ancient climates are the subject of paleoclimatology, its importance made clear by the glacials and interglacials of the Quaternary. To understand the Quaternary, geologists used to concentrate on the landforms and deposits left by glaciers or icecaps, but climatologists, oceanographers and geochemists have lately joined them, aware (partly because of the odd weather of the last few decades) that climate changes on a human time scale have major economic, political and social consequences.

The earth's climate is an engine whose parts are the land, the ocean and the atmosphere. The engine runs on heat from the sun. Some of this heat is reflected back into space, but much reaches the surface and warms the soil, the air and the water. Over the oceans the heat causes evaporation; clouds form, travel and condense, and rain falls somewhere.

Near the equator, the sun's rays strike the earth at a high angle, heating the surface more than do the rays that come in at higher latitudes. The hot equatorial air, being light, rises and flows north and south (Figure 3.4), distributing its warmth across the planet. Rising air expands, cools and so is less able to hold moisture, hence the equatorial rain belt. Aloft, the warm air gradually cools and between 25° and 35° N and S latitude some of it sinks and returns to the equator as a surface wind. The sinking air is compressed and creates a subtropical high-pressure zone. Because the compression gives the air a greater capacity to hold moisture, the subtropics are quite dry.

The upper air continues toward the pole, cooling and sinking gradually as it goes. From the polar region it returns as a cold surface wind. The seasons complicate this simple picture because in winter the temperate and polar zones are less or not at all heated, with the result that a sharp temperature gradient forms between 45° and 60° N and S, the circumpolar vortex.

53

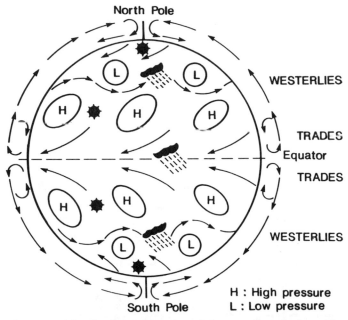

Figure 3.4. The planetary circulation of the atmosphere begins with warm air rising at the equator and spreading north and south. As it rises, the air cannot hold its moisture, and tropical rains fall. In the subtropics, some of the now cooler air returns to the equator, converted by the Coriolis force into easterly trade-winds. The sinking air is compressed and thus dry and the climate sunny. Beyond 45°C, a high-altitude westerly wind, the jetstream, carries low-pressure systems and their rains eastward. Easterlies prevail around the poles.

The airflow is modified by the rotation of the earth. The effect, called the Coriolis force, alters the direction of every current on earth, whether air or water. The force is proportional to the velocity of the current and to the latitude, increasing from zero at the equator to a maximum at the poles. Looking downstream, the Coriolis force turns a current to the right on the northern and to the left on the southern hemisphere. It can be visualized (but not fully understood!) if one imagines oneself standing at the North Pole and throwing a ball south with great force. While the ball flies, the earth turns eastward under it, and when it finally hits the ground, it does so far west or to the right of the intended target.

The Coriolis force turns the surface air flowing from the subtropics toward the equator into easterly trade-winds which blow all year between 10° and 25° or 30° latitude. In polar regions the return flow is similarly converted into the polar easterlies.

The northward flow between 6 and 20 km (20,000 and 70,000 ft) turns because of the Coriolis force into a westerly wind, the jetstream. Blowing between 45° and 60°, the jetstream derives its energy from the circum-polar vortex and is strongest in winter, because the vortex is at

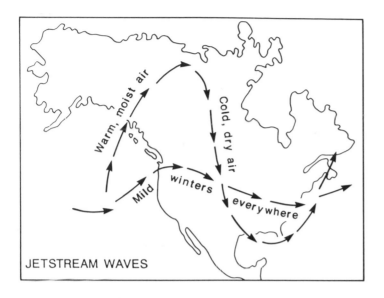

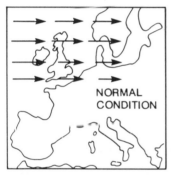
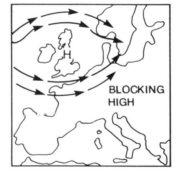

Figure 3.5. Changes in the path of the jetstream across North America greatly influence the weather of the temperate zone, especially in winter. Normal jet flow brings mild winters to the United States and rain to the west coast, but a weak flow with large waves causes drought in California, bitterly cold, dry conditions east of the Rockies, and snowy winters in the east as moisture is picked up in the Gulf of Mexico. Other long-lasting effects are caused by blocking highs that interrupt the normal west-to-east airflow in mid-latitudes. The usually rainy summer of northwestern Europe is then replaced by clear skies, consistently sunny months and drought.

a maximum then. It carries, like beads on a string, the endless series of depressions that bring rain to western Europe and the west coast of North America.

The jetstream does not flow straight on its path around the earth, but in three to six horizontal waves. The position and number of the waves depend on many factors: on major mountain ranges, for example, on the presence of ice packs or on large patches of cold and warm ocean water (Figure 3.5). The waves are also affected by blocking highs, islands of high pressure that force the jet to divert around them. Blocking highs drift slowly from west to east, but may remain stationary for long periods of time. They then bring fine, clear, cold weather in winter, and hot, dry weeks in summer. In 1972 a blocking high over Europe caused harvest failures in the USSR and raised the price of grain in the United States.

The strength of the jetstream depends on the temperature gradient from the equator to the poles. When the polar regions are cool, this gradient is high and the jetstream flows straight and fast. When the high latitudes warm up, the jetstream slows and its waves swing widely. This brings unusual weather patterns, because when the jet turns north across the ocean, it brings warm moist air from low latitudes; when it turns south over land, it is cold and dry. Thus a winter may be unusually severe in Britain and blessedly mild in Moscow at the same time, while the reverse may be true some years later if the waves change position and amplitude. An example of the impact of the jetstream on the regional climate is provided by the now fairly well-understood droughts and wet winters of the recent past on the Pacific coast of North America (Figure 3.5).

The global behavior of the atmosphere is known as the planetary circulation, because it depends only on the presence of a sun, an atmosphere and a spinning planet. Features of the earth from the beginning, trade-winds, temperate westerlies, a jetstream, and tropical and subtropical zones have always existed and will do so until the end.

If we put continents and oceans on the simple earth, the airflow is modified. Anyone living near the sea knows the sea breeze which comes up by mid-morning, vanishes in late afternoon, and is replaced at night with wind from the land. During the day the land heats more rapidly than the sea; warm air rises and is replaced by cool air blowing in from the sea. At night, the sea stays warm longer than the land, and the airflow is reversed. The same process, but with seasonal rather than daily reversals, and on a much larger scale, produces the monsoons (Figure 3.6). Over central Asia the air heats in summer and, in rising, draws a flow of moist air from the Indian Ocean into East Africa and southern Asia. The oceanic winds, their direction controlled by the Coriolis force, bring the much-needed rain of the southwest monsoon. In winter, Asia is cold, the ocean warm, and the monsoon, now blowing from the northeast, is dry and cool (Figure 3.6).

Caused by the contrast of land and sea, monsoons appeared as soon as sufficiently large continents had formed on the surface of the earth (Section 7.3), but their patterns have varied as continents drifted and ocean basins opened or closed (e.g., Chapters 10, 11).

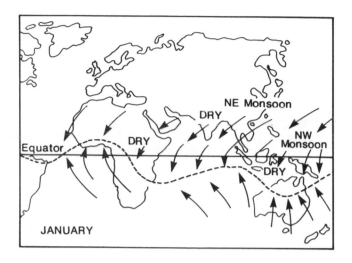

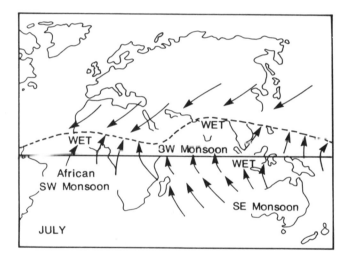

Figure 3.6. The climate of large continents and the adjacent oceans is dominated by monsoons. In winter, the heart of the continent is cold, and dense, dry air sinks and flows outward to the warmer ocean. In summer, rapid heating of the continent results in rising air and a landward flow of cool, moist ocean air. The Coriolis force alters the radial outward and inward flow directions, leading to northeast and southwest monsoons.

Water holds much more heat than air and releases it more slowly; the upper 10–15 cm of the ocean contain as much heat as the whole atmosphere. As a result, ocean currents flowing long distances are a powerful means of distribution of heat. Without them the climate would be much harsher, more like the interior of Asia, the local and seasonal contrasts much greater, and weather changes more abrupt. The size, shape and position of the ocean basins as they changed through time have strongly

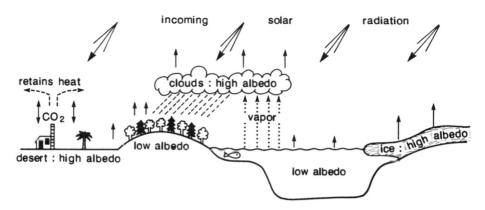

Figure 3.7. Not all incoming solar radiation goes to heating land, sea and air. A large part is reflected by clouds, snow and ice, and deserts, all of which have a high albedo. Forests, lakes and oceans have a low albedo and absorb much of the incoming heat. Carbon dioxide, generated copiously by our burning of fossil fuels and forests, allows the incoming heat to pass, but blocks the reflected radiation and so warms the earth. This is nowadays known as the green-house effect.

influenced the climates of the past and we must study them (Chapters 10–12).

3.4   FILTERS

Not all solar energy goes toward heating the air, the ocean and the land. Some is reflected back by clouds or by the earth's surface. The ability of a surface to reflect solar radiation is called its albedo (Figure 3.7). Clouds have a high albedo and return nearly all radiation; it turns much cooler when clouds move overhead. Snow and ice reflect 50–85 percent and deserts 25–35 percent of the incoming radiation. An earth with large deserts or vast icecaps would be much cooler than one that is mostly green or covered with water, because forest and sea reflect very little (5–15 percent). Variations in the earth's albedo due to changes in the cover of snow and ice, the vegetation, or the position of the continents, have had a large but not yet fully appreciated impact on climate.

A warm ocean evaporates more than a cold one and so has a thicker, more widespread cloud cover. This reduces the amount of radiation that reaches the water; the ocean cools, and the cloud blanket diminishes. This process, called a negative feedback, is a stabilizing influence. We shall later encounter its opposite, a positive feedback which accelerates change.

Not only clouds reflect radiation back into space; dust in the atmosphere does the same. A volcano sending a vast plume of ash up into the stratosphere can lower the global temperature measurably and for an

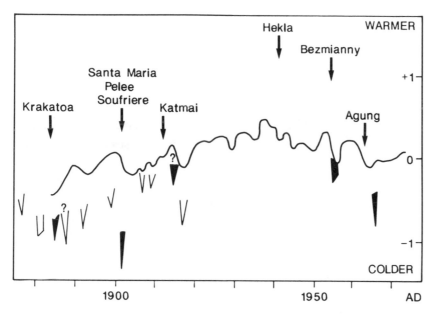

Figure 3.8. Large volcanic eruptions are sometimes followed by abnormal cold spells. The late 19th and 20th centuries provide many examples (arrows). Not every eruption brings a cold spell, however, nor are all cold spells caused by volcanic activity. The curve gives the average temperature of the northern hemisphere, while the V-shapes mark brief cold spells which are colored black when they follow immediately after major volcanic eruptions. The scale on the right shows the deviation in °C from the mean for 1870–1980.

appreciable time (Figure 3.8), until the ash has fallen back to the surface. A huge eruption of the Indonesian volcano Tambora was followed by the "year without summer" of 1816 in the eastern United States, and the explosion that created Lake Toba on Sumatra 73,500 years ago may have accelerated the onset of the last glacial. The volcanic products sulfuric acid and sulfur dioxide that form minuscule droplets (aerosols) suspended in the upper air have a similar or even greater effect. Dust and aerosols must be blown very high, however, or rain will wash them out before they have any effect. The burning of forests, the clearing of land, and many other human activities besides shuffling feet also stir up dust, but it does not rise high enough; the next rain clears the air in days or weeks. Only volcanoes, giant meteorites, or a cluster of nuclear bombs are candidates for a climatic role by virtue of the dust they raise.

In cold years, trees are stressed and grow thin rings. The tree-ring record, extending across thousands of years, shows many cold spells of which several coincide with the dates of major volcanic eruptions. But whether clusters of eruptions, spewing enormous quantities of ash and aerosols to great heights, do change the climate or merely modify it briefly, is not yet clear.

59

3.5   A GLOBAL GREEN-HOUSE

The opposite effect, a warming of the earth, is produced by gases in the atmosphere that allow the short-wavelength radiation of the hot sun to pass through, but trap the longer waves reflected by the cool earth. In effect, these gases turn the earth into a green-house (Figure 3.7). The green-house gas with the highest public profile is carbon dioxide (CO_2), a product of the burning of organic matter; even in small amounts it is an effective blanket. There are many other green-house gases. We are only just starting to appreciate the importance of methane released by the digestion of cattle and termites or when arctic soils thaw, or of many industrial chemicals, and especially of water vapor. In the form of clouds water cools the earth by reflection, but as a green-house gas it keeps us warm. Which it does best has not yet been established.

The green-house effect has been around from the moment the earth developed an atmosphere but for some three billion years it has been kept in check by life's processes (Section 14.6). In the 18th century, however, the industrial revolution began to burn coal in earnest and later oil. Since then the carbon dioxide content of the atmosphere has risen by one-fifth. At the moment it is only 0.03 percent, but as we continue to burn fossil fuels at a prodigious rate, cut our rain-forests, set fire to woods and savannas, and over-cultivate old soils rich in organic matter, the amount of atmospheric CO_2 will double in the next century. The impact of this man-made green-house will not be trivial and we should be deeply concerned about it.

How much do we know about the green-house of the future? First, it is beyond dispute that, as green-house gases accumulate, the earth will get hotter. How much hotter, however, is a hotly debated question. Since Charles Keeling, of the Scripps Institution of Oceanography in La Jolla, California, started measuring it in the 1950s, CO_2 has increased 7 percent (Figure 3.9). That does not seem much, but the figure does not include other green-house gases which began to increase more recently but are catching up rapidly.

At the same time, the mean global temperature has risen 0.5 °C, give or take 0.2 °C, in the course of this century. That number is not too solid because estimates of mean annual temperature are subject to many biases, and fluctuate so wildly from year to year that it is hard to spot an increase against this background. A thorough study covering the last 30 years has shown that the annual temperature rise and the increase in atmospheric CO_2 are correlated, but that does not necessarily mean that the increase of CO_2 is causing the small warming of the last few years.

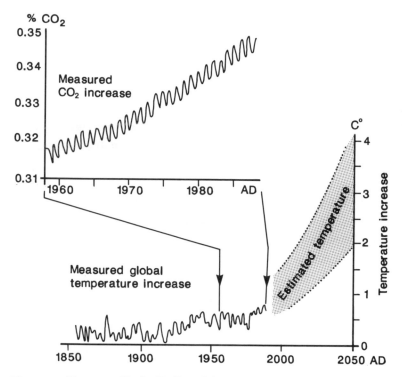

Figure 3.9. Since 1958, Charles Keeling of the Scripps Institution of Oceanography in California has measured the amount of carbon dioxide in the atmosphere. It has risen steadily, with curious oscillations that are due to the seasonal variation of the intensity of photosynthesis in the northern hemisphere. Although the mean annual global temperature varies a great deal from year to year (lower graph), a clear increase is visible from about 1920 onward. It seems set to continue, but how fast is not yet clear.

It might be coincidence or the other way around: if the earth is warming, so is the ocean, and a little CO_2 dissolved in its water will escape to the atmosphere. A lag of five months between the annual temperature increase and the rise of the CO_2, if not an artifact of the analysis (which it might well be), suggests just that. Still, the general rise of the atmospheric CO_2 content cannot result from climate. Whether we can yet see it or not, the green-house is a fact.

3.6 TOMORROW'S WORLD

What would the effect be of a doubling of the CO_2 content of the atmosphere and when might that come about? The latter question depends on our estimate of the amount of fossil fuel people will burn, how many forests they will cut, and so forth. It is made more insecure, because we do not know how much added atmospheric CO_2 will dissolve in the ocean; that is a terribly complicated process. But as a doubling is inevitable and almost certainly will happen in the next century, let us set the time between 2050 and 2075. What will the world be like?

The best guess, one might say the official wisdom, is that the mean global temperature will rise between 1.5 and 4.5 °C. We shall take a conservative 2.5 °C, but ask me again in ten years' time and I may have a different (and more confident) answer. That is well within the temperature range people tolerate with ease, and as it comes upon us slowly, most will not personally notice it very much. The rise will be less at the equator, but much more (4–7 °C) at the poles. What we would therefore notice is that the jetstream will be much weaker and more wavy because the equator-to-pole temperature gradient will be so much less. The climate is thus likely to be even more variable from year to year than it is already, hard to believe as that may be.

Crops will benefit from more warm days and more CO_2, but in some regions might be stressed by high heat. The current opinion is that it would make little difference for the harvest, but the demand for irrigation water would increase greatly. What would happen to the distribution of rainfall and runoff, climatic features that we understand less well than temperature, is not clear. Warmer oceans evaporate more and there ought to be more rain, but where it would fall is very uncertain, although the tropical rain belt will, of course, stay where it is. But even if the rainfall does not change, higher temperatures reduce runoff and hence the water supply. Add to this the greater need for water of a larger world population living in a hotter climate, and you have a problem that alone justifies much worry, although so far not enough to prod governments into action.

Something else that must change is the level of the sea. Ice and snow will melt when the polar regions get warmer, adding water to the oceans, possibly as much as 40 cm. Moreover, as the ocean itself warms, the water expands, raising sea level another 40 cm. Add to this the 10–20 cm the sea has already risen during the past century, and at doubling time its level might be almost one meter higher. That is an intimidating prospect for many major port cities (Figure 3.10), for the low countries by the sea, and for owners of beach houses. To drive the point home, consider that a 70 cm rise would flood 24 percent of Florida, containing 43 percent of the state population, and Louisiana would lose 27 percent of its land, housing 46 percent of the population. Coastal land is not very valuable in Louisiana, but it is in the Miami area, and also in New York. Only 0.6 percent of New York's area, housing a mere 2 percent of the population, would be under water, but much of that is in New

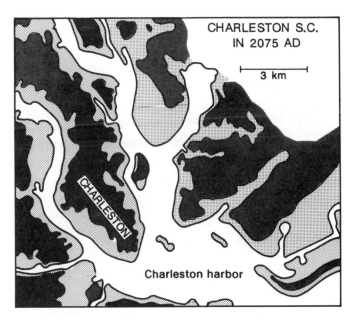

Figure 3.10. Charleston, South Carolina, and its harbor as they are now (light shading) **and** what will remain of it in 2075 (dark), if the sea rises according to a moderate estimate of 1.5 m (5 feet) above its present level.

York City. And let us not forget London, Holland, or Bangladesh, the last-mentioned already in serious trouble with floods.

So why is there no great flurry of activity to do something about it? Politicians dislike long-term decisions, "long-term" meaning beyond the day they hope to be re-elected, and avoid them as long as they possibly can. The only way to force the issue is the pressure of public opinion. So why does not the world demand action? That failure stems from the nature of scientific opinion and the lack of skill of scientists who present it. The warming is beyond doubt; it will surely come. The argument is about how much, when, and where, and science can only talk about probable outcomes, no certain yes or certain no.

Why we seem to be so confused and in disagreement can be illustrated with the coming rise of the sea. That ice and snow will melt and the sea will rise is a certainty, but when and how much are matters of probability. Recent work indicates that in high northern regions, say around Greenland, the temperature would rise first in the winter accompanied by more snow, because the warming adjacent sea will bring moisture. The increase in summer melting would come later. The amount of ice will at first increase, reducing the rise of the sea, but by how much? And for how long?

This sort of thing, especially if played as an adversary game as lawyers will do, provides plenty of excuse for stalling. Much uncertainty will disappear when we come to understand such key elements as how much heat and CO_2 the ocean can absorb and how fast, how quickly the albedo of the polar ice and snowcap will change, or what the regional patterns of temperature and precipitation are going to be. But the forecast will remain just that, a forecast, and that entails risk. If it is hard to accept the forecast, you can doubt it.

People accept this when dealing with the weather and still find forecasts useful, even where a wrong one may have serious consequences as in the offshore oil industry. But in the matter of climate change, pollution control or acid rain, the uncertainty of the predictions is a convenient reason to defer action. Proposing still more research is easier than making a politically unpleasant decision, and postpones that decision until at least after the next election.

It is impossible to forestall a doubling of CO_2, because the Third World depends on a major increase in energy consumption to gain much needed prosperity, while western economies would collapse from a reduction by one-third or more. Thus we must learn to live with it, and that should not be impossible if the necessary adjustments are properly anticipated and gentle rather than late and catastrophic. Life styles can be changed (without adding more air conditioning), and low-lying land protected if it is valuable (the Dutch have done it for centuries but, being already 600 cm below sea level, may have a tough task). New crops suited to higher temperatures and lower, more variable rainfall can be developed, and industry shifted to more suitable locations. And developing a better water economy is a pressing matter almost everywhere, even if no green-house threat were looming at all.

To talk about reducing green-house gas emissions is talking about a brief delay in doubling time. It is also a means to dodge pressure for action on pollution control, saving rain-forests, and protecting natural vegetation and wildlife. Above all, it conveniently distracts attention from making the polluters rather than the polluted pay. Much can be done that would make life more livable for almost everyone, often at a surprisingly low price.

The only major solution, the only one that would be really effective, is a large reduction in the number of people in the world, preferably by means other than disease, war and famine, and the best way, one of proven efficacy, is the education and raise in status of women; where that has been done, the birth rate has invariably dropped. While we are at it, cleaner economies that use fewer resources and energy of a better kind, and befoul the planet less, are a worthwhile goal too!

64

How does the study of the climates of the past bear on these issues? It does, because they are partial analogues of what is in store for us. Many lack the rapid change human action is causing, but ice cores taken in Greenland show that some major changes in temperature took only years, not even decades. The warm last interglacial, with its high atmospheric CO_2 content and sea level, was sufficiently similar to give us an inkling of what is coming and to test our omni-present computer models, and Greenland ice cores published in 1993 show that the temperature repeatedly changed from glacial to interglacial within decades. Happily, this demonstrates that we must go on with this kind of research.

4

Portrait of an ice age

The realization that we are living in an ice age is relatively new and did not come easily. In 1830 Louis Agassiz, a young Swiss geologist, resigned himself to the evidence he had seen with his own eyes that an icecap once covered the Swiss Alps, but it took him many years to convince others of the reality of an ice age. To us this seems odd, because the traces of the former icecaps of North America and Europe are so conspicuous. The walls of debris left where ice fronts once melted eloquently testify to processes that no longer shape the temperate zones. Once the concept of so large a climate change had been accepted, however, it proved fertile and occupies the minds of climatologists and geologists to this very day.

Progress in understanding the Quaternary ice age has not been swift. Until the early 1970s almost everyone believed that four or at most five glacial periods, separated by interglacials, adequately described the history of the ice age. This has turned out to be a major error that resulted mainly from a lack of appreciation of the damage done by successive ice advances to the record left by their predecessors. Only when a far more complete Quaternary record was found in the sediments of the deep ocean, did the truth emerge. The Quaternary ice age of the northern hemisphere has counted so far some 30 glacials, each followed by a brief interglacial, in the last of which we live and which we absurdly designate a separate unit, the Holocene. The oldest northern glaciations go back about three rather than two million years, thus starting well before the Quaternary, while Antarctica has had a full ice age for at least 15 million years.

As is often the case, these insights followed the adoption of new geochemical methods based on sophisticated instruments called mass spectrometers that are capable of measuring stable isotope concentrations accurately on very small samples.

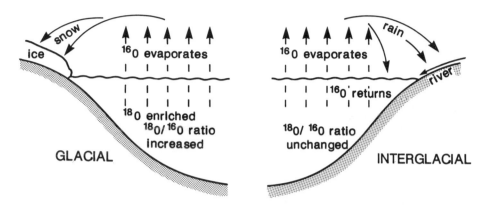

Figure 4.1. Of the two oxygen isotopes ¹⁶O is the lightest. When ocean water evaporates, more ¹⁶O leaves the sea than ¹⁸O, but at the present time all of it returns as rain or via rivers, and the oxygen isotope ratio of the ocean does not change. During a glacial period, however, ¹⁶O is stored in snow and icecaps for the duration and the ocean is enriched in ¹⁸o. As a result the oxygen isotope ratio of the water and of the calcareous shells of organisms rises.

4.1   A KEY TO MANY DOORS

Just as the use of radioactive isotopes brought a revolution in our notion of geological time, so stable isotopes have greatly increased our understanding of the history of climate and the oceans. Thanks to the two isotopes of oxygen the coming and going of glacials and interglacials can be seen in fine detail in ocean sediments, while carbon isotopes clarify the key role of organic matter in the history of atmospheric carbon dioxide and oxygen (Section 10.5).

Of the two oxygen isotopes, the lighter one, ^{16}O, is by far the more abundant. When a marine organism takes up oxygen in the form of carbon dioxide or carbonate and builds it into its tissue or calcareous shell, it acquires ^{16}O and ^{18}O in proportion to their abundances in the water. When it dies, the $^{18}O/^{16}O$ ratio is preserved in its fossil shell where it can be measured with great precision. Because ratios are not very convenient, the values are expressed as the deviation $\delta^{18}O$ (delta ^{18}O) from an arbitrary standard in parts per thousand (per mil). A Cretaceous fossil that happened to be handy when Harold Urey (who later got his second Nobel Prize for this work) looked for a suitable candidate, serves as the standard.

The light isotope evaporates from the sea more readily than the heavy one, so that atmospheric moisture and rain are enriched in ^{16}O. This would leave a tiny excess of ^{18}O in the sea (Figure 4.1), except that rain and rivers return the evaporated water to the sea where waves and currents mix it so well that the oxygen isotope ratio is everywhere the same.

During a glacial period, however, part of the evaporated water is locked in ice together with its excess ^{16}O, leaving the ocean with a higher $^{18}O/^{16}O$ ratio than in an interglacial. For every 10 m of water removed from the oceans and stored as ice on land, the $\delta^{18}O$ rises by 0.1 per mil. When the icecaps melt, the ^{16}O returns to the sea and the original ratio is restored.

The fossil shells of marine organisms deposited on the ocean floor thus record glaciations and deglaciations as oscillations of the oxygen isotope ratio. All we need to do is take a core of ocean sediment that encompasses sufficient time, analyze it, date its layers, and we are able to contemplate the climatic history of the ice age.

Nothing is, of course, ever quite so simple. The core, for instance, must not contain any hiatuses, and some organisms are fastidious about their isotopes and do not accurately reflect the isotope ratio of their environment. More important than these minor inconveniences is that the oxygen isotope ratios of fossils also depend on the temperature at which the organisms prepared their shells. Those that live in cold water, for example, take up more ^{18}O than the inhabitants of a warm sea. Temperature affects the ratio in the same direction as evaporation, and a 1 C° drop in temperature causes a 0.25 per mil increase in $\delta^{18}O$, the same as the removal of 25 m of water. Therefore, the $^{18}O/^{16}O$ ratio for microfossils from a glacial ocean is higher than that from an interglacial one not only because of water storage as ice but also because the waters were colder. We can correct for this temperature effect by analyzing the shells of bottom-dwelling organisms where the temperature changed little, and are rewarded by obtaining both ice volume and temperature.

Suppose we have analyzed a core and have found the temporal variation of oxygen isotope ratios shown in Figure 4.2. How do we date the many events it displays? The calcareous fossils can be dated with radiocarbon, but only up to 40,000 or 50,000 years of age. The magnetic polarity reversal time scale (Section 2.4) comes to mind, but its first usable reversal is 730,000 years old, leaving much time uncalibrated. We may get around that by interpolation if we assume, a bit recklessly, that the rate of deposition did not vary, but the answers are probably unreliable. An ingenious solution for this problem appears in Section 5.2.

4.2 GLACIAL–INTERGLACIAL CYCLES

The many oscillations between warm and cold climates of the Quaternary are vividly displayed in the oxygen isotope record of oceanic sediments (Figure 4.2), the only continuous record of the Quaternary we possess.

68

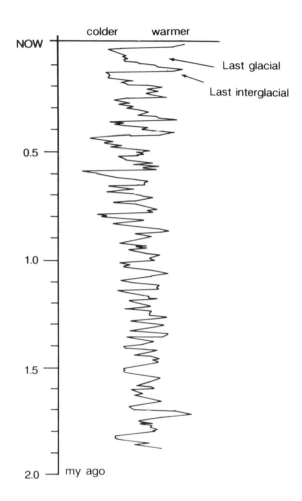

Figure 4.2. Oxygen isotope ratios measured on the calcareous shells of small oceanic organisms are an index for the amount of water stored as ice on land. The curve shows the fine details of Quaternary climatic change that oxygen isotope data are able to reveal. This includes some 60 major fluctuations since the beginning of the northern hemispheric glaciation more than two million years ago.

Sharp transitions from a cold state to a warmer one, known as terminations, delimit oxygen isotope stages which are true global time boundaries. This oxygen isotope stratigraphy underpins our understanding of the current (Quaternary) ice age, although many problems remain in correlating the ocean-based history with the record on land.

The global record shows a distinct evolution with time. At first the glacial and interglacial intervals were of roughly equal length, each pair lasting 40,000–50,000 years, and the difference between the warmest and coldest states was not large. Around a million years ago, however, a drastic change rang in the second half of the Quaternary. It is marked

69

by fewer but longer glacial phases, each 100,000 years or so long and separated by much shorter interglacials. The contrast in ice cover between glacials and interglacials indicated by the $^{18}O/^{16}O$ ratio rose to nearly twice what it had been before. The pattern also became much more complex, because after each interglacial the glacial increased step-wise in severity, reaching its maximum just before a new interglacial arrived.

It is fair to add that the oxygen isotope ration curve for the last 800,000 years (Figure 4.2) does feebly suggest four major glaciations and helps us to understand why a four-stage ice age was accepted for so long and was given up so reluctantly.

The oxygen isotopes mainly record changes in ice cover and, because the Antarctic icecap did not change much in the last two million years, what we see is primarily a history of the icecaps of the northern hemisphere. This history is supplemented by the record contained in the thick, windblown loess deposits of China which say that throughout the Quaternary the cold periods were windy and dusty and the plant cover thin, but, except for the last glacial, we know little about the locations and limits of the icecaps.

Our knowledge of the last interglacial–glacial–interglacial cycle, on the other hand, is fair, because no new ice advance has yet come to wipe out the traces of the old one. For a glimpse of the glacial world, we therefore turn to the last 125,000 years.

4.3   ICECAPS ON THE WORLD

The world at the height of the last glacial vividly illustrates what a major climate change is capable of. From 25,000 to 17,000 years ago, the cover of ice was extensive over much of North America and Europe, but patchy in Siberia (Figure 4.3). A dome of ice perhaps 3 km high rested on Canada, extended south to New York and the Ohio River valley and was bordered in the west by a smaller ice blanket on the northern Rocky Mountains and Cascades. Another icecap sat on Scandinavia from where it spread southward into Denmark, Germany, Poland and Russia, while smaller ones lay on the crests of the Pyrenees, Alps and Carpathians. The British Isles had their own icecap based on Scotland and Wales, but it may not have been connected to the North European one. South of the ice margin vast, treeless plains covered with a low, hardy tundra were traversed by innumerable meltwater streams.

In the southern hemisphere, on the other hand, the Antarctic icecap itself was not much larger than it is today, although its ice shelf extended much farther north. The icecaps of southern South America were greatly

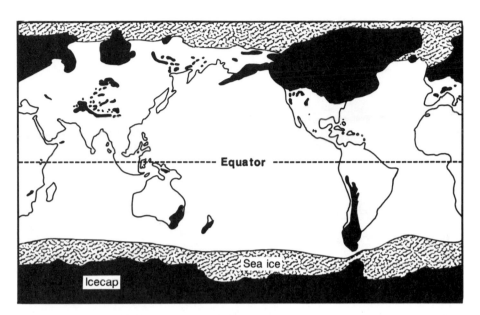

Figure 4.3. The world 18,000 years ago during the last glacial maximum. On the northern hemisphere icecaps covered much of the land, and the North Atlantic had become an extension of the Arctic Ocean. On the southern hemisphere, Antarctica was surrounded by a wide belt of sea ice, but on the southern continents the ice cover was limited. Glaciers crowned many subtropical and tropical mountains.

enlarged, as were the glaciers of the higher mountains of Australia, Africa, central South America, and even Hawaii. In all at the height of the interval some 40 million km³ of ice were piled on land, and one imagines polar conditions in the central United States and middle and western Europe that are today familiar only to Eskimos, Alaskan oil drillers and Russian coal miners.

Actually, conditions beyond the ice were not so severe. In North America, forests grew not far beyond its edge, the summer days were long and bright as they must be at that latitude, and the winters, although harsher than now, were not truly polar. The edge of the ice extended so very far south not because conditions were so arctic there but because a high mound of ice flows like molasses under its own weight (Figure 4.4). As long as the ice grew upward and moved faster than it melted, it continued on its way south, coming to a halt only where ice supply and summer thaw balanced. Under those conditions, a few slightly warmer summers at the ice margin could cause a quick retreat, whereas increased snowfall thousands of kilometers to the north would thicken the ice and, with some delay, push its edge farther south, sometimes even into the boreal forest.

A floating ice sheet, on the other hand, does not flow away from its

71

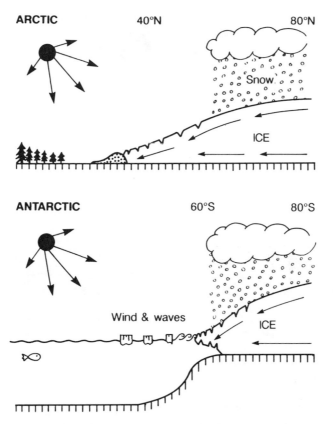

Figure 4.4. Because ice resting on a solid base flows under its own weight, the icecaps on the northern continents were able to advance far south into regions that even then did not have an arctic climate. In the southern hemisphere that was not possible, because the deep ocean surrounding Antarctica, where tides and waves erode the floating ice edge, presents an impassable barrier.

source; it merely sinks deeper when it thickens. That is why the Arctic Ocean, although frozen, was not the nucleus of the icecaps of the north. The Antarctic ice, on the other hand, rests mainly on land, although from its edges a floating shelf extends over an ocean too deep for the ice to touch bottom. In contact with the warmer sea, the shelf, attacked by waves and daily raised and lowered by the tide, is contained by melting and the breaking off of icebergs which then drift away. Even the largest, thickest icecap can be stopped by a narrow sea as long as it is more than a few kilometers deep. The Southern Ocean has always protected the southern continents from the Antarctic ice.

Some northern parts of Siberia, seemingly well situated to accommodate an icecap, did not have one and neither did northern Alaska. To form glaciers and icecaps, ample snowfall in winter is as necessary as a low temperature. Icecaps form where the snowfall in winter exceeds the thaw of summer. Storms reach most of Siberia only after passing across the width of Asia where they drop rain and snow on the mountains of the far south. The Arctic Ocean, ever-frozen, does not evaporate and furnishes almost no precipitation. In North America, winds from the Pacific brought then as now snow to southern Alaska and British Columbia, but were dry when they reached the northern slopes. The icecaps of eastern and central North America and northern Europe, on the other hand, were fed mainly by wet Atlantic storms.

4.4   GLACIAL OCEANS

Surprisingly, no one gave much thought to the nature and climatic role of the glacial oceans until recently, when oceanographers, climatologists and geologists banded together to find out how the two-thirds of the world that are under water fared during the Quaternary ice age. In the 1970s, under the flag of the CLIMAP Project, they clarified many issues and obtained many answers, although, as is often the case, other questions that seemed simple at the start have remained elusive or become more complex.

Deep-sea sediments consist mainly of the tiny calcareous and siliceous shells of planktonic organisms. These fossils record, and under the microscope reveal, the condition of the surface sea in which they lived and died. The information can be extracted if we assume, as is reasonable for a past no more distant than a few hundred thousand years, that most species have not changed their environmental habits very much. Because years of ocean exploration have put at our disposal many ocean sediment cores, we can obtain from the species assemblages estimates of salinity, temperature, and even nutrient levels in the surface waters of glacial oceans. If we use only species that flourish seasonally, we can map the temperature and salinity distributions of the sea surface for summers or winters now long gone. Dating the core samples is, of course, critical and often tricky (Section 5.2), but various methods are at our disposal.

Consider (Figure 4.5) the sea surface temperature in the North Atlantic for the summer of the year 18,000 BP (before present). The ocean was frozen from Greenland to Great Britain and full of pack ice as far south as a line from Cape Hatteras to Spain. The Gulf Stream, which today warms the shores from Great Britain to Spitsbergen, went directly across from Florida to the Azores and warmed only Africa. A week at a Carolina or Portuguese resort in water of 6–8 °C would hardly have qualified as a summer holiday.

Compared to this, changes in the North Pacific were trivial (Figure

73

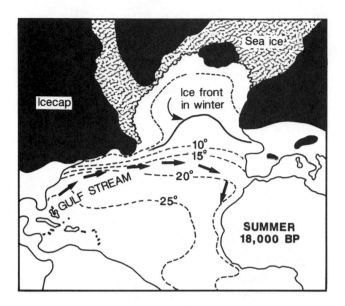

Figure 4.5. At 18,000 years ago, the North Atlantic was an arctic sea. The edge of permanent sea ice ran from Britain to Nova Scotia, with drifting pack ice farther to the south. Even in summer, the temperature between Spain and the Carolinas did not exceed a frigid 10 °C, while the Gulf Stream, today responsible for the mild climate of northwestern Europe, went straight from Florida to North Africa.

4.6) except near Japan, but the equatorial ocean was much cooler than it is today, and many a subtropical sea was rather like the nearshore waters of the present North Sea in August. At mid-latitudes, on the other hand, there was little change in temperature; Haiti and Indonesia, California and the Azores were much as they are now.

4.5 THE WORLD BEYOND THE ICE

What was the rest of the world like while ice covered much of the northern hemisphere? What were the conditions in glacial Greece, California, Brazil, or central Africa? The question is not easily answered, because when we compare evidence from different parts of the world, we must be sure that we look at the same instant in time. It simply will not do to assume that a glacial deposit in Maine was simultaneous with a swamp in Colombia if their ages are uncertain by several thousand years, because the fluctuations of the glacial climate were so numerous and so often brief.

Obviously, the large climatic swings of the high and middle latitudes could hardly have failed to have repercussions in the subtropics and tropics. Indeed, we have long known that near the equator the present

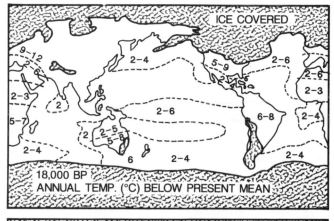

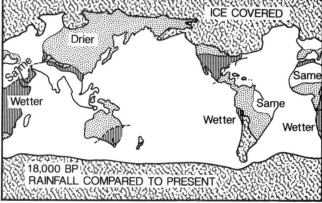

Figure 4.6. Not everywhere was the last glacial period a time of extreme cold. Over most of the world, except near the edge of the icecaps, temperatures were only a few degrees lower. Ocean surface waters were, as might be expected, lower in the extreme north and south but, surprisingly, also near the equator. The pattern of precipitation changes, showing for instance a much drier Amazon region, is complex and not fully understood. Note that icecaps and sea ice are represented by the same pattern.

climate was preceded by a different, much wetter one. Lake levels in East Africa and the Middle East once were well above those of today. Prehistoric rock-paintings in the Sahara bear witness to a flourishing population in a lush land, rich in game. Logic has suggested to many geologists that the wet periods, called pluvials, ought to correspond with the glacials of higher latitudes, but logic has told others that they should coincide with warm and hence wet interglacials. Unfortunately, hypotheses resting only on logic, no matter how compelling, often lead us astray, because logical arguments are no better than the premises they rest upon.

Neither concept turns out to be entirely true. Because the temperate, subtropical, and tropical climate zones were compressed toward the equator by the expanding cold zones, the northern Sahara benefitted from

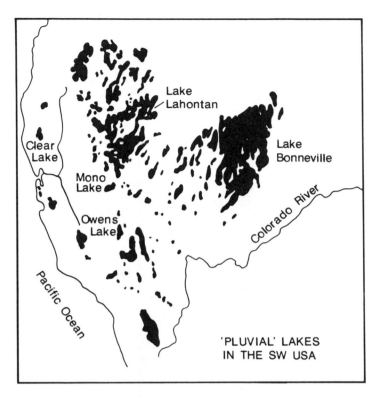

Figure 4.7. During the last glacial maximum, high air pressure over the North American icecap drove the storm tracks far south. This gave California, Utah and Nevada a much wetter climate than today. The many enclosed basins of the interior were occupied by lakes without outlets, of which the old shorelines can still be seen. A few lakes remain, such as Great Salt Lake, a remnant of glacial Lake Bonneville.

the winter rains that now fall in the Mediterranean. California also received more rain during glacial times, because the high pressure over the icecap to the north deflected Pacific winter storms southward. With a climate much like that of Oregon today, many lakes (Figure 4.7) fringed with pine and juniper and vast piñon woodland filled interior valleys, while the coastal zone was green with redwood forests. Today the lakes are mostly gone, but the groundwater stored during those wet millennia still supplies the needs of Los Angeles and permits for a while the profligate semi-desert farming that enriches the state due to a brief reliance on the rains of long, long ago.

Most of the world, however, was drier then (Figure 4.6), because a cooler ocean means reduced evaporation and low evaporation means a lower average global precipitation. On land it was cooler everywhere, but the effect was felt most strongly at higher latitudes. After all, a drop of 2–3 °C below a yearly average of 27 °C is felt less keenly than one of

5–10 °C below a present mean of only 15 °C, as was the case in central North America.

These temperatures derive from the study of the late glacial vegetation in many parts of the world. We possess a technique called pollen analysis or palynology which is nowadays very sophisticated, and a valuable means for the study of the plant cover of the past. Flowering plants produce pollen grains, often in great abundance, that are usually diagnostic for the species or at least the genus, and they disperse the pollen by wind or through helpful insects. Small as the grains are, always of microscopic size, they are surprisingly resistant against decay and are preserved especially well in the organic-rich sediments of lakes and bogs. Because plants respond sensitively to changes in environmental conditions, in particular climate, they provide us not only with a better image of the glacial and post-glacial landscape, but also with a climate record that in many instances can be correlated with the oxygen isotope record of the oceans.

Pollen tells us that on the northern continents, south of the tundra, grew a boreal forest of birch, pine and spruce as today in Canada (Figure 4.8). In the eastern United States the temperate broadleaf forest that would cover most of the land today if human beings had not cut it down had retreated south to favored climes in Florida and the Gulf Coast. In Europe, where the E–W-trending mountain ranges formed a glaciated barrier, the retreat had been to refuges in southern Europe, North Africa and the Near East, so distant that many species never returned.

A cooler climate and less precipitation also decreased the area covered by tropical rain-forests in South America, central Africa, India and southeast Asia. The Amazon region, for example, may have been as much as 4–6 °C cooler than today. This caused trees that now grow on mountain slopes to descend to the plains, while an open, steppe-like vegetation changed much of what is (or was) green jungle today to the seasonal brown of savannas. The complex communities of the tropical rain-forest, often thought to be the product of a very long evolutionary history, have been established in their present areas for at most six or seven thousand years, and throughout the Pleistocene tropical rain-forest existed in its vastness only during brief interglacials, receding into small refuges during glacials but always spreading again.

4.6 THE LEVEL OF LAND AND SEA

One after the other, glaciations came and went throughout the Pleistocene, each drawing enough water from the ocean to alter the shape of

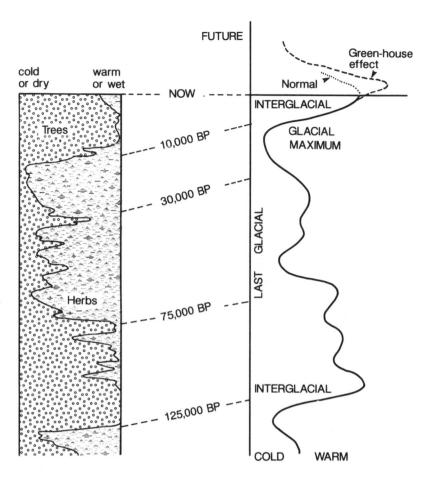

Figure 4.8. Pollen grains buried in peat or wet mud resist decay and can tell us much about the response of the vegetation to climate changes. Even something as simple as a graph of the ratio of tree to herbaceous pollen (left) clearly defines cold and warm (or wet and dry) spells. Combined with other data, pollen records can be converted into a climatic history of the last 100,000 years. A speculative extrapolation into the future by John Imbrie of Brown University suggests that the return of another glacial period will be delayed by the green-house effect of our excessive burning of coal and oil.

the continents. Each time as the ice melted, the sea rose, inundating the former coastal plains. Such changes in sea level, if they are simultaneous across the world, are called eustatic. If all ice remaining in Greenland and Antarctica were to melt today, the sea would rise about 70 m. The next to last glaciation, a large one, accumulated some 75 million km³ of ice, removing a layer of water more than 160 m thick from the oceans. During the last glacial maximum, 18,000 years ago, the level of the sea fell between 100 and 150 m. Why are these estimates so crude? Could we, should we, be more precise?

A glacial drop or rise of the sea can be estimated in many ways. We might, for instance, calculate the total volume of ice from the known area and thickness of the icecaps. The area can be determined from the terminal moraines, the walls of debris left at the outer fringes of the ice, but the thickness presents more of a problem. Mountain tops piercing a glacier are splintered by frost and look jagged, whereas the flanks, once covered by ice, are worn smooth. Unfortunately, in Scandinavia and Canada the ice covered all but a very few mountains and our thickness estimates lack precision. In Antarctica, the former extent and thickness of the ice are uncertain, because the edges of the icecap, extending as they do into the sea, left only diffuse traces on the seafloor, and few mountains rose above the ice.

One might look for shores of the right age, now submerged on the continental shelves, the shallow platforms that surround each continent. Shores often have a distinctive relief and deposits, and radiocarbon dates on buried wood or shell can tell us their ages. Given dates and present depths for many old, now-submerged shores, the maximum lowering of the sea and the history of its subsequent rise can surely be ascertained?

Unfortunately, when this obvious idea was pursued further, it produced a startlingly wide range of estimates and only slowly have we learned why this was so. As it turns out, the problem is the behavior of the earth's crust when a large load is placed on it or taken away. Because we do not possess a fixed tide gauge to measure sea level, we must define it relative to the adjacent land, and both land and seafloor refuse to maintain fixed levels.

Vertical movements of the earth's crust due to loading or unloading are called isostatic, in contrast to tectonic movements that accompany earthquakes and mountain building. Icecaps are heavy and we have long known that they depress the land which rises again when the ice melts. Yet curiously, very few people realized until recently that adding or subtracting ocean water has an analogous effect on the seafloor.

Continents and ocean floors, the components of the earth's crust, are rafts floating on a dense fluid, the mantle (Figure 4.9). Continents obey Archimedes's buoyancy law and, if weighted down with an icecap, sink until a mass of mantle material equal to the added weight has been displaced. Conversely, if the ice melts, the raft rises. The seafloor responds in the same way to the removal or addition of water. As a rule of thumb, a 100 m layer of ice will depress a continent by about 27 m, and every 100 m of water removed from the sea will cause the seafloor to rise 30 m. Isostatic compensation also applies to other kinds of loads: a new

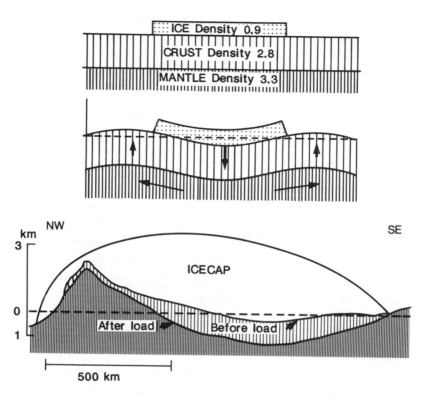

Figure 4.9. The earth's crust floats on a dense mantle which behaves as a very viscous liquid. If we place an icecap on land, the crust bends, displacing enough mantle material to equal the mass of the ice. The excess mantle material flows outward, raising a low, broad welt around the icecap, but because it is so viscous and flows so slowly, equilibrium in either direction will be much delayed compared with the growth or melting of the ice. The cross-section of the present Greenland icecap shows the deep depression, in part even below sea level, of the underlying bedrock.

volcano or a large delta depress the crust, and the trimming of a mountain by erosion causes uplift.

There are, as always, complications. Because the crust has considerable strength, it bends like a beam and subsides for some distance beyond the edges of the load (Figure 4.9). As it sinks, displaced mantle material flows away to either side and raises a low welt around the depression. When the load is removed, the mantle flows back, but because it is viscous, the compensation, which depends on its flow rate, is a slow process. As a result, subsidence or uplift continues for thousands of years after a load was emplaced or removed. Although their icecaps vanished some 8,000 years ago, Canada and Scandinavia have thus far risen only a little more than half of the ultimate 500–800 m of compensation. Well before the once ice-covered regions have returned to equilibrium, thousands of years from now, Hudson Bay will be dry land and Stockholm raised above the sea.

It is evident that if we attempt to measure the rise of the sea near formerly glaciated regions, we must be prepared for messy corrections, but that is not all. Oxygen isotope ratios of marine fossils show that the amount of water removed from the sea and stored as ice during the last glacial maximum was equivalent to a layer 165 m thick, considerably more than even the deepest submerged shores would suggest. A considerable weight was thus removed from the ocean floor, and we must make a correction for the isostatic effect. If we do that, we find that the seafloor rose some 50 m in compensation, and the net drop in sea level was only 115 m, a value comfortably within range of other estimates. A further complication is the attraction exercised by a high land mass on the sea. The combined result of all these forces is that the rise or fall of the sea can be quite different from place to place.

Let us accept then that the late glacial sea was about 120 m lower than the present one. The continents are surrounded by areas of shallow water, the continental shelves. The shelves vary in width but are today generally less than 100–150 m deep; at that point the bottom begins to slope steeply down to the floor of the ocean. With the sea more than 100 m below its present level, the coastal geography of the world differed greatly from that of today, except where the continental shelf is narrow, as for example on the west side of the Americas or in the Mediterranean. In the Gulf of Mexico, a coastal plain more than 200 km wide emerged, the North Sea was dry, and the Indonesian archipelago was an extension of southeast Asia (Figure 4.10), large enough to cause its equatorial climate to become much drier.

When the last glacial interval came to an end about 17,000 years ago and the ice began to melt, the sea rose, slowly at first (Figure 4.11), but then accelerating until the first major melting phase ended around 12,000 years ago with a flourish that raised sea level 24 m in less than a thousand years. A rapid return to a colder climate during the so-called Younger Dryas, about 10,500 years ago, briefly slowed the rise, but it resumed less than a millennium later with another swift spurt that brought the shores of the world close to their present positions (Figure 4.10 C: North Sea).

This signaled the demise of the northern icecaps except for Greenland, and at 8,000 BP, with the sea still 25 m below its present level, the rise slowed drastically but did not stop, not even today (Section 3.6). It is not clear where that water is coming from; perhaps the West Antarctic icecap is unstable and continues to melt. If so, another 7–10 m rise may be in store.

To assess that distressing possibility, we need more precise knowledge of the most recent part of the post-glacial sea level rise. Unfortunately,

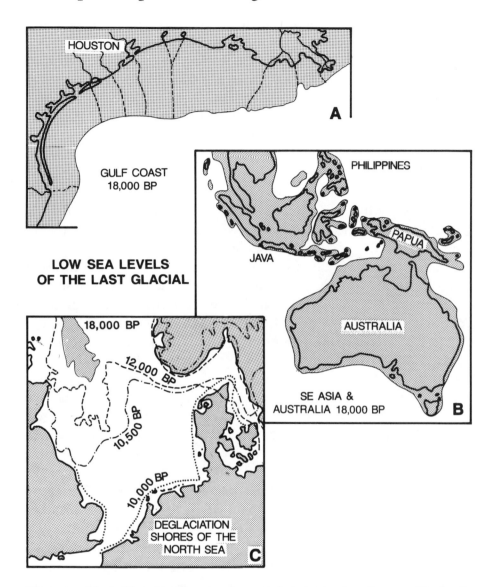

Figure 4.10. The buildup of icecaps removes a great deal of water from the ocean. During the last glacial maximum this caused sea level to fall by about 120 m, exposing the continental shelves and in many places greatly altering the geography of the world. The subsequent rise took place in several stages as the map of the deglaciation of the North Sea shows. Some of these steps were rapid; note the large area lost between 10,500 and 10,000 years BP.

the data for the last several millennia are conflicting and opinions on the subject are sharply divided. Some believe that the present level was reached several thousand years ago, others that the sea is still rising, while a third group thinks that it has for some time been oscillating by one or two meters around the present mean. The problem is that, when we deal with a few meters' difference in level and a few centuries in time, we are

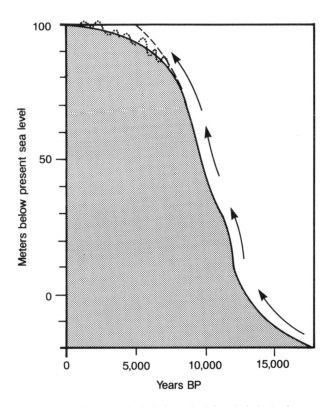

Figure 4.11. At the end of the last glacial period the ice began to melt and the sea rose again. The best documented history of this rise (solid line) rests on fossil corals off Barbados Island studied by R. G. Fairbanks. Controversy, especially regarding the last 5,000 years, has not subsided fully, and by way of example the diagram shows three popular curves for the late rise.

pushing the limits of the data. More research is in order, research that would have very real value in devising ways to protect low-lying cities and lands, but also research that, because the rise or fall of the sea is such a complex process, would intimidate the most intrepid investigator.

The rate at which the shoreline moved landward during the two main melting phases was far from trivial. Between 16,000 and 9,000 years ago, the shores of the Gulf of Mexico proceeded landward at some 3 m/year, too rapid for coastal deposits of any consequence to form. Even the Mississippi River, large as it is, was unable to build itself a delta; it merely retreated within its own valley ahead of the rising sea. Only when the rise slowed to almost today's value of about 20 cm/century, did the landward march of the shore decrease enough to allow the construction of beaches, barrier islands, marshes and lagoons that grew upward as fast as the sea rose. The present position of the shore is often close to where it was 8,000 years ago, even though sea level was then some 15–20 m lower. As time passes, the lagoons and estuaries trapped behind sand barriers by the rise of the sea will fill with sediment and vanish, and the few beautiful coastal wetlands that have escaped conversion to condominium villages and yacht harbors will be converted into fertile but dull coastal plains, fronted by straight beaches.

4.7 ONSET AND DECLINE OF A GLACIAL PERIOD

Anyone examining in detail the last interglacial and glacial intervals is caught immediately in a deep bog of stratigraphic terminology designed to reflect events in local terms. The last glacial, for example, is called Wisconsin in the USA and Canada (but with different starting dates), Würm in the Alps, Weichsel in northern Europe, Devensian in England, Valdai in Russia, and Zyryanka in Siberia, and that is just the beginning. In general discussions we avoid this by using the oceanic oxygen isotope stages (Figure 4.12), but the practice has not fully sunk in, partly because local specialists tend to be conservative and attached to their own stratigraphies, and partly because we cannot yet correlate the many small ice advances and retreats on land with the marine record. Here I shall use isotope stages; the odd numbers stand for warm times (interglacials or interstadials) and even numbers for the cold ones (glacials or stadials).

Most people believe that the last glacial maximum is typical of the whole interval and even of some 80 percent of the last two million years. This is not so; most of the time conditions were a good deal less severe, a fact important to remember, especially for archaeologists who are fond of setting early human beings in an "ice-bound" Europe of blizzards, tundras and mammoths that fails to reflect the reality of the environment in which modern human beings evolved.

The oxygen isotope record (Figure 4.12) shows that the warm last interglacial arrived about 125,000 years ago, followed at 110,000 BP by two cooler intervals separated by slight warming. The first major glaciation of isotope stage 4 arrived some 75,000 years ago. At this time forests vanished from middle latitudes to be replaced mostly by tundra and steppe (Figure 4.8). Milder conditions returned during stage 3, followed around 25,000 years ago by the true glacial maximum of stage 2 that peaked at 18,000 BP. Only stages 4 and 2 qualify as truly severe.

At the height of the last interglacial, the sea stood 5–7 m higher than today, bearing witness to a somewhat warmer climate, perhaps roughly equivalent to what green-house warming will eventually present us with. It then fell, fluctuated around −40 m until the start of stage 4 some 75,000 years ago, when it dropped briefly to −75 m. Once again it rose

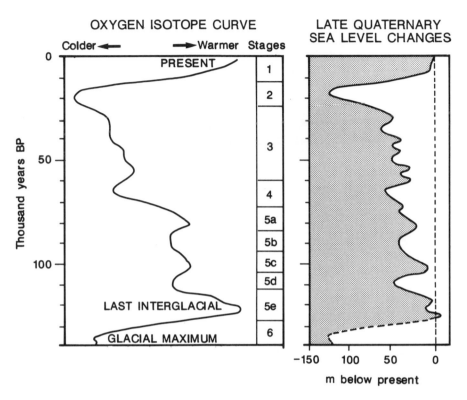

OXYGEN ISOTOPE CURVE — Colder ← → Warmer — Stages

LATE QUATERNARY SEA LEVEL CHANGES

PRESENT — 1

2

3

4

5a

5b

5c

5d

LAST INTERGLACIAL — 5e

GLACIAL MAXIMUM — 6

Thousand years BP — 0, 50, 100

−150 100 50 0

m below present

Figure 4.12. Oxygen isotopes and sea level changes for the last 140,000 years, from late in the previous glacial period to the present. The isotope curve is based on oceanic sediment core data, while the sea level curve rests on a combination of raised coral reef terraces in New Guinea and sea level inferences drawn from oxygen isotope data. The central bar presents the oxygen isotope stages now widely used to subdivide the Pleistocene.

and oscillated around −50 m until the final maximum. Therefore, most of the time, the exposed continental shelf was much smaller than during the maximum, and the ice volume ranged from less than half to two-thirds of its peak value.

Where was the ice margin during the 80,000 years that the icecaps were building to their ultimate glory? This is hard to say, because the final advance destroyed or rendered unreadable the deposits of earlier positions, and dating ice margins older than 40,000 years, the practical limit of radiocarbon methods, requires much luck. Still, a plausible hypothesis for the icecaps of isotope stage 3 (Figure 4.13) shows how different the "average" last glacial was from its better known final phase.

A mere thousand years after reaching its peak the ice in North America began to melt and the ice fronts withdrew at speeds ranging from a few hundred to, in some places, more than 1,000 m/year. The eastern ice sheet withdrew from Ohio to Hudson Bay, a distance of 2,000 km, in

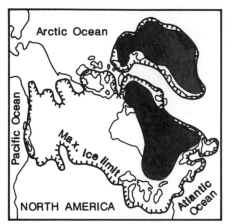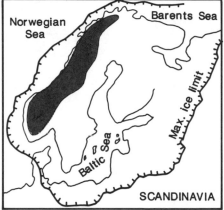

Figure 4.13. Many people, especially archaeologists, regard the glacial maximum around 18,000 years ago as typical for the entire last glacial period. This is not so! The Scandinavian ice sheet, for example, occupied for much of the time only the central Scandinavian highlands (shaded). The USA, except for its northeastern part and extensive mountain glaciers, was more ice-free than generally believed.

less than 6,000 years and was gone 7,000 years ago. In the west, the Cordilleran ice sheet left Puget Sound 14,000 years ago and disappeared 4,000 years later. In Europe the ice cleared Britain, Denmark and northern Germany before 11,000 BP. Retreating into the mountains of Scandinavia, it vanished 8,000 years ago.

Compared to the great length of the entire glacial, its demise in much less than 10,000 years was strikingly abrupt. The development of the first truly cold stadial 75,000 years ago was also sudden; the ice may have thickened at a rate of 20–60 cm/yr and locally advanced as much as one kilometer per year.

How sharp major climate changes can be, is illustrated by a brief event about 11,000 years ago. At that time the ice had been long gone from northwestern Europe, and birch and pine woodland had replaced the ice and the mossy, shrubby tundra beyond its fringe. But around 10,500 BP the sharp cold spell of the Younger Dryas wiped out the woodland, brought back the tundra from southern Sweden to Britain and the Low Countries, restored the herds of reindeer and their Paleolithic hunters to their old domain, and created a new icecap on Scotland, all within a century. This dramatic event lasted only 700 years and ended as abruptly as it had begun. Recent data from Greenland ice cores bring it in focus; oxygen isotopes and dust concentrations show that the improvement of the climate occurred in less than two decades, and that south Greenland warmed about 7 °C in half a century. Even more striking is the case of the last interglacial: these same Greenland ice cores clearly illustrate sharp, almost instantaneous changes from warm to almost glacially cold. The cores destroy the hope that climate changes are slow on a human time scale and that therefore we need not fear the green-house effect too much, but encourage those who believe that much can be usefully learned from studying the past.

5

Explaining glaciations

Why should the earth's climate have oscillated for so long between very cold and reasonably mild, between long glacials and brief interglacials? Why should each change from one extreme to the other have been so sudden? Why, during each glacial interval, did less cold interstadials alternate with cold stadials, and why was the warm interglacial climate interrupted by little ice ages? How do we explain this flip-flop response of the climate system and what are the forces that make it behave so? We shall see in Chapter 11 that the Quaternary ice age came to us on soft feet, gently announcing itself some 30 million years ago, then advancing in distinct steps of which the last one, less than a million years ago, is the most striking feature of Figure 4.2.

The normal state of the world appears to be one of ice-free poles and a small equator-to-pole temperature gradient, but ours is not the only ice age the world has seen. The earliest major glacial deposits turn up between 3 and 2 by ago, but we know little about them and they may have been local. The ice ages of the late Precambrian and Paleozoic, however, are not in doubt; the wide distribution of their characteristic deposits permits no other conclusion. The late Precambrian ice age seems to have been one of the largest (Section 14.7), and descended to curiously low latitudes, while the Paleozoic ones left extensive marks on the continent of Gondwana around the South Pole of that time (Section 7.3). Each was in some way different from all others, but all had large polar icecaps which advanced far into normally temperate lands. Why ice ages happen is an unsolved mystery (Section 12.4), but we know enough about at least the last one to speculate why it displays such a fascinating rhythm of cold and warm phases.

5.1 CAUSES OF GLACIAL AND INTERGLACIAL CLIMATES

Before we consider some of the current explanations, it is well to take note of just what must be explained. We need a cause that provides oscillations between harsh and mild conditions. We must allow for the possibility that these oscillations return with predictable regularity. Because variations of short duration are superimposed on the rhythm of interglacials and glacials, we must seek causes that operate on several time scales, from centuries to hundreds of thousands of years. And because of the evidence for a sudden onset and decline of many events, any explanation should include sharp transitions from one state to another. Each model, whatever force it invokes, must deal with the atmosphere and the ocean and with interactions between the two.

The hypotheses are legion, but they divide into just three categories. Some seek the cause in variations of the sun's output. Others look to heat filters interposed between donor sun and receptor earth. Cosmic dust clouds, volcanic ash and a carbon dioxide blanket are examples of such filters. A third class rests on variations in the orbit of the earth around the sun.

Many propositions are difficult to test, dealing as they do with past conditions not likely to leave an unambiguous signal in the rock record. Variations in solar output or passages of the earth through a cosmic dust cloud are detectable only through their climatic consequences, leading us on a merry path of circular reasoning. Such ideas are not fruitful.

Other hypotheses rest on firmer ground. Volcanic eruptions, proposed long ago as a cause for ice ages, intermittently return to center stage. The effect on the global temperature of a large plume of ash and aerosols blown to high altitude is real enough as the 1992 eruption of Pinatubo in the Philippines showed, but it demands that ash and gas are ejected very high, and even then the effect lasts only a few years. Do we expect clusters of giant eruptions lasting millennia to have occurred at the start of each glacial interval?

The geological record is inconclusive. Data from boreholes in the ocean floor have convinced a few that cold intervals in the Middle Miocene, Pliocene and Pleistocene came after times of intense volcanism, but the evidence is fragile. Besides, if one wants it badly enough, a correlation between two sets of events that are frequent and poorly fixed in time can usually be made. It is also true that coincidence does not mean that one brought about the other; both may be due to a third cause. And there is no evidence that these eruptions were of exceptional violence, nor that they had the striking periodicity of the Pleistocene record. It is perhaps no surprise that blithe spirits once suggested that we have it the wrong way around; isostatic compensation for ice growth and decay fractures the earth's crust and causes volcanoes to erupt. For now it is probably best to reserve judgment on any correspondence between fire and ice.

Might the heat-preserving blanket of carbon dioxide and other greenhouse gases thin out during glacials and be restored in interglacials. Ten years ago, when writing the first edition of this book, I discarded that notion for lack of evidence and because, even if atmospheric CO_2 were to vary much with time, the process seemed hard to reconcile with the rhythm of glacials and interglacials. Now better data allow us to return to the role of atmospheric gases later in this chapter, but we shall still find them wanting.

On an entirely different tack, Maurice Ewing and W. F. Donn proposed many years ago that glaciations and deglaciations are due to a basic instability of the arctic climate. Begin with an ice-free Arctic Ocean: seawater evaporates, snow falls on the adjacent land, and an icecap forms. The higher albedo causes the temperature to drop and the Arctic Ocean freezes, cutting off the supply of moisture. Deprived of snow, the icecap shrinks, the temperature rises, and the Arctic Ocean once more becomes free of ice. The cycle might have started when the Arctic basin drifted into a cold polar position. Today, this hypothesis seems less plausible, because the Arctic Ocean did not freeze until the Pliocene, long after it had become centered on the pole, and its ice cover has been permanent for more than two million years.

5.2   REVIVAL OF AN OLD IDEA

None of the explanations considered above suits all of the requirements imposed by the ice age conditions, failing above all to account for the frequent alternation between cold and warm conditions. That aspect is addressed by a different explanation that links the apparently periodic climate variations to periodic changes in the way the earth circles the sun. First conceived by the Scot James Croll in 1867, soon after the existence of an ice age had been accepted, the idea was given its final form nearly 70 years later by the Yugoslavian mathematician M. Milankovitch. It was long disregarded, because it could not be matched to the traditional scheme of four glacials separated by interglacials, but was resurrected in 1976 in a now famous paper in the journal *Science*, "Var-

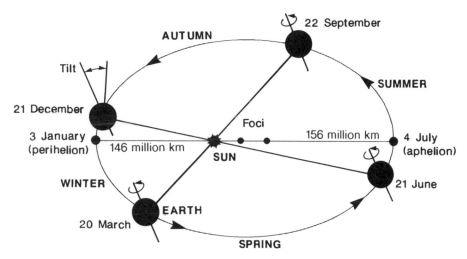

Figure 5.1. The orbit of the earth is an ellipse but the sun is not one of the focal points. As a result the earth is farther from the sun (at aphelion) at one end of the long axis and closer (at perihelion) at the other. The length of the long axis of the ellipse varies over time, but at present the earth is closest to the sun in December. Thererefore the northern winter is somewhat warmer than it would be if the winter solstice came at the opposite end of the orbit.

iations in the earth's orbit: Pacemaker of the ice ages," by James Hays, John Imbrie and Nicholas Shackleton. The astronomical theory argues as follows (Figure 5.1).

The earth's orbit is an ellipse with a small eccentricity that varies by 6 percent over a period of 100,000 years (Figure 5.1). The distance between sun and earth varies accordingly. At present the earth is closest to the sun (at perihelion) on January 3, thus reducing the wintry cold, and farthest away (at aphelion) on July 4, thereby cooling the summer. At the other end of the cycle, summer warmth and winter cold are enhanced, but the effect is slight. Furthermore, the axis of the earth is tilted, currently at 23.4°, with respect to the orbital plane (the ecliptic). Because of the tilt (obliquity), the northern hemisphere receives more heat during the one half of the year when it faces the sun, the southern hemisphere for the other half. The obliquity varies from 21.8° to 24.4° over a period of 41,000 years, and with it vary the length of day at high latitudes and the heat received there.

The earth's axis is not just tilted, but like a top it wobbles with a period of about 22,000 years. The wobble shifts the dates of the winter and summer solstices and the equinoxes clockwise around the orbit. This is called precession (Figure 5.2). Nowadays, the winter begins just before the earth is closest to the sun, but 11,000 years ago it was as far away as possible on that date.

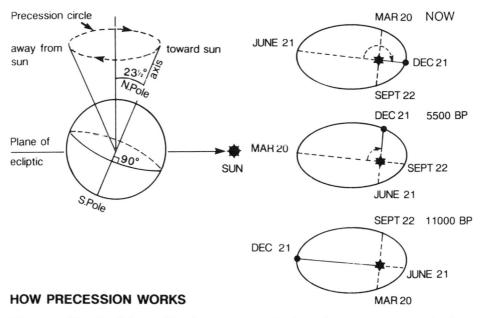

HOW PRECESSION WORKS

Figure 5.2. The tilt of the earth's axis, at present 23.4°, also varies over 41,000 years, thereby varying the degree of seasonality from smaller when the axis is closer to vertical to larger when it is tilted more. In addition the axis wobbles like a top over a period of about 22,000 years. This causes the dates of the equinoxes to travel along the orbit. At present the winter solstice happens when the earth is closest to the sun, but 11,000 years ago mid-summer came at that point.

Only the eccentricity has influence on the total amount of heat the earth receives and it is a small one, only 0.3 percent. That seems too small to take it seriously, but to it we must add the changing heat distribution with latitudes and seasons that is caused by the obliquity and the precession. At high latitude, for example, temperature changes in winter have little impact because it is anyway cold enough for snow to accumulate. The summer temperature, on the other hand, counts, because in cool summers less snow melts, so setting the stage for more accumulation in the following winter. The earth is thus in a position favorable to the forming of ice when the sun is distant in summer and the tilt is small (Figure 5.3).

Summed together, the periods of the eccentricity, obliquity and precession form a complicated curve from which the variations over time of the amount of solar heat received at any latitude in any month or season can be computed (Figure 5.4). The curve is sound, but the changes are small and seem hardly adequate to have major climatic consequences.

To test Milankovitch's proposal we must show that the periodicities of his curve match variations of temperature or ice volume indicators in

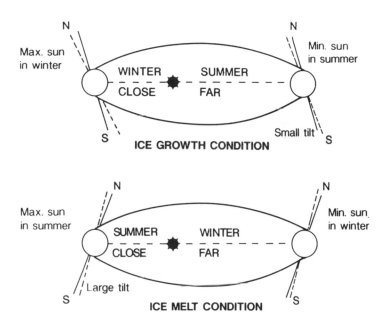

Figure 5.3. The change with time of the orbital behavior of the earth has an impact on climate that is illustrated here for two extreme conditions. The first is a state conducive to the growth of icecaps, the second to their melting.

the rock record. The oceanic oxygen isotope data (Figure 4.2) are an obvious candidate. The test, carried out by Imbrie and collaborators in the paper cited above, showed that the oxygen isotope ratio varies over time as the sum of three curves with periods of 100,000, 41,000, and 23,000–19,000 years. So close a match with the Milankovitch curve cannot be dismissed as mere coincidence, and control by orbital forces of the rhythm of glacial and interglacial stages has been widely accepted.

The confirmation has yielded a valuable by-product. The dating of oxygen isotope records is rendered uncertain by a small number of radioactive dates and their uncertainties. Normally, rates of sedimentation are used to interpolate between dated points, but the rates vary and small hiatuses are common. A distorted chronology is the result. But the variation with time of the orbital geometry can be calculated without such flaws, and provides us with a clock. The oxygen isotope record reflects the ticking of this clock and can be calibrated with its aid. The details of the technique, which we owe again to John Imbrie of Brown University and his co-workers, are beyond us here, but the technique is comparable to tuning a violin. The years appearing alongside Figures 4.2 and 4.12 have been obtained in this way.

Although the astronomic theory is now generally regarded as sound, it has its own oddities. Why should the 100,000-year eccentricity cycle,

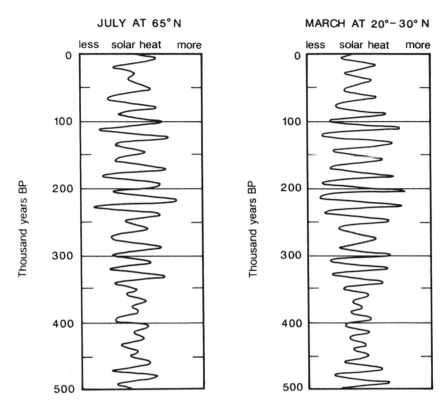

Figure 5.4. Because of the orderly change in eccentricity, obliquity and precession, the variation over the last 500,000 years of the amount of solar heat received at various times and latitudes can be calculated. It is shown here for July at 65° and for March at 20°–30° N latitude. The similarity with the oxygen isotope curve (Figure 4.2) is obvious.

the smallest of the three effects, dominate the climate record of the last 800,000 years, while obliquity and precessing cycles controlled it before (Figure 4.2; Section 12.3)? Equally puzzling is that, while orbital variations extend indefinitely into the past (Section 12.3), ice ages and their interglacials and glacials appear only occasionally.

5.3 BEYOND MILANKOVITCH

The orbital hypothesis relies only on the amount, time and place of solar heating as they relate to the change of seasons, and it clearly fits the chronological evidence. But seasonality alone seems too simple a mechanism to force climate to shift so rapidly from one state to its opposite, and yet that is what we see. To achieve these abrupt transitions may require a self-reinforcing process, a positive feedback. T. C. Chamberlin, noted for facing down Lord Kelvin regarding the age of the earth (Section 2.1), recognized already around the turn of the century the role of positive feedback in the glacial/interglacial cycle, but with this insight, as with many others, he was too early.

Let us try the following reasoning. Start when the sun is distant in summer and the tilt is small. On land not all snow that fell during the winter melts because the summer is too cool. The area of white ground is enlarged and the albedo raised. More heat is reflected, even less snow melts next summer, and the albedo increases again. During the 1960s, a cooling trend reduced the annual temperature of the northern hemisphere by a mere 0.5 °C, but the snow cover increased disproportionately; after the miserable summer of 1971 it rose a full 20 percent. More snowfall would do it too, but several cool, overcast summers have a larger effect than the same number of very snowy winters.

Assuming the climatic trend does not reverse, the thickening blanket of snow could become an icecap. Every 22,000 years the precession cycle would cause a setback as summers became warmer and some of the ice melted, but each time the next round would start from a higher level than before, a larger white area with a larger positive feedback. The cumulative effect would repeat for four precessional cycles until, after 100,000 years, the maximum tilt coincided with summer perihelion, and the ice melted.

If we carry out this thought experiment quantitatively, we find that it is too simple. It helps explain the curious prominence of the 100,000-year eccentricity cycle which by itself has such a small primary effect, but other components of the climatic system need to be taken into account, the albedo, the moisture supply, green-house gases, or planetary and monsoonal wind systems. And above all the ocean, that gigantic reservoir and carrier of heat, because it operates so much more slowly than the atmosphere, on a scale of months or years, rather than hours and days. A source of moist air in a nearby, warmer ocean, for example, would undo our tidy result.

An instructive example is the North Atlantic Ocean which, when the last glacial began, remained as warm as it is today for some time. As a warm ocean in an already cold world, it furnished ample moisture to storms that took it north and dropped it as snow. In this manner, the time lag as the oceans adjust to global cooling may have been a major factor in the rapid growth of the icecaps. But when the ocean finally cooled, its role as a snow machine diminished as the low level of evaporation cut back the supply of moisture, perhaps even to the point that the ice cover began to shrink.

As we consider the ocean–atmosphere system, we should also think again about the fundamental instability of climate invoked almost half a century ago by Ewing and Donn (Section 5.1). Complex systems like climate need not always respond to stimuli in proportion to their strength. Some may react so slowly to the pressure for change that the conditions have changed again before the adjustment is complete. Others, more stubborn, may not respond at all until the push exceeds a certain threshold value; then they flip fast and far. The sudden decline so typical of glacial periods would make sense in this light.

There is growing evidence that the influence of changes in the earth's orbit is not limited to interglacials and glacials, and that the monsoon, controlled year after year by the seasonal cycle of the sun, is also affected by the geometry of the earth's orbit. Between 12,000 and 5,000 years ago the earth was tilted about one degree more than it is now and was at perihelion in mid-summer rather than in December (Figure 5.2). Consequently, more heat was received in summer and over a broader band of latitudes than today, and less in the winter. The seasonal difference in temperatures increased by 7 percent, enough, as computer models showed, to produce a much stronger monsoon.

Computer modelling is a useful tool and the results are impressively quantitative, but they are no better than the data that are entered and the premises that underlie the model. At the present time, both often leave something to be desired, and the results should always be verified. In this case, that was easy, because it had long been known that the levels of many African and southwest Asian lakes were much higher between 12,000 and 5,000 years ago than before or after, pointing to higher rainfall and/or less evaporation (Figure 5.5). Pollen data from as far away as Siberia indicate a northward shift of warmth-loving vegetation and there was a sharp reduction of Saharan dust-storms, all changes that one would expect from an intensified monsoon. The great Harappan civilization of the Indus Valley, which flourished 4,500 years ago in what is now a desert, was able to take advantage of the tail end of this great monsoon period, but the change to the present climate may have contributed to its demise.

While this shows that there is more to orbital forcing than glacials and interglacials, other reductions in subtropical lake levels 10,000 and 7,500 years ago remind us that atmosphere and ocean respond to Milankovitch forcing in complex ways that are as yet far from well understood.

5.4   ARE THE OCEANS INVOLVED?

Orbital forcing works mainly by altering the seasonal temperature contrasts that affect the extent and volume of ice sheets, but it does not

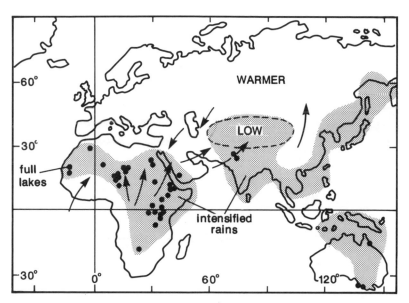

Figure 5.5. Computer models based on the Milankovitch theory predict a deep low-pressure area over central Asia in summer for the interval from 12,000 to 5,000 years ago. This strengthened the monsoon winds (arrows) and intensified and extended the monsoon rainfall (shaded area; compare with Figure 3.6). Ancient lake levels confirm this: lakes now standing low were filled with water (black dots) and areas now dry offered a decent living to prehistoric farmers and pastoralists.

explain two other important features of glacials: that they arrive and depart precipitously, and do so in the northern and southern hemispheres simultaneously.

The oceans store much heat that they only yield after a long delay, a property that explains, among other things, the contrast between mild maritime and harsh continental climates. They act as the "flywheel of the climate system," a nice phrase coined by Francis Bretherton. Consequently, the northern oceans, and especially the North Atlantic with its warm Gulf Stream bringing moisture to high latitudes (Figure 4.5), deserve our attention.

The study of sediment cores from the deep North Atlantic and ice cores from Greenland does not confirm the gradual transition from glacial to interglacial and back again that is implied by the astronomical theory and partly documented by the oceanic oxygen isotope record. Instead, as on land, the response of the North Atlantic ocean–atmosphere system increasingly looks like a series of abrupt flip-flops from one state to the other.

The end of the last glacial was marked by several such sharp climatic oscillations. The most striking is the sudden return to near-glacial con-

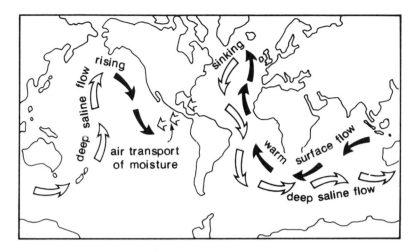

Figure 5.6. In the North Atlantic, the salinity of the surface waters increases slowly over time because, as the warm water evaporates, the winds carry the atmospheric moisture to the Pacific. When the salinity has risen so high that the surface water is dense enough to sink, the conveyor belt turns on and exports the saline water to the Pacific where it returns to the surface. In the North Atlantic it is replaced by warmer, less saline water coming from the Indian Ocean. After some time, the salinity is again low enough and the conveyor turns itself off.

ditions in the North Atlantic 11,000 years ago in the Younger Dryas (Section 4.7). It happened at a time when the main flow of meltwater from the North American icecap to the Gulf of Mexico came to a virtual halt. The obvious inference is that the meltwater found a new outlet through the St. Lawrence Seaway, but if that is true, the inflow of cold freshwater into the North Atlantic might have brought on the Younger Dryas cold.

That looked like a good explanation until a new study of the sea level rise resulting from the melting of the icecaps proved something that had been suspected before. The Younger Dryas cold event was accompanied by a major slowing of the rise of the sea, surely a result of a reduction of the production of meltwater.

At this point Wallace Broecker of Columbia University came up with an elegant idea. The global ocean circulation from the Atlantic to the Pacific, which takes place partly at the surface and partly in mid-depth, is driven by salinity oscillations (Figure 5.6). At the present time, warm and therefore light surface water moves north in the Atlantic until near Iceland it meets the cold, dry, polar air. It cools and evaporates, so increasing its salinity. The now heavier water mass sinks, flows southward through the Atlantic at mid-depth, crosses the Indian Ocean and enters the Pacific. In the northeastern part of that ocean it has become warm enough to rise and continues south as a cool surface current, gradually becoming less saline by rain.

Broecker proposed that the "salt conveyor" has two modes, on and off. In the off mode, the salinity in the North Atlantic rises slowly, because the wind carries Atlantic water vapor to the west coast of the Americas where it falls as rain and drains into the Pacific. When the salinity reaches a level that allows the water to sink, the conveyor turns on. At the surface, heat and salt are transported northward in the Atlantic, and salt is exported in deep water to the other oceans. Eventually, so much salt will have been subtracted that the density of North Atlantic water becomes too low and the system turns itself off. A pulse of meltwater would help here. With the conveyor on, the heat transported north causes ice to melt, with the conveyor off a net ice growth is possible. The late Quaternary record on land and at sea suggests that this process oscillates with a period of roughly 2,000 years, reminding us of little ice ages which return, also roughly, every 2,500 years (Section 3.2).

Whether this is the right idea remains to be seen. Data on such a fine time scale are still sparse, the mechanism surely is also influenced by localized events, the quantitative aspects are uncertain, and the impact of retreating or advancing ice sheets on the atmospheric circulation is not taken into account. But it does offer us a new way in which the ocean and the atmosphere may modify orbital forcing and, as a bonus, illustrates the stumbling course of progress in science.

5.5   GLACIAL AND INTERGLACIAL GREEN-HOUSE GASES

In Chapter 3 much was made of the human-induced climate change that will bring us the green-house of the 21st century. Is this green-house a human invention or has the past also seen it come and, ones hopes, go? If so, one expects green-house gases to be abundant in the interglacial atmosphere and low in glacial times. Could variations in the amount of green-house gases account for part or all of the climatic change we see at each transition from warm to cold and cold to warm?

Useful information regarding atmospheric carbon dioxide is contained in ocean sediment cores (Section 10.5), and over a decade ago Nicholas Shackleton of the University of Cambridge observed that the CO_2 content of the atmosphere was high in the last interglacial, dropped to low in the glacial period, and rose again in post-glacial time.

Atmospheric gases can be measured in cores of Greenland and Antarctic ice. The cores have yielded a detailed record of carbon dioxide,

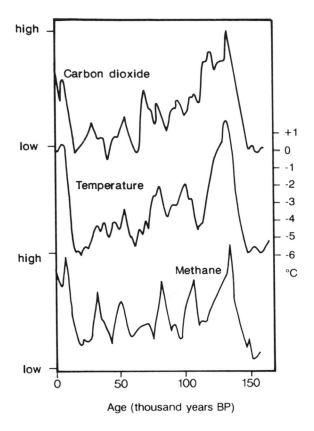

Figure 5.7. The percentages of carbon dioxide and methane in the atmosphere above Antarctica have been measured in air bubbles enclosed in a long Antarctic ice core taken by the Russians. Over the past 150,000 years, the variations have closely paralleled the temperature record.

methane (another green-house gas), and temperature back 160,000 years into the end of the penultimate glacial period. And indeed, the interglacial CO_2 values were only a little below the present level, then dropped to half that concentration during the glacial interval (Figure 5.7). The gas and temperature curves are strikingly parallel and anyone equipped with the appropriate mathematics can verify that green-house gases and Milankovitch cycles are well correlated.

Does this show that the green-house effect was a major force of glacial–interglacial climate changes? The answer, alas, is not so unambiguous. The solubility of CO_2 in the ocean depends on the ocean temperature. If the atmospheric CO_2 content changed before the temperature did, a green-house effect may be inferred, but if the temperature change came first, one suspects uptake or release of dissolved CO_2 by the oceans. In other words, is the change in atmospheric CO_2 concentration cause or effect? Alas, the chronology is not quite good enough for us to be certain of the answer here.

Moreover, plants in the ocean and on land use CO_2 to make organic matter and they are sufficiently abundant to have a major impact on the partition of CO_2 between ocean and atmosphere. We do not know enough about this phenomenon and its interaction with the circulation of ocean and atmosphere to evaluate it. And so we are caught once more in this intricate net of interrelations where, it seems, no problem can be solved first or by itself, but where all issues must be addressed simultaneously on all fronts.

The opposite effect of cooling by atmospheric aerosols and dust should not be forgotten. At present they keep us cooler by about 2 °C than we would be if the air were pure. Ice cores again tell the story and it is what one would expect; glacial periods, due to thin vegetation and much wind, are dusty, and interglacials are not. Another piece of the jigsaw puzzle, but for the time being we must leave it unfinished.

PERSPECTIVE

Climate is a variable condition, and even small changes, a few degrees cooler, a slight increase in rainfall, may alter the balance of erosion and deposition, modify the setting of life, and leave a distinct imprint on the rock record. Ice ages are extreme among the many climate changes the world has seen, but they do illustrate the remarkable degree to which we can read the sedimentary archives. As we descend deeper into the past, details inevitably fade and speculation assumes a greater role, but not even the Precambrian entirely lacks a decipherable record.

The pattern of glacials and interglacials appears to have been triggered by changes in the orbit of the earth, modified by the interaction of ocean and atmosphere; that much seems assured now. For the precise mechanisms, we must wait awhile, wait until we have a better understanding of the global climate, wait until we can date the geological record with greater refinement, wait until we have a better grasp of the conditions on the globe as a whole, wait until we can connect better the events recorded at sea with those on land.

What about the future? Will ice once again cover New York and London? Will huge lakes inundate the southwestern United States? Scholars have ventured various guesses regarding this interesting question, but guesses is all they are. The most reassuring is John Imbrie of Brown University who believes that the next round will be delayed by the man-made carbon dioxide blanket and is still some 20,000 years away. Others, using essentially the same data, recommend that we brace ourselves for a cold year 3000.

FOR FURTHER READING

Oddly, books on ice ages and climate change are still few, but some come to mind. For lay readers, the following offer a fair choice, the latter more at college level: Battan, L. J. (1974). *Weather* (Englewood Cliffs, NJ: Prentice-Hall); and Critchfield, H. J. (1983). *General Climatology* (Englewood Cliffs, NJ: Prentice-Hall).

On the inconstant climate consult: Crowley, T. J. & North, G. R. (1991). *Paleoclimatology* (New York: Oxford University Press); Grove, J. M. (1988). *The Little Ice Age* (London: Methuen); and Roberts, N. (1989). *The Holocene: An Environmental History* (Oxford: Blackwell).

On climate change and human affairs the classic work is: Lamb, H. H. (1988). *Weather, Climate and Human Affairs* (London: Routledge); a light but nice overview is Roberts, W. O. & Lansford, H. (1979). *The Climate Mandate* (San Francisco, W. H.

Freeman); while Leggett, J. (1990). *Global Warming: The Greenpeace Report* (Oxford: Oxford University Press) provides an up-to-date foundation for all serious students of the issue, although it contains some advocacy.

Readers keen on economic consequences should start with Maunder, W. J. (1989). *The Human Impact of Climate Uncertainty: Weather Information, Economic Planning and Business Management* (New York: Routledge). For humans and environment see Goudie, A. (1986). *The Human Impact on the Natural Environment* (2nd ed.; Oxford: Basil Blackwell).

SPECIAL TOPICS

Bradley, R. S. (1985). *Quaternary Paleoclimatology* (Boston: Allen & Unwin).

Broecker, W. S. (1987). Unpleasant surprises in the greenhouse? *Nature*, **328**, 123–26.

Broecker, W. S. & Denton, G. H. (1990). What drives glacial cycles? *Scientific American*, **262**, 43–50.

Chen, R. S. & Parry, M. L. (1987). *Climate Impacts and Public Policy* (Vienna: International Institute for Applied Systems Analysis).

Dawson, A. G. (1991). *Ice Age Earth: Late Quaternary Geology and Climate* (New York: Routledge).

Detwiler, R. P. & Hall, C. A. S. (1988). Tropical forests and the global carbon cycle, *Science*, **239**, 42–47.

Houghton, R. A. & Woodwell, G. M. (1989). Global climatic change, *Scientific American*, **260**, 18–26.

Imbrie, J. & Imbrie, K. P. (1979). *Ice Ages, Solving the Mystery* (Garden City, NJ: Enslow Publishers).

Jones, G. A. (1991). A stop–start ocean conveyor, *Nature*, **349**, 364–65.

Jones, P. D. & Wigley, T. M. L. (1990), Global warming trends, *Scientific American*, **263**, 66–73.

Kutzbach, J. E. (1987). The changing pulse of the monsoon, in *Monsoons*, J. S. Fein & P. L. Stephens (eds.), pp. 247–67 (New York: Wiley).

Newell, R. E. (1979). Climate and the ocean, *American Scientist*, **67**, 405–16.

Newell, R. E., Reichle, H. G. & Seiler, W. (1989). Carbon monoxide and the burning earth, *Scientific American*, **261**, 58–64.

Rampino, M. R., Self, S. & Stothers, R. B. (1988). Volcanic winters, *Annual Reviews of Earth and Planetary Science*, **16**, 73–99.

Repetto, R. (1990). Deforestation in the tropics, *Scientific American*, **262**, 18–24.

Schneider, S. H. (1989). The Greenhouse effect: Science and policy, *Science*, **243**, 771–81.

Stothers, R. B. & Rampino, M. R. (1983). Historic volcanism, European dry fogs and Greenland acid precipitation, 1500 B.C. to A.D. 1500, *Science*, **222**, 411–13.

Vellinga, P. & Leatherman, S. P. (1989). Sea level rise: consequence and policy, *Climatic Change*, **15**, 175–90.

Webster, P. J. (1981). Monsoons, *Scientific American*, **245**, 109–18.

Wigley, T. M. & Kelly, P. M. (1990). Holocene climate change, ^{14}C wiggles and variations in solar irradiance, in *The Earth's Climate and Variability of the Sun over Recent Millenia*, J.-C. Pecker & S. K. Runcorn (eds.), pp. 547–60 (London: The Royal Society).

Drifting continents, rising mountains

Our views have constantly to be revised and adjusted, but occasionally new aspects of far-reaching consequence shed such unexpectedly different light on existing problems that the effect might be compared to that of a revolution. All that had hitherto been sacrosanct crumbles to the ground; hardly anything is left untouched.

J. H. F. Umbgrove, *The Pulse of the Earth*

A GEOLOGICAL REVOLUTION

The sciences, like children, alternate brief periods of rapid growth with long intervals of semi-stability. Explosive renewal came in physics with the introduction of the Bohr–Rutherford atom model and in biology when the genetic code was deciphered by Crick and Watson. Consolidation follows and the revolutionary ideas of yesterday become the conventional wisdom of today, until once more the intuitive and sometimes irrational processes of a revolution replace the orderly march of traditional scholarship. Each time, the discipline emerges from the turmoil refreshed and matured.

The triumph of uniformity over catastrophism early in the 19th century and the victory of Darwinian evolution half a century later were geology's first revolutions. Long stability followed; my geological education in the 1940s was not so very different from that of early years of the century. Today, another revolution lies behind us; tempers have cooled, legends of the heroic days are told, and the new concepts are well enshrined. But the calm is deceptive, the fires of another and equally important revolution have been lit, and we shall soon have to acknowledge its achievements.

The uprising of the 1960s, known as the plate-tectonics revolution, had a precursor, and its fate is informative. Early in this century, the German meteorologist Alfred Wegener noted, as others had done before him, the similarity of the opposing coasts of the Atlantic. Being a geophysicist, he desired to get rid of sunken continents and vanished land bridges, then often invoked to explain trans-oceanic fauna connections but known to be incompatible with isostasy. In 1915, in *Die Entstehung der Kontinente und Ozeane*, Wegener tested, with a large set of geological matches across the oceans, the notion that a single great continent had shattered and drifted apart. In Europe he received a surprised but sympathetic hearing, but on the other side of the Atlantic the ambience was distinctly chilly. In 1926, at a meeting in, of all places, Atlantic City, American geologists to a man read him out of the company of acceptable thinkers. Only on the southern continents, where solid evidence stared everyone in the face, did geologists continue to support his ideas, but their remoteness denied them proper attention.

As an idea whose time had not yet come, continental drift languished for decades. The reluctance of geologists to consider it seriously was instinctive and emotional rather than factual, and it did not help that Wegener was not a certified geologist. His failure to provide a suitable driving force for drift was also severely criticized, but when the great British geologist Sir Arthur Holmes presented a feasible one, no one listened. Clearly, the flaws and cracks in the existing scientific structure still seemed small to the geological profession. Having just learned to live with the abyss of time, they were not yet prepared to deal with the uncertainty of place that is inherent in a world of drifting continents. It was not until the middle of this century that startling results from the study of the magnetic properties of rocks and our growing knowledge of the ocean floor culminated in the plate tectonics theory and brought about a belated victory for Wegener's ideas.

6

Continental drift and plate tectonics

6.1 WANDERING POLES OR WANDERING CONTINENTS?

Certain igneous rocks are magnetized when they cool according to the magnetic field of the earth at that time. Basalt preserves the field especially well, but many other rocks and some sediments do so too. The measurements yield the direction, the azimuth to the magnetic pole; with judiciously spaced samples, we can determine its past position by triangulation.

The rock also retains a magnetic record of the paleolatitude at which it formed. If one strides poleward carrying a bar magnet suspended on a string, the bar remains parallel to the lines of magnetic force and will thus point more steeply downward the closer we approach the magnetic pole (Figure 6.1). The angle, the inclination, depends only on the latitude. Alas, the azimuth to the pole and the paleolatitude are all we obtain; we know the direction of the meridian and the sample's position on it, but not the paleolongitude. With enough samples we can orient two ancient continents with respect to the pole and restore them to their initial latitudes, but we cannot know whether they were adjacent to one another or separated by a vast ocean.

By the late 1950s, curious facts had begun to emerge from a large number of measurements of this kind. If, for example, one obtained a set of magnetic pole positions for different moments in time from rocks on a single continent, the data would not cluster around a single point, but traced a line on the globe, a polar path. Either the pole was fixed and the continent drifted or a wandering pole was observed from a stationary continent. The data did not permit a choice between the two options, but to most geologists a wandering pole seemed easiest to accept.

Once polar paths had been obtained for several continents, however,

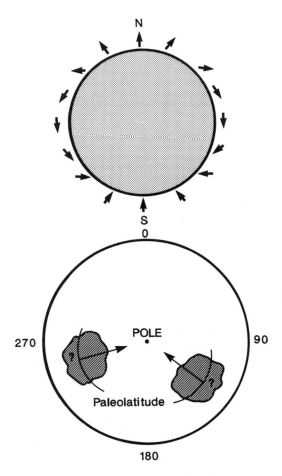

Figure 6.1. The magnetic minerals of many igneous and some sedimentary rocks line up with the earth's magnetic field when the rock forms. This permits us to determine the azimuth of the magnetic pole as it was at that time. Moreover, because a magnetic bar dips progressively more steeply as we take it from the equator to the pole (top), we can also obtain the paleolatitude. Because the paleolongitude cannot be determined, we are unable to say whether two continents, correctly oriented and placed at the proper latitude, existed side by side or were separated by an ocean (bottom).

it was clear that it could not be so. Consider the polar paths for Europe and North America (Figure 6.2). Far apart in the early Phanerozoic, they converge with time on the present pole. Similar separate but converging polar paths exist for all continents, and if we join the continents in approximately the arrangement proposed by Wegener, all polar paths coincide. This agrees much better with continental drift than with the concept of multiple wandering poles.

Notwithstanding the simple, compelling logic of the magnetic data, they persuaded only a few, and those mainly geophysicists, that Wegener

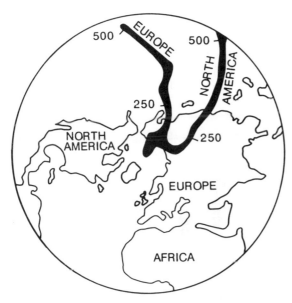

Figure 6.2. The positions of the magnetic pole for various times in the past, measured on a single continent and plotted on a world map seen from the pole, delineate a band, a polar (wandering) path. Individual continental polar paths (shown here for Europe and North America) converge on the present pole for the present time. The numbers alongside are millions of years. The paths for past times can be made to coincide by reuniting the continents as proposed by Wegener.

deserved another hearing. The more conservative geologists took refuge behind whatever paleomagnetic measurements failed to fit this simple picture. Later the discrepancies were largely eliminated by a better understanding of rock magnetism and how to measure it properly, but by then oceanographers had reopened the matter of continental drift from a different angle.

6.2 CONTINENTS AND OCEAN BASINS

Because of their sheer size, continents and ocean basins have an aura of permanence. They consist of different kinds of rocks, and the distinction between those of continental origin and others formed on the ocean floor is usually clear even for the oldest strata. Oceanic rocks are sometimes found on the continents, but as far as we can tell no ocean basin has ever been raised to become a continent, nor is it physically possible for an entire continent to founder and become an ocean.

Together with their submerged margins the continents occupy slightly more than 40 percent of the earth's surface. Their mean elevation is only a little above sea level (Figure 6.3), while the ocean floor lies about 4,000

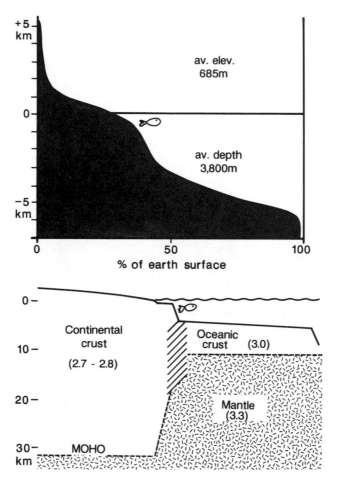

Figure 6.3. Most of the surface of the earth lies either a few hundred meters above sea level or several thousand meters below (top). This two-level earth makes sense if we assume that a thick, light continental crust and a much thinner but slightly denser oceanic crust float like rafts on a mantle denser than either of them (bottom: numbers in parentheses are rock densities in grams/cm³). Isostatic compensation then produces the two dominant levels.

m deeper. Isostasy explains this remarkable two-level earth. The continents, consisting in large part of rocks resembling granite in composition, are light and thick. The ocean floor, made of oceanic basalt, is heavy but thin. Like rafts on a swimming pool, both float on the underlying mantle which is denser than either. The boundary between the mantle and the overlying oceanic or continental crust, named after its discoverer the Mohorovicic discontinuity or Moho, can be easily detected with artificial seismic waves. With continents and oceans in isostatic equilibrium, the difference in buoyancy (which depends on density times thickness) causes the difference in mean elevation.

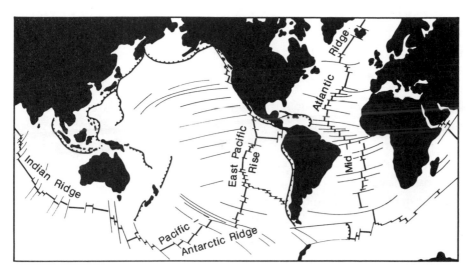

Figure 6.4. The most prominent features of the ocean floor are broad mid-ocean ridges and deep, narrow trenches. The mid-ocean ridges, in many places offset at right angles by fracture zones, form an almost globe-girdling system, but most trenches are in the Pacific Ocean.

The Moho has long been regarded as a sharp boundary between two reasonably homogeneous layers of very different composition. The US National Academy of Sciences, dreaming of glory, once planned to spend much tax-payers' money by drilling a deep hole in the ocean floor, the so-called Mohole, to see what the Moho and the mantle might be made of. Cooler heads cancelled the plan at the last minute, and fortunately so, because we now know that the Moho varies greatly over short distances, reflecting not only the origins of continental and oceanic crust but also their history.

Curiously, the rock record on the continents reflects the presence of adjacent oceans from the very beginning, whereas the age of the oldest oceanic rocks does not exceed 200 my. Also, the sediment blanket on the ocean floors is far too thin to represent a long career as the world's ultimate sanitary landfill. Was there or was there not an early ocean? A most intriguing paradox.

Among the most striking features of the ocean floor are the mid-ocean ridges. They girdle the earth like the seam on a tennis ball (Figure 6.4) and have unusual properties. They always rise to about 2,500 m (*c.* 8,000 ft) below sea level and their crests are usually cleft lengthwise by a narrow, fault-bound rift valley, the site of frequent shallow earthquakes of the sort that indicates tension. At intervals the crest is also offset by sets of linear ridges and troughs called fracture zones; at right angles to the crest they lend the ridge a startlingly orthogonal pattern.

Mid-ocean ridges are built of basalt flows and intrusions and are dotted with submarine volcanoes and scattered volcanic islands, such as the Galapagos, Ascension or Iceland. On the ridge crest the lavas are fresh and seem young, but as one draws away downslope, they and the volcanic islands increase in age. Similarly, the sediments in the rift valley are thin and patchy, but thicken down the flanks of the ridge. Because open-ocean sedimentation rates do not vary in this way, this implies that the age of the crust increases with distance from the axis. Also, the flow of heat from the earth's interior, which is very high on the floor of the rift valley, rapidly decreases downflank and away from the axis.

The shallow, hot, mid-ocean ridge contrasts sharply with another major set of features of the deep-sea floor, the trenches; narrow, elongated troughs 5–7 km (17,000–23,000 ft) deep. Most trenches occur in the Pacific which they encircle almost completely (Figure 6.4). They have a very low heat flow, active volcanoes are rare, and they generally contain little sediment. Trenches bordering on continents may have thick deposits, but because a prolific source is nearby, this does not imply great age. Like mid-ocean ridges, trenches are associated with numerous earthquakes, but in this case they are of a kind that is due to compression of the crust rather than tension. The earthquakes occur along a plane, the Benioff zone, which descends steeply to a depth of 300–700 km.

Chains or arcs of islands accompany most trenches. The first lies close to the trench and consists of deformed, mildly metamorphosed sediments. Farther back, active volcanoes make up a second arc. Instead of the oceanic basalt typical of mid-ocean ridges, these volcanoes pour out andesite lavas that are intermediate in composition between oceanic and continental rocks. Where a trench borders on a continent, the sedimentary arc is part of the continental edge and the volcanic arc lies well inland, as for example in the Andean ranges of western South America landward of the Peru–Chile trench.

6.3   A DAISY CHAIN OF HYPOTHESES

These features, most of them discovered or at least imprinted on the minds of geologists after World War II, make of the ocean floor a simple, orderly domain but one where much demands explanation. The first attempt at a universal hypothesis used a suggestion made in 1929 by Sir Arthur Holmes but long buried in the depths of libraries: the idea that the mantle is stirred by convection currents. Suppose that, far down, the mantle is heated. It expands, becomes lighter and, now capable of flowing, rises. Below the Moho the flow spreads sideways and, when it has cooled

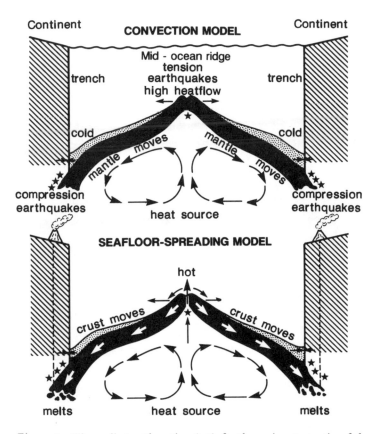

Figure 6.5. The earliest explanation (top) for the curious properties of the ocean floor involved convection in the upper mantle below a stationary crust, but failed to explain why the seafloor increases in age away from the ridge crest. Hess's seafloor-spreading model (bottom) solved this problem by allowing the ocean floor to ride with the mantle. This requires the formation of new crust on the ridges and the disposal of old crust in trenches. This model fails to account for the lack of trenches in the Atlantic Ocean which does have a mid-ocean ridge.

sufficiently, sinks once more. Such convective flow can readily be demonstrated in a pan of boiling soup. Above the rising column (Figure 6.5) a mid-ocean ridge forms, because the overlying crust heats, expands and forms a bulge on the seafloor. Stretched by the flow that diverges underneath, it cracks, to the accompaniment of tension earthquakes and volcanic eruptions. Away from the rising column the crust is cooler and the depth of the ocean gradually approaches its normal value. Where the cold limbs of the convection current sink, they drag the crust down, causing compression earthquakes and forming trenches. The 300 km depth of the Benioff zone seems excessive, but that is a detail.

This model explains many things, but it fails to deal with the change in age away from the ridge crest. It is also useless for the details of the trench-and-arc complex, nor does it say anything about fracture zones.

That last failure especially is bothersome, because these features are so large, so common, and clearly so important. They also sometimes behave strangely, as in the eastern Pacific where they offset the crest of the East Pacific Rise by hundreds of kilometers but terminate against the coast of California without distorting it (Figure 6.4). Patently, they are not ordinary faults with simple horizontal slippage.

In 1960, H. H. Hess of Princeton University improved on the basic convection model by changing it just a little (Figure 6.5), but enough to make his paper a true classic. He allowed the crust above the Moho to ride passively on the convecting mantle, like a raft on a river. At the ridge crest, this creates a gap that must continuously fill with lava which solidifies, then cracks again. Excess crust accumulates at opposite ends; it is dragged into the mantle by the descending limbs, melted and re-cycled. Now we need the whole Benioff zone, the sedimentary arc be-comes the crumpled edge of the opposing slab, and the volcanic arc is fed where the sinking slab melts from the heat of friction. Andesite, in fact, is like basalt adulterated with oceanic sediments and seawater. The model also accounts for the apparent age increase away from the crest that is implied by the ages of the volcanic islands and the thickening of the sediment cover.

Hess's seafloor-spreading hypothesis was well received by marine ge-ologists because it explained so many different things at once. Continental geologists, on the other hand, noting that the whole cycle took place in the oceans, were not much stirred by it. Soon, however, the hypothesis received impressive support from two independent sources and could no longer be ignored.

The first confirmation came from an explanation for fracture zones invented by the Canadian J. Tuzo Wilson. Traditionally, the fracture zones had been regarded as faults with horizontal slip that cut an initially continuous ridge and gradually increased the distance between the seg-ments (Figure 6.6), as strike-slip faults do on land. Wilson suggested that the ridge never was continuous, but consisted from the start of segments offset by what he chose to call transform faults. As the crust is carried away from each segment, it moves in the same direction on both sides of the fault and there is no need for earthquakes. Between the crest segments, however, the pieces moves in opposite directions with a sense that is the reverse of what a strike-slip fault does, and earthquakes must occur there.

The relative directions of motion along a fault can be determined from earthquakes and it would be a simple matter to validate Wilson's sug-

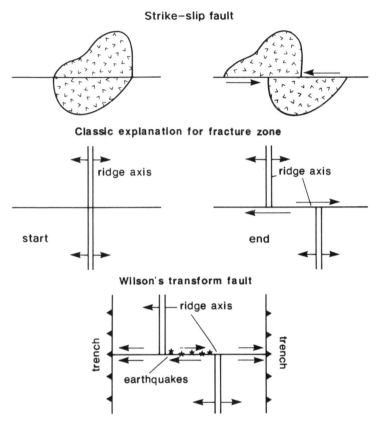

Figure 6.6. Faults (top) along which pieces of crust slip horizontally past each other are called strike-slip faults. At first, the fracture zones that offset mid-ocean ridge crests were also regarded as strike-slip faults (center), by assuming that the ridge crest was initially continuous. If seafloor-spreading is true, however, motions exist that are fundamentally different. These motions (bottom) define a transform fault, and are confirmed because earthquakes occur only between ridge crests.

gestion. Lynn Sykes, a seismologist at Columbia University, did so right away, and the credibility of seafloor-spreading rose considerably.

6.4   MAGNETIC ANOMALIES AND POLARITY REVERSALS

Almost simultaneously, support for the seafloor-spreading model came from an entirely different side: the data collected during years of patient towing of magnetometers behind research ships. The magnetic field of the earth is simple in principle, but is locally distorted by the magnetization of rock formations. On land, the distortions tend to be irregular, but at sea there are patterns of alternating high and low values. If we color the values above the average earth field, the positive anomalies, black and leave the negative anomalies white, the map resembles a zebra skin with stripes parallel to the mid-ocean ridge crests that are distorted by the fracture zones (Figure 6.7). Invariably a high positive anomaly perches on the ridge axis, and the anomalies to either side are usually markedly symmetrical.

The first magnetic anomaly map was made in the late 1950s for the northeastern Pacific, and presented a bold enigma that provoked various extravagant explanations. The correct answer was found by Drummond Matthews and his research student Fred Vine, two Cambridge University geophysicists. Put forward quietly in 1963, it rested on magnetic polarity reversals (Section 2.4). Vine and Matthews reasoned as follows (Figure 6.7). Suppose the seafloor spreads and a fissure opens. The lava that fills it cools and adopts the present normal polarity. If we tow a magnetometer across this new crust, its magnetization and that of the earth's field, having the same direction, sum and we see a positive anomaly. If, on the other hand, some time ago an earlier crack had filled with lava while the earth's field was reversed, it would have acquired a reverse polarity. Measuring today across the older block, the strength of its field, being opposite to that of the earth, would subtract and a negative anomaly would be seen. Thus, like a magnetic tape recorder, the spreading ocean crust tracks the reversals of the earth's field through time as positive and negative departures from the present one.

When the crust formed at the ridge axis splits in two, the pieces, bearing symmetric halves of the latest magnetic anomaly, move apart and a new block is inserted between them. Thus, the central block always has the present magnetic polarity and the anomalies on the flanks, each being half of the former central block, are mirror images of each other. If the seafloor spreads at a constant rate, the widths of successive anomalies depend only on the length of time between reversals.

As soon as a reasonably dated polarity reversal time scale became available, Vine tested his idea by plotting the widths of the anomalies against reversals dated on land. The fit was fine, and Vine's hypothesis and Hess's seafloor-spreading model were simultaneously confirmed. One needs luck for any discovery, and Vine had it; there was no a priori reason why the seafloor should spread at a constant rate, and it does by no means always do so, but it happened to be true for the data that he used. As a bonus he obtained seafloor-spreading rates ranging from 2 cm/year in the North Atlantic to more than 15 cm/year in the central South Pacific.

Luck was what L. W. Morley, a Canadian, did not have. He had

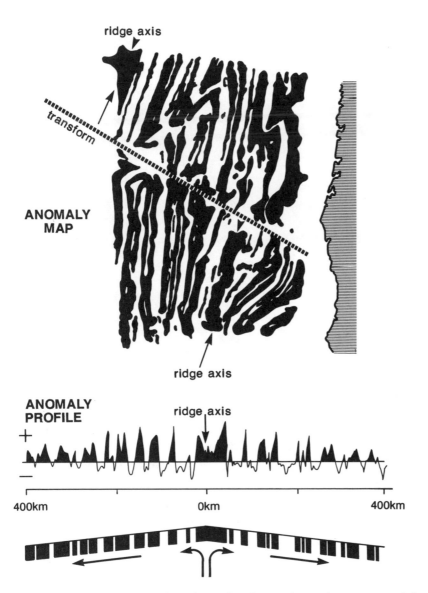

Figure 6.7. Observations at sea have shown that the oceanic crust has a pattern of alternating high (black) and low (white) magnetic values that resembles the skin of a zebra (top). The stripes are distorted along transform faults/fracture zones. A magnetic traverse across a mid-ocean ridge (center) shows a pattern of positive anomalies above and negative anomalies below the average earth's field. The anomalies are symmetric to either side of a broader central anomaly (bottom). They can be explained as the result of seafloor-spreading (see text).

independently arrived at the same conclusion as Vine, but the publication of his article was delayed and it finally appeared in a less well known journal. When he emerged from obscurity, credit had already accrued to others. A bitter fate.

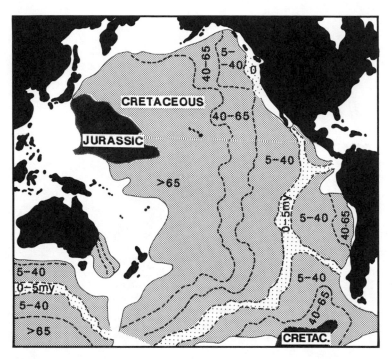

Figure 6.8. This partial map of the ages of the ocean floor based on marine magnetic anomaly data shows that, as a result of subduction, oceanic crust becomes more scarce with increasing age. The numbers are the ages of the Cenozoic crust in millions of years.

Vine's article made believers out of many, and the next logical step was soon taken. If we assume that, on some ridges at least, the spreading rate has always been constant, reversals older than those found on land can be dated by extrapolation from magnetic anomalies observed at sea. Whether we are really dealing with a constant spreading rate can be verified by comparing many spreading ridges. Later the dates can be calibrated with basalt samples obtained by drilling in the ocean floor. In this way, the polarity reversal time scale has now been extended back to almost 150 my. Soon after that we run out of crust; the oldest so far identified in the oceans occurs in the western Pacific and is of middle Jurassic age (*c.* 170 my).

Once reversal times had been fixed and numerous magnetic profiles obtained, a map of the age of the oceanic crust could be constructed (Figure 6.8). It shows clearly how young the ocean floor is compared to the billions of years of the continents, and how little remains even of oceanic crust formed as recently as 100 my ago. The seafloor-spreading hypothesis, by demonstrating that ocean floor is continuously destroyed and renewed, has neatly solved the paradox of old continents and young oceans.

6.5   PLATE TECTONICS

For many, Vine's article confirmed seafloor-spreading beyond any further doubt. All at once, mid-ocean ridges, trenches, Benioff zones, fracture zones, ocean crust ages, heat flow distributions, and magnetic anomaly patterns fitted into a single system. The continents, and therefore continental drift, were not part of the hypothesis, however.

Still, serious problems remained. Seafloor-spreading demands a separate convection cell between each pair of fracture zones, and each rising column must be offset with respect to its neighbors as the ridge axes themselves are offset. As the number of known fracture zones grew and their spacing shrank, the narrow, elongated shape and large number of the cells became improbable. Puzzling also was that, while the hypothesis demanded that each mid-ocean ridge have a pair of trenches to discard the excess crust, the Mid-Atlantic Ridge, best known of all, had no trenches (Figure 6.4). The same is true for large parts of the ridge in the Indian Ocean.

In retrospect, it is clear that the solution to these and other problems needed only the removal of a single mental block: the conviction that the Moho was the principal physical boundary underneath continents and oceans. As long as one shares this conviction, continental and oceanic crustal blocks are the major structural units and mantle flow below and passive drift above the Moho are the logical consequence. Getting rid of such mental blocks can be very difficult, but is invariably rewarded with new insight.

Simultaneously, three young scholars, perhaps less burdened by tradition than their elders, discarded preconceived tectonic units, went to see where the main seismic and volcanic activity in the world was located, and discovered that those zones of activity are narrow, sharp, and continuous (Figure 6.9). Together they define a small number of large segments of the earth's surface now known as tectonic plates. Plates are delineated by trenches and mid-ocean ridges linked by transform faults such as the dangerous San Andreas in California or the faults of northern Anatolia, the Caucasus and Iran. Some plates contain only ocean crust; others carry both oceans and continents. This insight was presented, separately and almost simultaneously, by Dan McKenzie at Cambridge University and Jason Morgan at Princeton, and elaborated by Xavier LePichon. It was soon confirmed with seismic data by Lynn Sykes and Jack Oliver, then at Columbia University, and plate tectonics was born.

In a world subdivided as in Figure 6.9, ocean basins and continents are clearly secondary elements, and that means that the plates must in-

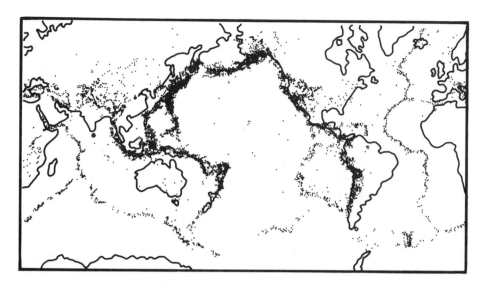

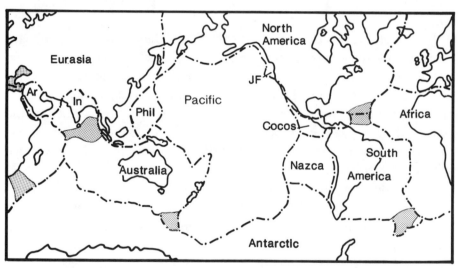

Figure 6.9. The geological activity on the earth's surface in the form of earthquakes and volcanic eruptions is concentrated in narrow but distinct zones which are usually not located at the boundaries of continents and oceans. These zones divide the earth's crust into a small number of tectonic plates (bottom). In: Indian plate; Phil: Philippine plate; JF: Juan de Fuca plate; Ar: Arabian plate. The shaded areas are examples of anomalous boundaries where earthquake activity occurs across a broad region; they are now known to be quite common.

clude more than just the crust. They are pieces of the lithosphere, the outer shell of the earth. The lithosphere is about 120 km thick (Figure 6.10) and consists of a thick slice of mantle with a thin topping of crust. It is rigid and floats on the asthenosphere which is ductile, i.e. it deforms without breaking and behaves in many ways like a very viscous fluid.

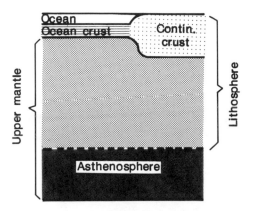

LITHOSPHERE

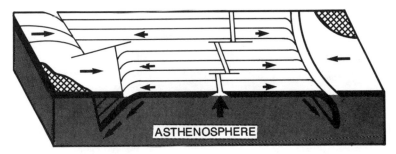

Figure 6.10. Recognition of the active zones of Figure 6.9 led to the plate-tectonics model which regards the earth as covered with slabs of lithosphere (plates) floating on a weak asthenosphere (top). The 120-km-thick plates diverge at mid-ocean ridges, converge in trenches now recognized as subduction zones, and slide harmlessly past one another along transform faults. Continents and oceans ride passively on the backs of the plates.

Lithosphere and asthenosphere were not new inventions; the evidence for their existence had been available for decades but had been judged to be of little importance. What exactly they are is a matter of debate even now, more than 30 years after they were recognized as key elements in the dynamics of the earth.

Plates meet at three kinds of boundaries. They drift apart, converge and collide, or slip past each other on transform faults when their boundary happens to be parallel to the direction of their relative motion (Figure 6.10). At a divergent plate boundary part of the mantle melts because the pressure is relieved by rifting and creates new lithosphere. Consequently, this lithosphere is always basaltic in composition, has a high density, and forms oceanic, never continental crust. Even when a divergent boundary opens within a continent, the melting mantle soon begins to create oceanic crust in its own new ocean basin.

When plates collide, one is forced underneath the other, is subducted. If both are oceanic, either one may be subducted, but if one is continental and the other oceanic, the continental edge will remain on top, because it is the more buoyant one. When two continents collide, neither can be subducted. The edges crumple, but ultimately all movement ceases. Because the surface of the earth is covered entirely with tight-fitting plates, the movements of the entire set may be modified when that happens. There is growing evidence that this adjustment, notwithstanding the enormous size and mass of the plates, is often geologically speaking swift, something of the order of a million years.

It is now clear why the Mid-Atlantic Ridge does not need trenches of its own. As long as the total amount of lithosphere created at all mid-ocean ridges equals the amount destroyed in all subduction zones, it is not important where the recycling takes place. The new crust formed in the Atlantic is compensated where the Pacific plates are being subducted under the west coast of the Americas. The American plates grow larger, the Pacific plates shrink, but a worldwide equilibrium is maintained.

An important rule in plate tectonics states that all plate activity occurs at their boundaries and that the interiors are not deformed. Obviously, this is an oversimplification, because earthquakes do shake the interiors of plates and volcanoes do erupt there. The activity to which they testify, however, is not enough to alter the size and shape of the plate significantly. At least, so we hope, but our confidence in the simplicity of nature is waning. Take, for instance, the Indian and Australian plates (Figure 6.9). Long thought to be a single plate, it has become clear that there are two and that the Australian plate is closing on the Indian plate at a rate of more than 1 cm/year. If everything went according to the rule, we would see a sharp collision zone somewhere, but we do not (Figure 6.9). The earthquakes testifying to the collision occur in a zone many hundreds of kilometers wide and do not define a plate boundary in the classical sense. Does this simply mean that a new subduction zone, not yet sharply defined, is being born? Or do plate interiors deform after all? There are other cases of a diffuse state of affairs. Anatolia pushes west into the Aegean Sea to get out of the way of Arabia which moves north, and here also the motions are distributed over a large area in the Aegean and Greece. The last word is not in.

6.6 MEASURING PLATE MOTIONS

The axiom that plate interiors do not deform is important because it allows us to deal with plate motions as exercises in spherical geometry,

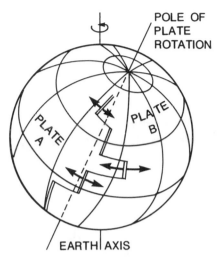

POLE OF
PLATE
ROTATION

PLATE
A

PLATE
B

EARTH AXIS

Figure 6.11. The relative motions of two plates A and B on a sphere can be described as a rotation around an axis which exits from the sphere at two poles. The plates are separated by a mid-ocean ridge (double line) and transform faults (single lines); the heavy arrows indicate directions of relative motion. The poles of plate rotation should not be confused with the poles of rotation of the earth itself.

if we use Euler's theorem. This theorem, having emerged from obscurity into the clear light of the geological revolution, says that any movement on a sphere can be regarded as a rotation around a properly chosen axis through the center of the sphere. The movement of one plate relative to another can be similarly described as a rotation around a joint axis (Figure 6.11).

The points where the axis exits the sphere are the poles of rotation of the two plates; they must not be confused with the poles of rotation of the earth or catastrophic misunderstandings will result. Transform faults are parallel to the relative motion of two plates, and define small circles perpendicular to the axis of rotation, like lines of latitude (parallels). Having found the pole of rotation of two plates, we obtain their relative velocity from magnetic anomalies by dividing the distance between opposing reversal boundaries by their age. Then, proceeding pair by pair around the globe, South American to African plate, African to Indian, Indian to Australian, Australian to Pacific, etc., we can define the entire system in terms of poles and rates of rotation. It is clear that this exercise would lose most of its rigor if the axiom that plates deform only at their edges were found to be often and seriously wrong.

When this chapter was written for the first edition of this book, magnetic anomalies and transform faults were all we had to estimate plate motions, but none had been measured directly. To do that seemed difficult or even impossible, because it meant measuring changes of centi-meters per year between points tens or even hundreds of kilometers apart. But it turned out to be feasible after all, and with considerable precision too. The main tools are the Global Positioning System that uses satellites for triangulation, and a more exotic method that takes advantage of the time difference between radio signals emitted by quasars that are received at stations that are very far apart.

And so we know that the North American plate moves southward relative to the Pacific plate at a speed of 46 mm/year, a motion spread across several subparallel fault systems. Each year the North Atlantic Ocean widens by 10 or 20 millimeters, quite slowly compared to the East Pacific Rise which, in the South Pacific, grows larger at more than 150 mm/year. South America rides over the Nazca plate to the west at 84 mm/year, and Africa approaches Europe by just less than one meter/century. These short-term measurements are close to estimates obtained from magnetic anomaly patterns over much longer intervals. This is pleasing, because it shows, as one would expect for such enormous masses, that the motions are steady.

6.7   WHAT DRIVES THE PLATES?

A good, although obvious question this, especially if we remember that Wegener failed to persuade the world of continental drift in part because he was unable to say how continents could make their way through ocean basins. It is ironical that even now we are not entirely sure of the answer, although Sir Arthur Holmes's mantle convection remains the favorite. Since the plates cover the whole surface of the earth, they impede each other's freedom of motion and so tell us little about the rates and directions of flow in the upper mantle that supposedly pushes them around somehow.

Other forces might drive or help drive plates, all of them feasible if, as is widely believed, the mantle is capable of flow at a rate of 10–20 mm/year. The pull of a cold and therefore heavy subducted slab sinking into a warmer, lighter mantle might do it, or the push of magma rising at a mid-ocean ridge (Figure 6.12). Because the friction between lithosphere and asthenosphere is small, the plates might even simply slide down the gentle slopes created at the base of the lithosphere by heating and ex-pansion of the underlying asthenosphere.

In theory these possibilities can be tested. Is the speed of a plate related to the size and age and therefore to the weight of an attached subducting slab? Are plates without any subduction zone or mid-ocean ridge sta-

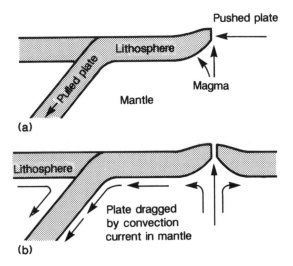

Figure 6.12. Various forces such as push at the divergent boundary and pull by a subducting slab might, alone or in combination, drive the plates. The so far most widely accepted driving force, however, is drag by convection in the mantle. The relative contributions of the various forces to the movements of the plates are not known.

tionary or at least very slow-moving? Trying for the answers to these and other questions has shown that plates with long subducting boundaries move faster than those not so endowed, but that neither the length, the age, nor the angle of a sinking slab seem to be very important. It is also odd that plates carrying a large continent move more slowly than those that do not have that burden. The Pacific and Indian plates which are wholly or mostly oceanic move faster than the entirely continental Eurasian plate. This confirms an old suspicion that the lithosphere is thicker under continents than under ocean basins, like a ship with a deep keel, but why one does not know.

Unfortunately, we are short of plates, having only a dozen or so, and that set lacks many of the combinations we should like to test in this experiment. It also should be no surprise that the search achieved less than was hoped because of the demon of empirical science who causes the differences one seeks to find to be very close to the uncertainties of the measurements.

6.8   POSTSCRIPT TO A REVOLUTION

Revolutions tend to be affairs of the young, and the revolution in the earth sciences was no exception. This stands to reason where the overthrow of conventional wisdom is at stake, but young revolutionaries need courage and persistence to face the stubborn resistance of the older generation. As a reward, their names become scientific household words overnight. Such instant fame is not an unmixed blessing, however, because after setting an entire discipline on its ear in a single lecture, what does one do for an encore? The anxious wish to live up to impossible expectations has at times led to extravagance in the search for new ideas, bt more often to boredom as one returned to the long and often tedious task of checking and consolidation.

Naturally, alternatives should be and were considered. Many suggestions were downright silly, but a much debated one was the expanding-earth hypothesis, stubbornly defended by S. Warren Carey of Tasmania. Take a small earth, equipped 250 my ago with a single supercontinent, and make it expand in the Mesozoic. As it swells, the supercontinent is torn and the pieces scatter wide and far. At first sight this makes sense. It does not require subduction nor can it account for the compression of the crust that builds mountains, but small changes in the model might accommodate these points. More serious is that it cannot be reconciled with the well-documented drift of the continents that caused them to join the supercontinent in the first place.

The authors of these and other propositions, although unable to generate better models than plate tectonics, served science well by forcing the big thinkers to maintain high standards of critical scrutiny and intellectual honesty that otherwise might have been temporarily relaxed.

Now, after a quarter century, the paradigm is starting to lose some of its simplicity and elegance. Violations of what the late Norman Watkins called the principle of minimum astonishment are turning up. Attempts to reconstruct by strict geometry the plate configurations of the past 50 my have occasionally run into difficulties that could only be solved by inventing new plates or by abandoning the rule that plates do not deform internally. Complex convergent plate boundaries with unorthodox histories have been identified; some seem to have tiny plates that behave like roller bearings between the big ones. There is a growing tendency to assign importance to differences between continents and ocean basins. And special events and *ad hoc* assumptions have been invoked for episodic intensifications of mountain building that do not follow the stately progress of crustal creation and destruction. In short, the "new global tectonics" has had to learn to live with large local deviations.

To geologists who learned long ago that the earth is always more complicated than at first it appears, this did not come as a big surprise, but it has been disconcerting to others who prefer their earth to be orderly. Also the need to accommodate a growing number of anomalies has loosened some useful constraints that plate-tectonic models once possessed.

Is this serious? Does it spell the demise of plate tectonics or, even worse, the impossibility of a theory of the earth? I do not believe so. Notwithstanding some defects, and an overzealous search for anything startlingly new, plate tectonics has settled many ancient controversies, clarified many major issues, and proved its worth by raising profound new questions. Its impact continues in many areas, including that story of stories, the evolution of life, and it has been of help in practical matters such as finding oil and minerals. The "flaws" are in part merely a recognition that nature is not perfect in a mathematical sense, but they invite us to build further upon the first simple model.

Plate tectonics has fundamentally altered our view of the earth and we shall not return to the old ones. Few are the scientific theories that survive intact forever, and our concepts of the dynamics of the earth and the behavior of its surface will change as new information comes to light, but the continents will continue to drift, and we shall persist in seeing the earth as a dynamic whole, not a disorderly set of unrelated events.

7

Continental breakup and continental drift

The theory of plate tectonics states that the earth's surface is seriously deformed only at plate boundaries. When a continent breaks up, divergent plate boundaries form, and new oceans and more but smaller continents take shape. At a convergent boundary, oceanic crust is lost, island arcs form and mountain ranges rise. Collisions between continents combine them into supercontinents. New oceans are evidence for divergence, while the scars of ancient collisions, called sutures, testify to the existence of former oceans closed by collisions.

7.1 TRACKING THE DRIFT OF CONTINENTS

Plate tectonics has presented us with a world where the shapes, sizes and positions of continents and ocean basins forever change. Luckily, it also offers the means to reconstruct their past arrangements. To this end we use magnetic anomalies and transform faults to rotate the plates by simple geometric operations into their proper positions at any moment in the past (Section 6.6). It is essential that we have enough old ocean crust at hand to provide us with the data. Because of subduction, forever swallowing the ocean rock record like an anaconda ingesting a pig, this procedure limits us to the last 100 my.

For most of the Mesozoic and the entire Paleozoic, not to forget several billion Precambrian years, we must use other, less satisfactory means. A good set of paleomagnetic measurements will suffice to give a continent its proper orientation and latitude, provided we do not mix data from normal and reversed polarity intervals, but it will not provide its longitude (Figure 6.1). We cannot know whether this continent was surrounded by a vast ocean or close to other continents. Means to find longitude are much sought after, but so far little has turned up that is widely applicable.

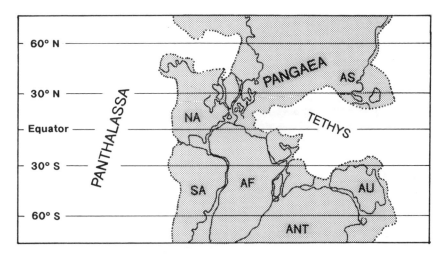

Figure 7.1. The supercontinent Pangaea, surrounded by the superocean Panthalassa in the early Triassic, 220 my ago.

To sort out the paleogeography of the ancient world demands the skill of a jigsaw puzzle addict together with sophisticated, computerized use of geological data. Even at their best, these procedures involve some subjective judgment, and what is a fine solution to tectonicists may be anathema to those who lean toward paleontology. This is best understood by looking in some detail at Wegener's supercontinent; it will then be no surprise that a widely accepted paleogeographic atlas of the pre-Cenozoic world is not to be expected very soon.

7.2   PANGAEA AND PANTHALASSA

So Wegener was right. Some 250 my ago a single supercontinent romantically named Pangaea, the "all-land," sat in the superocean Panthalassa, the "all-sea" (Figure 7.1). Ever since, swarms of enthusiasts have tinkered with Wegener's assembly of Pangaea. The present coast is far too ephemeral to have any meaning in this regard, and Wegener put the boundaries of each piece at the shelf edge. Even better is the place where the continental crust makes way for the oceanic one at a depth of 2–4 km below sea level. The perfect fit remains elusive because, when continents split, their edges are severely mangled, rendering it difficult to find the exact position of the break. For the broad overview we need, however, such uncertainties are of little consequence.

Pangaea itself was assembled during the Paleozoic from many pieces, the largest of which, Gondwanaland (Figure 7.2), consisted of Africa, South America, India, Antarctica and Australia, and was the main sur-

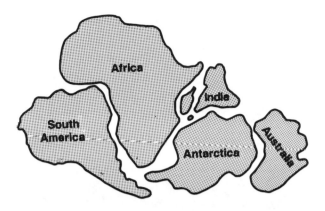

Figure 7.2. Gondwanaland, much as it was first reconstructed by Alfred Wegener, but modified when it became obvious that the 2–4 km depth contour is a better indicator of the edges of the continents than the coasts or the edges of the continental shelf.

vivor of a late Precambrian supercontinent. Around this hulk numerous smaller pieces, each with its own history, drifted in ever-changing arrangements like colored chips in a kaleidoscope. North America, another child of the late Precambrian, looked much as it does now, but the smaller fragments that now make up Europe and Asia seem very unfamiliar. Some were small indeed; for much of the Paleozoic, for example, Britain managed, as it persists in trying today, an existence quite independent of Europe.

The construction of Pangaea was completed early in the Mesozoic, around 220 my ago (Figure 7.1), but the supercontinent, in its full glory almost three times as large as Eurasia, our largest continental mass, did not survive intact for very long.

7.3   THE FACE OF PANGAEA

Maps of the Paleozoic continents as they drifted across the planet challenge us to contemplate the great land masses, those innumerable islands, those unknown seas and climates, and life in the Paleozoic in general. It is for future geologists to know these things with confidence, but we do deserve a glimpse of Pangaea, the largest land of the Phanerozoic and maybe of all times.

The story begins with Gondwana's march across the South Pole, nicely documented with paleomagnetic data (Figure 7.3). When the Sahara crossed the pole, the bedrock was scratched by glaciers and glacial deposits were laid down between 460 and 420 my ago. Then, for the next

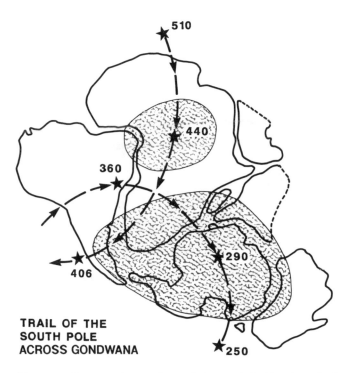

Figure 7.3. Between the late Precambrian and the Permian, the South Pole traversed Gond-wanaland. The numbers along the track mark its progress in millions of years. Several times during this voyage, icecaps formed around the pole (shaded areas), but at other times the polar regions were free of ice. The polar path shown here is only approximate, awaiting more data.

50 my, there is no sign of icecaps, although the climate remained cold. Late in the Devonian, another series of icecaps began to build, probably intermittently. Extensive, well-studied glacial deposits were laid down (Figure 7.4) as South America, southern Africa and Antarctica passed across the pole, until it cleared the edge of Antarctica in the late Permian, 250 my ago. Besides the unmistakable deposits, there is also growing evidence that the innumerable sea level changes observed in the Silurian and Pennsylvanian (Figure 9.4) of North America reflect Milankovitch-style climate changes and their impact on the waxing and waning of the icecaps.

This short paragraph suffices to convince us not to think of the Paleozoic world as unchanging over the many millions of years of its history. During this long, long time the climate switched back and forth from glacial to warm conditions, while Gondwana, the main continent, continuously changing its orientation with respect to the planetary winds, grew ever larger. Mountain ranges rose, creating deserts in their rain shadow, and the circulation of the ocean was profoundly altered as more

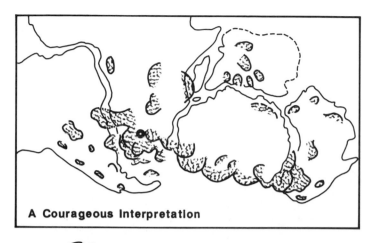

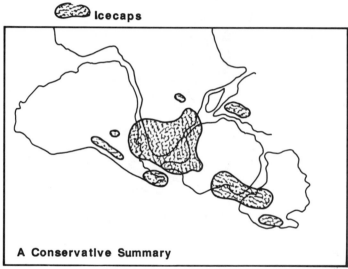

Figure 7.4. Two interpretations of the south polar icecaps of the Carboniferous about 320 my ago. The upper one has much detail and therefore seems more realistic than the lower one which, however, in its conservatism is more commensurate with the evidence.

continents joined, until the supercontinent extended from the far north to the far south. We can no more visit Pangaea briefly and come away with a fair impression of what it was like than a visit to Hudson Bay in the summer would make us understand the Arctic. Only snapshots can be taken, but those have provided geologists with rich food for controversy.

The nature of paleomagnetic data and the ambiguities of continental edge-fitting permit the construction of more than one plausible Pangaea (Figure 7.5). It is also troublesome that the drift of the continents was often rapid compared to the precision with which we can put an age to

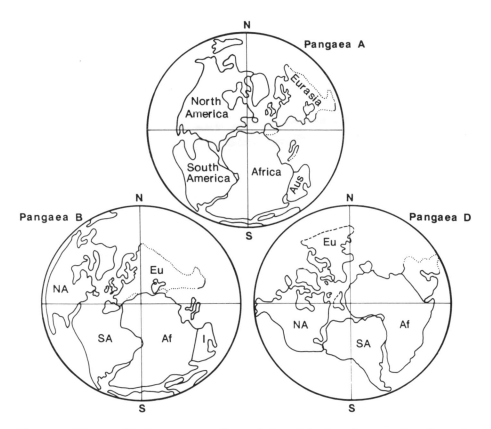

Figure 7.5. Three possible Pangaeas out of several that all fit the paleomagnetic and global tectonic information. Pangaea A is a compromise between paleomagnetic and geological data, Pangaea B a fit for a specific time (Early Permian), and Pangaea D a less probable visual fit based only on paleomagnetic data. The odd and varying shapes of the present continents are due to the projection.

the paleomagnetic and geological data. Therefore, all reconstructions need to be tested against the rock record.

The early Paleozoic sea was high and it was a watery world, rich in warm, shallow seas studded with reefs and full of life. This is helpful, because the fossils enable paleontologists to define many faunal provinces among the shallow marine communities of the time. Shallow marine faunas do not easily cross oceans. If seen to occur on opposing shores, they imply that those shores were not separated by wide and deep oceans. In consequence, we can test with faunal provinces reconstructions of Pangaea that rest on edge-fitting and paleomagnetic data.

Sediments that reflect climate add a great deal of information, as Judith Totman Parrish of the University of Arizona has shown. Salt deposits (evaporites) form in the hot, dry coastal zone of the subtropics between *c.* 15° and 35° N and S, and so do deserts with their windblown sands.

135

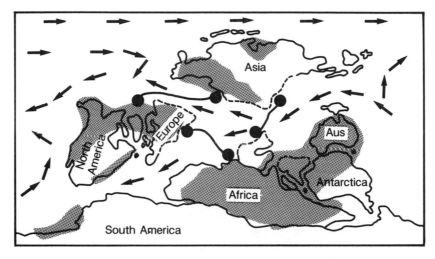

Figure 7.6. The late Silurian world about 415 my ago. Land is shaded, the arrows indicate possible main ocean currents, and the black dots mark opposing areas which according to the fossil evidence ought to be connected (black lines), and could not be separated by major ocean currents. The odd shape of the present continents is again due to the projection.

Many climate zones also have typical soils. Most are easily eroded, but those of the subtropics tend to form a calcareous hardpan known as calcrete which, like the limestone it resembles, strongly resists the ravages of time. Glacial deposits are restricted to high latitudes and even during ice ages usually stay above 45°–60° (but see Section 14.7). Coal, on the other hand, is mainly a product of warm climates and high rainfall, conditions best fulfilled in the equatorial rain belt.

With a great deal of this sort of information already in the literature, we can fill our paleogeographic maps, put on equators and poles that are independent of the paleomagnetic data, check that the pieces fit geologically, make sure we do not have oceans where the fauna says we should not and, as a bonus, get a better idea what Pangaea looked like. All this for the price of reading several thousand geological descriptions from all over the world, often not available in our library or written in a language we cannot read, a daunting task.

Still, attempts have been made. Art Boucot, a paleontologist at Oregon State University, compared Silurian and Devonian faunal provinces with paleogeographic maps and was not pleased with what he saw, because his shallow water faunas crossed too many oceans (Figure 7.6). I, because of my background, would rely more on soils and sediments which fit the paleomagnetic reconstructions better, and wonder whether there might not be ways to structure Boucot's data differently. Newer studies

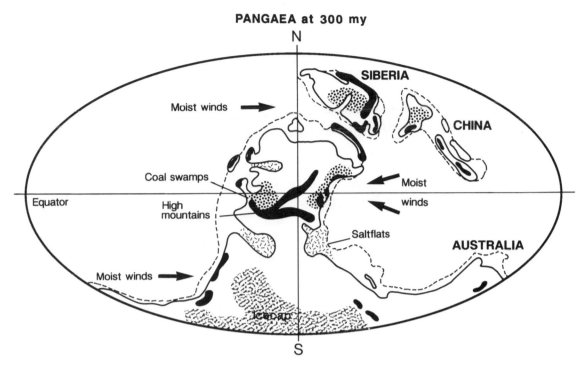

Figure 7.7. Late in the Carboniferous (Pennsylvanian) Pangaea was already a supercontinent, although incomplete, with Siberia and China plus southeast Asia still waiting in the wings. The south polar icecap, the lush equatorial swamps in the east, the subtropical deserts and the evaporites fit a simple and plausible climatic pattern (see Figure 3.4) with a strong influence of the monsoon. The shape of the south polar region is a result of the map projection.

indicate that this is indeed so and it is necessary to deal separately with different components of the fauna because they tell different stories.

Consider once more Pangaea 300 my ago, almost but not quite complete (Figure 7.7). The only icecap was located appropriately in the far south, there being no land at the North Pole, and in the subtropics evaporites deposited on desert flats and coastal salt pans were abundant. Reaching almost from pole to pole, the supercontinent barred the way to ocean currents and planetary winds. Our understanding of climate suggests wet west coasts in the temperate zones and much rain in the equatorial region. Coal beds formed in swamp forests do bear out this conjecture, but a few occur at surprisingly high latitudes. Perhaps they were the Paleozoic equivalent of Scottish peat bogs and Canadian muskeg. Or are they telling us that our understanding of the climate of today does not suffice to understand that of the Paleozoic? Or perhaps not all is right with our map!

Elsewhere, this huge land, here and there bordered or traversed by lofty mountains remaining from the collisions of its birth, was bone-dry in its interior or dominated by monsoons of a strength not seen even in present India. The widespread redbeds and salt deposits of the desert and the dune sands of the Permian show that during its brief existence much of Pangaea was a harsh, arid world not unlike present central Asia.

The availability of a reasonably good paleogeographic base and the lure of a world so different from our own have tempted John Kutzbach, a climatologist at the University of Wisconsin, to model the late Pangaean climate and Panthalassan ocean currents. Using several different sets of basic assumptions, the answer was always that the continent had hot summers, cold winters, and a grand monsoon circulation. Much would have been arid all year, the rest seasonally, while rain would have poured down only in the eastern coastal regions and in the tropics and beyond 40° N and S on the west coast, in fair agreement with Figure 7.7. The area of the coal swamps in the far north had enough rainfall, *c.* 700–800 mm/year, for a temperate rain-forest, as in coastal Washington State today.

7.4   PANGAEA DISMEMBERED

Already in the Triassic, before all pieces had come together, the super-continent began to show signs of stress. In New England and elsewhere on the eastern seaboard rift valleys formed as blocks of crust subsided along faults, accompanied by copious lava flows that covered large areas, the plateau basalts. The Palisades on the Hudson River are a fine example. Shortly afterwards, such rifts appeared elsewhere in Pangaea too, an-nouncing that all was not well, but it would be hasty to conclude that breakup and drift were about to begin.

These Triassic rifts were similar to those that now scar East Africa from Mozambique through Tanzania and Kenya to the Red Sea, forming fault-bound valleys on a broad, gentle rise a few kilometers high and up to 1,000 km wide. This African rift system, more than 20 my old and famous as the birthplace of the earliest human beings, has so far produced a beautiful country, spectacular volcanoes and grand wildlife, but has failed to sunder the continent. Only in the Red Sea and Gulf of Aden has separation begun, and that just a few million years ago.

Clearly, rifting and drifting are different things. Although cracks ap-peared in Pangaea as early as 200 my ago, marking the trace of future oceans, the North Atlantic did not begin to open until 20–30 my later, and the South Atlantic waited another 50 my (Figure 7.8). For a while, the great eastern embayment of the Tethys, growing westward, divided

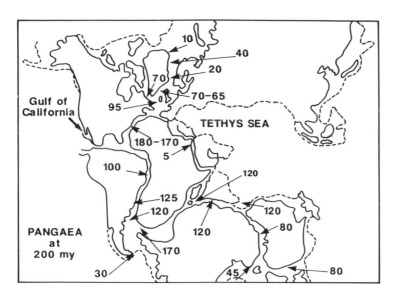

Figure 7.8. When Pangaea broke apart, drift began at different times in different places. The earliest new oceans were the southern North Atlantic and the one separating Africa from Antarctica; the South Atlantic followed considerably later. The process is not complete; the Red Sea began to open 5 my ago and the Gulf of California is even younger; neither is finished. In 50 million years Baja and Southern California will be a long, narrow continent somewhere in the Gulf of Alaska, separated by an ocean from North America.

the supercontinent into two parts, Laurasia and ancient Gondwana, but then Laurasia itself split and Gondwana, shattered like a pane of glass, sent pieces every which way to form new continents or to attach themselves to Asia and Europe. It all took time, and even now new oceans are forming in the Red Sea and the Gulf of California.

What, precisely, happens when a continent disintegrates? The African rift valleys and the embryonic Atlantic and Indian oceans tell us much about the sequence of events. It all begins (Figure 7.9) when the crust forms a "hot" dome, probably when it drifts over a heat source well below the lithosphere, a hotspot. I note parenthetically here that scientific jargon is shifting from bad Greco-Latin to sometimes excessively colloquial English; the fundamental particles of physics have "flavors," and the earth has hotspots.

As the brittle crust bulges up, it stretches and fractures, and keystone blocks sink to relieve the stress and accommodate the increase in surface area. Three rift valleys meeting at the top of the dome are an efficient and apparently common way to accomplish this. Faulting and stretching of the hot rock thin the lithosphere, and lava rises through fissures to build volcanoes. At this stage the rift valley is high above sea level; the land slopes away from it, and the rivers drain outward. As a result, the

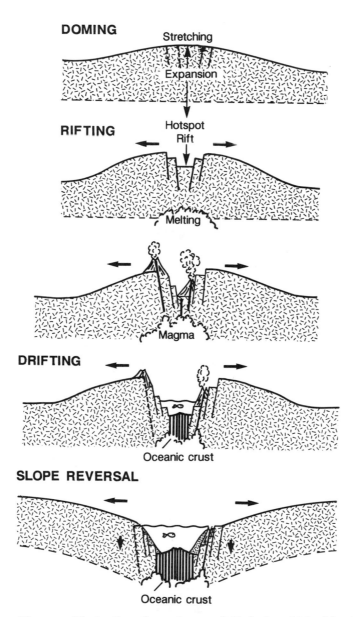

Figure 7.9. The breakup of a continent probably begins with local heating by a hotspot beneath the lithosphere. The expansion creates a surface bulge which cracks, forming a rift valley lined with volcanoes. Eventually, but not always, the stationary phase of rifting is followed by drift, oceanic crust is intruded, and the sea invades the rift valley. The margins cool as the distance to the mid-ocean ridge increases, the relief of the continent reverses, and major continental drainage now enters the sea.

140

first sediments in what will be a new ocean are laid down in patches in lakes, river plains and deserts, or as blankets of volcanic ash. Although the sides of the rift valley diverge as the crust stretches, the plate has not yet sundered and there is no drift.

As East Africa shows, continents may remain in this stage for millions of years, and sometimes go no further. The Pangaean crust, however, was ultimately stretched five-fold and, now very thin at the bottom of the rift valley, sank below sea level. An ocean formed, shallow and narrow to begin with, its edges turned up like those of the Red Sea or the Gulf of California, and oceanic lavas from the mantle began to create oceanic crust. Drifting had begun. The distance between the new continents and the hot mid-ocean ridge gradually increased, the continental margins cooled, became more dense, and subsided. As a result the slope of the continents reversed toward the ocean and large rivers began to carry sediment into the sea.

When Pangaea came under stress, domes with triple rifts may have formed so close together that their branches connected to become continuous rift valleys, forerunners of new oceans (Figure 7.10). In each triplet the third branch was a failure, pointing away from the new ocean and ending blind. Although of no further significance to continental drift, these failed rifts had, as we shall see below, a splendid future in oil.

7.5   DOMES AND HOTSPOTS

Continents join because they happen to be on a collision course, but why they disintegrate is less obvious. There is good reason to believe that hotspots are a key element and that the process can be seen at work today, but what causes hotspots?

Most of the world's active volcanoes are associated either with mid-ocean ridges or with subduction zones, but some, remote from plate boundaries, are not so easily pigeonholed. In the heart of the Pacific plate lie the Hawaiian Islands, a string of active and recently extinct volcanoes sitting atop a broad rise of the ocean floor (Figure 7.11). The easternmost volcanoes, those of the island of Hawaii, are active and a mere few hundred thousand years old, but those of the adjacent island of Maui are barely alive or just extinct. The farther west we go, the longer the volcanoes have been dead. Beyond the last island there are other, even older ones, truncated by waves, reduced to pinnacles or even entirely submerged beneath the sea. Where the chain, now deeply sunken, reaches an age of 45 my, it turns sharply north, but the age progression continues.

Drifting continents, rising mountains

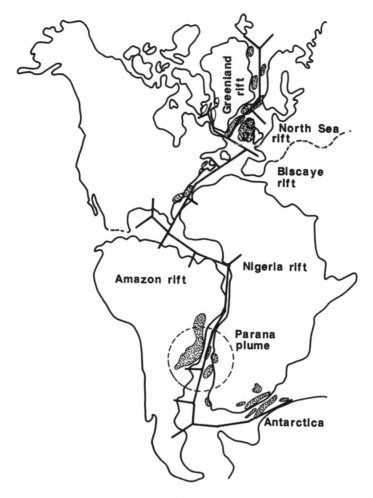

Figure 7.10. The dismemberment of Pangaea may have begun with rows of domes above sublithospheric heat sources. Such domes often crack in the pattern of a three-pointed star (heavy black lines). Two rifts of each dome ultimately connect with their neighbors to form the new ocean, in this case the Atlantic, whereas the third, an abortive rift, dies. The growth of some domes was accompanied by copious outpourings of lavas called plateau basalts (stippled).

Other linear volcanic chains in the Pacific, although not all, have the same trend and are also only active at their eastern end.

To account for this clearly meaningful pattern, assume a heat source deep below the lithosphere, due to causes not yet fully understood. A hot column of mantle material rises until it reaches the base of the lithosphere and heats it, creating a broad swell. The top of the column melts partly, and the magma breaks through the crust, where a volcano grows, then dies as the plate moves past the hotspot. The dead volcano sinks, erodes away, and a new one rises behind it, until a whole chain

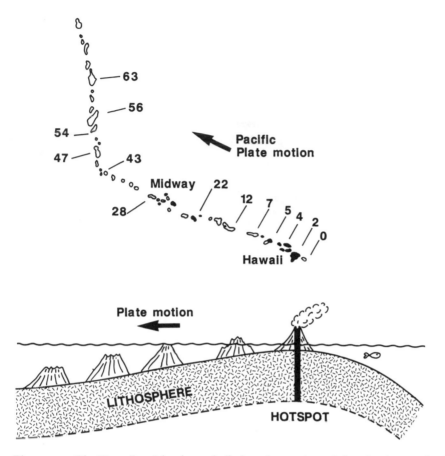

Figure 7.11. The Hawaiian Islands are built by a hotspot located deep in the mantle. The lithosphere fissures and volcanoes form, then die when they are beyond their source. A new volcano then forms upstream; a small one is already forming under the sea southeast of Hawaii. Emerged volcanoes are black. The result is an island chain parallel to the direction of plate motion. The age progression of the volcanoes in millions of years measures the speed of the movement. It is clear that more than 45 my ago the Pacific plate moved in a more northerly direction.

is formed, young at one end and old at the other. Although we do not yet know why there should be hotspots, nor from what depth they rise, the evidence for their existence is decisive.

The perceptive reader has probably noticed that the way to determine plate motions described in Section 6.6 has one regrettable flaw. Because all determinations, carried from one plate to the next, are relative, the whole lithospheric shell could slip around the earth, plates and all, but we would never know. The stationary hotspots, being anchored well below the lithosphere, come in handy because the lines of volcanoes give us the direction of plate motion and the age progression measures the velocity. If the plate changes direction, the volcanic chain also turns.

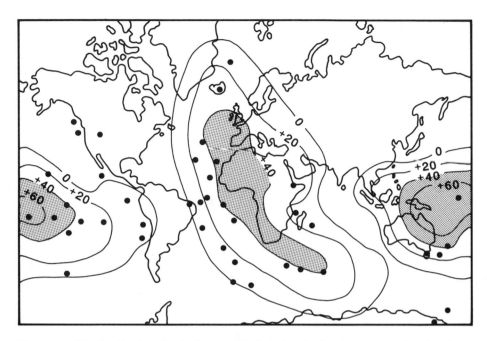

Figure 7.12. The distribution of active hotspots (black dots) is closely related to vast, low elevations of the earth surface shown here by contours in meters above the geoid, the average surface of the sea if no land were in the way. Seismological evidence indicates that the bulges are due to rising flow in the mantle.

Thus hotspots relate the entire jigsaw puzzle to the standard frame of reference based on the earth's spin axis and the equator.

Whether a hotspot is indeed fixed can be tested if there are enough of them on a single plate, each marking it with a linear island chain. If it survives the test, the plate becomes a point of reference to establish whether hotspots under other plates are also fixed. The Hawaiian island chain suggests, and new evidence confirms, that most if not all hotspots have fixed positions.

Hotspots are not common on continents nowadays, although geologists with a broad definition of a hotspot would disagree, but that was different in the past. At present they are clustered in an east–west zone across the Pacific and a north–south zone across the Atlantic, Africa and Europe (Figure 7.12). Because all of them are located on two large swells that rise above the geoid, the smooth surface the earth would have if there was only ocean, they probably mark the sites of huge masses of convecting mantle that have raised the top of the asthenosphere.

We have mentioned the enormous lava flows, the so-called plateau basalts, that accompanied the initial rifting. A classic case is the giant volcano which, 65 my ago, spread nearly one million cubic miles of lava

(2,500,000 km³), now known as the Deccan Traps, across central western India in less than half a million years. The eruptions, being immensely larger than the Hawaiian hotspots or the gentle mantle upwelling on mid-ocean ridges, surely had a dramatic impact on the regional and perhaps even the global environment at the time.

Bob White and Dan McKenzie of the University of Cambridge have shown that all three kinds of mantle upwelling, mid-ocean ridges, hot-spots, and the mantle plumes that pour out plateau basalts, are in principle identical. When tectonic forces stretch the lithosphere, it fractures, the mantle rises, part of it melts and the melt erupts at the surface. Whether this happens gently or catastrophically depends on the temperature of the mantle.

If the mantle is 100–150 °C hotter than usual, as is often true for mantle plumes, and the plume encounters the lithosphere, it mushrooms underneath it and heats and expands the lithosphere into a flattish bulge 1,000 to 2,000 km in diameter (Figure 7.13). The partial melting is especially copious at that high temperature, and produces a huge flow of lava at the surface, while a large volume of magma also consolidates against the base of the crust and thickens it, adding from five to ten million cubic kilometers to the crust. Eventually, however, the ageing plume settles down to a more steady flow, the mushroom decreases in size, and less and less lava erupts at the surface. As the plate moves on, a trail of ordinary hotspot volcanoes connects it with the original plateau basalts (Figure 7.14).

Still, it does not seem that Pangaea could have been set on the road to ruin only by rows of mantle plumes. The rising plume does not melt and erupt until the pressure of the overburden is reduced; it follows rather than causes the stretching. Under the thick continental lithosphere this means a long time of heating and thinning the lithosphere from below. This domes the crust up and fractures it at the top, until the pressure is reduced enough to allow copious melting to take place. A wet lower lithosphere would help to melt it. In short, the plume must incubate. Some of the rifts that fragmented Pangaea failed to produce plateau basalts (Figure 7.10); presumably they did not incubate long enough. Once the breakup is well under way, however, hotspots may help plates to slide down the gentle slopes of the asthenosphere domes (Figure 7.12). And there the issue rests for now.

Why should Pangaea have fallen victim to so virulent a case of the hotspots, whereas the present ones, common as they are, do not appear to threaten the integrity of our continents? To answer this, it has been suggested that the continental crust acts as a blanket. If the blanket is

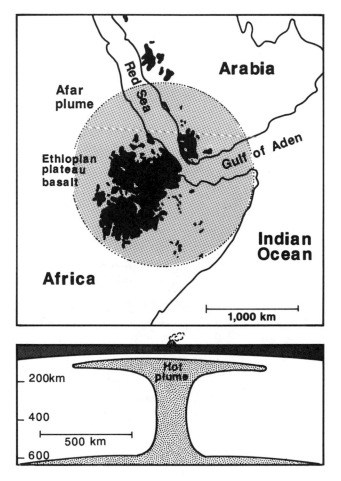

Figure 7.13. The lavas of Aden and Ethiopia are examples of White and McKenzie's hot plume model (bottom). The dome produced by the Afar plume, pouring out vast plateau basalts, straddles the Red Sea which here forms a classical triple rift with a failed branch pointing southeastward into Africa.

small, the accumulating heat escapes around the edges, but if it is large, the heat flow is impeded, and stretching and thinning of the lithosphere occur. This might render the lithosphere more vulnerable to the action of hotspots and to the lateral pull of mantle convection as well. Is this true? We do not know. Computer models show that the thinning of the lithosphere would be rapid, a matter of only a few tens of millions of years, and that the heat flow would focus on the thinnest spots. This makes it more plausible, but models need information and there is much we do not know about the mantle.

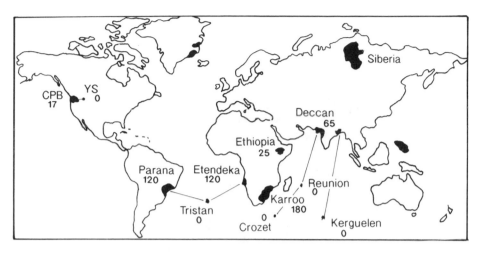

Figure 7.14. Plateau basalts produced by mantle plumes of many ages occur on the continents and in the oceans. When the plumes grow older, the copious flow of lava decreases to that of ordinary hotspots, and as the plate moves on, a hotspot trail connects the later volcanoes with their more impressive origins (thin lines). The numbers are plume dates in millions of years. YS: Yellowstone hotspot, the present location of the plume that formed the Columbia River Plateau Basalt (CPB).

7.6 EDGES OF RIFTS AND MARGINS OF CONTINENTS

About half the present continental margins of the world began as edges of Pangaea's rifts. Called passive margins because of their present (but by no means past!) lack of earthquake and volcanic activity, they have an interesting history and considerable economic importance.

When the embryonic rifts began to sink below sea level, their flanks were stairs of fault blocks descending to the axis (Figure 7.15). The first invasions of the sea were hesitant; wherever closed basins existed and the climate was sunny and dry, as in the early South Atlantic, thick evaporites were laid down intermittently on top of the earlier continental deposits. But in time the shallow sea became permanent and in its clear waters reefs flourished on the raised edges of the fault blocks, with quiet lagoons and beaches behind. Such narrow seas are often fertile, as the present Gulf of California illustrates, and organic matter in abundance settled to the bottom where it was buried in the mud.

Eventually, the sea deepened below the sun-lit level where coral reefs can flourish, and the reefs died. At the same time large rivers, following the reversal of the slope (Figure 7.9), began to bring sediment into the sea in ever larger quantities, burying reefs and organic mud under thick

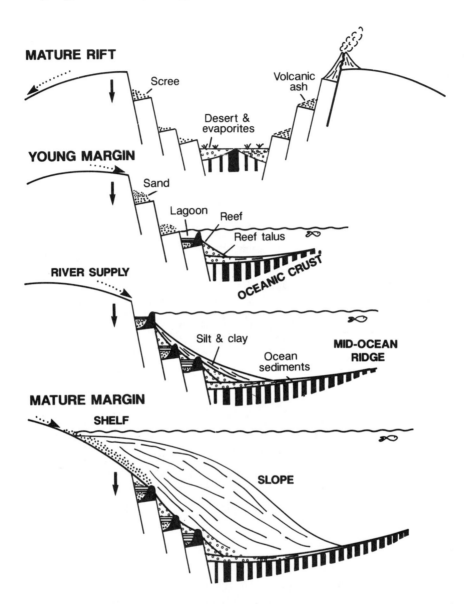

MATURE RIFT

Scree

Volcanic ash

Desert & evaporites

YOUNG MARGIN

Sand

Lagoon   Reef

Reef talus

OCEANIC CRUST

RIVER SUPPLY

Silt & clay

Ocean sediments

MID-OCEAN RIDGE

MATURE MARGIN

SHELF

SLOPE

Figure 7.15. A mature continental rift forms an elongate, shallow trough near sea level where evaporites form together with volcanic sands and sediment washed down from the slopes. With the onset of drifting, the rift widens, and the margins cool and subside. The sea invades, coral reefs grow, organic matter accumulates in the sediments, and beaches form along the shores. When subsidence begins to reverse the direction of continental drainage, large quantities of silt and clay bury the reefs and black organic shales, creating the continental shelves of a passive margin.

layers of silt and clay. If the sediment supply was large, as on the east coast of North America, deposition kept up with the subsidence, and a prism of shelf sediments formed that now measures from 6 to 10 km thick. On other continental margins, for example those of West Africa or Western Australia, there was less runoff, and the reefs and fault blocks of the early rift are not so deeply buried and can be more easily examined.

Taken together, the passive margins of the world are the largest storehouse of sediments on earth. They also contain a great deal of oil. About two-thirds of all giant oil fields, those yielding more than half a billion barrels, lie buried in modern or ancient passive continental margins. There they hold among themselves more than half of the world's oil reserves. Most of that good oil is found not on the margins of rifts that became real oceans, but in the sediments of rifts that failed.

To extract oil efficiently and profitably several conditions must be met (Figure 7.16). There must be abundant organic matter in a source bed of some kind, decayed marine plankton preserved in dark shale or limestone. This precursor of oil is not usable; it must simmer at moderate temperature and under the pressure of a moderate overburden of sediment before it converts into oil and gas. If the temperature is too high or the time too long, natural gas will form rather than oil, or the organic matter may turn into useless carbon dioxide. If the time, depth, or temperature are too limited, the organic matter stays in the shale, and the shale oil, although it might be mined some future day, is not now a particularly valuable commodity.

The oil must migrate out of the shale where it is too finely dispersed to be extracted, and travel to a porous and permeable reservoir before we can tap it with the drill and pump it out. Sandstones and reef limestones make good reservoirs. In the reservoir the oil, being lighter, separates from the water that is always there, and floats on top. If there is gas, it gathers above the oil. The reservoir must be sealed tightly, a clay will do, or oil and gas continue upward to the surface and seep away.

Rifted margins satisfy all these conditions. They have much organic matter for an oil source, the reefs and beach sands for reservoirs, and younger clays for a seal. In their early history they are warm but not too deeply buried; oil and gas can form and accumulate. Failed rifts are even better. They stay longer near the source of the heat, and there is less risk that their oil will be destroyed by excessive pressure, because they do not subside so far. Their reservoirs are not so deeply buried and so are less costly to drill and, even better, many of them are now on land, where recovery is easier and much cheaper than at sea. On mature margins like the Atlantic seaboard, the early reservoirs are very deep, and even if their

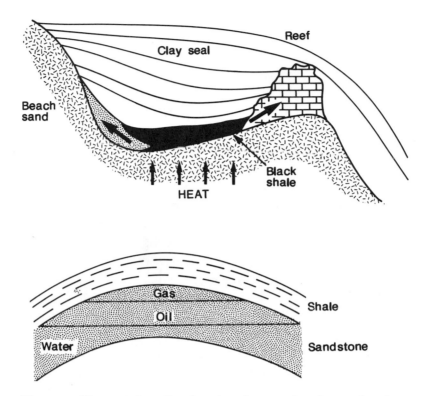

Figure 7.16. To accumulate oil and gas in such a way that they can later be extracted with profit, one needs a source bed rich in organic matter. The organic matter must mature by gentle but prolonged heating and the oil and gas must be able to migrate into permeable and porous reservoirs, reefs for example or beach sands. In the reservoir the oil separates from the omnipresent water and floats to the top; gas accumulates above it. A clay or silt seal is required to prevent the oil and gas from escaping.

contents have not already been converted into useless carbon dioxide, they will be expensive to find and exploit. The overlying shelf wedge of silts and clays is not rich in organic matter, has few good reservoirs and did not benefit from adequate heating. Oil accumulations do occur there, but they are not as worthwhile as the failed rifts of the North Sea, Venezuela, Nigeria, or the Persian Gulf.

8

Converging plates and colliding continents

Where plates converge, subduction recycles the oceanic crust and erases the evidence for the existence of old oceans, although not always completely. Thus it is in former subduction zones and collision sutures that the only remains of ancient ocean floors are preserved. Together with paleomagnetic and paleontological data, these are our only sources of information regarding the vanished oceans of the past.

Narrow strips of deformed, metamorphosed and often deeply eroded rocks, called mobile belts (Figure 8.1), are common on all continents where they surround stable cores of great antiquity, the cratons. The oldest mobile belts are early Precambrian in age and their significance is not yet fully clear (Section 13.5), but those of the late Precambrian and Phanerozoic represent converging plate boundaries, either extinct like the Appalachian, Alpine and Scottish–Norwegian ranges, or active like the Andes or Himalayas. All contain andesitic volcanic rocks and intensely metamorphosed and folded sediments, often intruded by granites. Therefore continental margins bordered by subduction zones are known as active margins.

Plate-tectonic theory says simple things about what must happen at converging plate boundaries, but as we have gone on to apply the theory to ever more real mobile belts, the difficulties have increased and reality has overtaken theory.

8.1 SCENARIOS OF SUBDUCTION

The simplest case of convergence is a collision between two oceanic plates. Being of equal density, either may be subducted, and a volcanic arc forms on the overriding plate at the point where the subducted slab is deep enough to begin to melt. If an oceanic plate meets a plate bearing a continent, only the heavier oceanic plate can be forced down, and

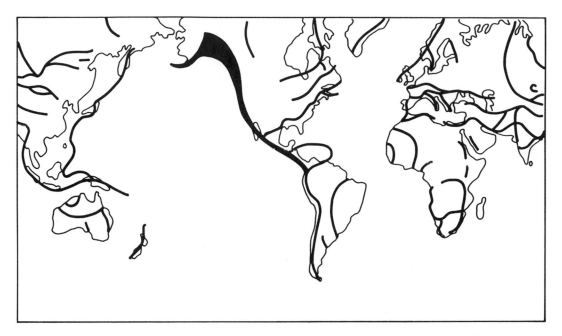

Figure 8.1. Subduction zones and sutures mark places where plates converge or did once converge, the so-called mobile belts. The oldest sutures shown are of Precambrian age and lie in the interiors of the continents, for example Africa and North America. In Asia, the sutures are mainly late Precambrian and Paleozoic in age. The youngest, such as those on the periphery of the Pacific, are still active.

volcanoes will always grow on the continental side. If the plate boundary should happen to lie a little seaward of the continental edge, once again either plate may be subducted, but if the subducting slab is the one attached to the continent, the end will come as soon as all oceanic crust has been consumed (Figure 8.2). The direction of subduction must then be reversed, and a new chain of volcanoes will form, this time on the opposite side. The old slab becomes detached, is absorbed in the mantle and, because it no longer drags the plate down, the crust above it rises. Such a reversal, involving abandonment of a slab of lithosphere 100 km thick and sometimes up to 700 km long, with the simultaneous creation of a new one, is a mind-boggling event, but theory demands it because continents cannot be subducted, and the evidence is there that it actually does happen.

Let us take a closer look at the subduction zone (Figure 8.3). As one approaches the trench from the ocean, a series of small fault blocks usually step down into it, the result of the bending and stretching of a brittle upper crust. On the opposite side of the trench, where the plates meet, the fore-arc builds up from deformed slivers of the upper part of the subducting crust, sheared off the way a carpenter's plane pulls wood

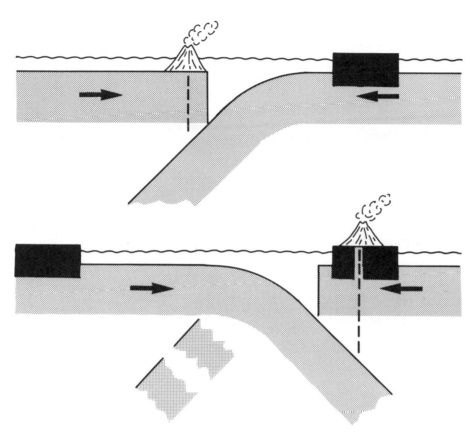

Figure 8.2. When plates collide, subduction consumes one of them. If that plate should happen to carry a continent, the continent will in time find itself at the edge of the subduction zone. Being too light, it cannot go down, the subduction must reverse its direction, and the other plate descends. Continental crust is shown in black.

curls off a piece of lumber. The oldest slivers are on top, and successive later ones are thrust under the wedge that is already there, each new one raising the pile a little more. Eventually, parts of the fore-arc appear above sea level to form islands of deformed and metamorphosed sediments, sometimes mingled with fragments of oceanic crust. Barbados in the Caribbean is an example, as are the Coast Ranges of Oregon and Washington, a sedimentary arc piled against a continental plate edge. Erosion then sets in and sediments are deposited in the trench. Behind the island arc lies a fore-arc basin filled with sediment washed in from both sides, and behind it a volcanic arc, with sometimes a back-arc basin on the far side of the subduction complex. This basin has a young crust, a high heat flow, and a set of magnetic anomalies, all showing that it is involved in seafloor-spreading. The cause and manner of back-arc spreading are still being debated. As with the forces that drive the plates, we

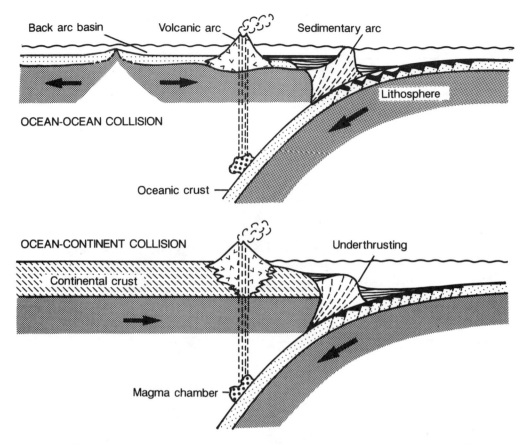

Figure 8.3. A subduction zone is wider than just the trench. An ocean–ocean collision zone begins where the plate bends down, often with small step faults compensating for the bending of the brittle upper crust. For clarity their size is exaggerated here. In the trench, the oceanic crust (stippled) and pelagic sediments (black) are buried under trench deposits entering in from the opposite side. Sediments and at times parts of the oceanic crust are scraped off by the edge of the overriding plate and thrust under previous slivers, creating a deformed sedimentary arc. A sediment-filled basin and volcanic arc lie beyond. Behind the volcanic arc spreading occurs in a back-arc basin. Continent–ocean collisions are similar, except that the deformed arc usually lies directly against the edge of the continent without a string of islands. It also normally lacks clear back-arc spreading.

are handicapped here because we lack enough cases of back-arc spreading to test all possible hypotheses.

The volcanic arc consists of andesite lavas with prodigious amounts of volcanic ash. Unlike basaltic mid-ocean volcanoes such as Hawaii or Iceland, which tend to be relatively well behaved, the volcanoes of subduction zones are nasty, violently explosive, and unpredictable. Krakatoa, Vesuvius and Mount St. Helens are well-known examples of this behavior. The volcanic arc is usually situated where the subducted slab reaches a depth of 100–120 km. At that depth its volatile components, the sediments with their enclosed seawater and the altered uppermost basalt, melt in a horizontal, tubular chamber, where they generate a magma rich in water vapor, carbon dioxide and other volatiles. The magma rises, melting its way upward out of the tube as plumes, each crowned at the surface by a volcano. This model is simple in principle, but leaves some questions unanswered. Why, for example, do volcanoes erupt episodically instead of continuously, while the subduction itself is continuous?

The fore-arc wedge consists of sediments and basalt scraped from the surface of the subducting plate by the edge of the other one. This is plausible enough, because the sediments are soft and the top of the basalt, deeply altered as it flowed out into cold seawater, is not very solid either. Still, a good part of the friable layer is subducted, unlikely as that may seem to anyone who has ever held a handful of oceanic ooze, to be melted at great depth so that it may feed andesitic volcanoes. How much is dragged down and how much remains at the surface to end up in the fore-arc wedge is a key question. Under the west coast of Mexico about 200 km^3 of sediment per kilometer of trench length have been subducted in the Cenozoic alone. That should be enough to build a fore-arc wedge some 50 km wide, but the existing one is much narrower, implying that most sediment was subducted.

Occasionally there is no wedge at all, or one so small that it must be temporary. Sometimes also the edge of the overriding plate appears to be sinking rather than being pushed up. One explanation for such behavior is that the subducting plate grinds away the bottom of the overriding one. This thins the crust and reduces its buoyancy enough for even a continental plate edge to subside, in sharp contrast with the more usual subduction process that results in the growth of the continental edge by accretion. The evidence for this tectonic erosion is still meager, but given the importance we attach to the idea that the continental crust is indestructible, it is a troubling suggestion.

The sutures of ancient collisions often contain substantial slabs of oceanic crust made of oceanic basalt, its alteration products and metamorphic equivalents, and associated oceanic sediments. Collectively these immigrants from another world are known as ophiolites. How do ophiolites escape the usual fate of the oceanic crust of being subducted into the depths of the mantle, a fate that would seem inevitable except for a few scrapings that end up in the fore-arc wedge?

Imagine a plate with a continental edge advancing on a mid-ocean ridge (Figure 8.4) as, for example, the South American plate does. In time, the mid-ocean ridge itself is subducted and the subduction zone

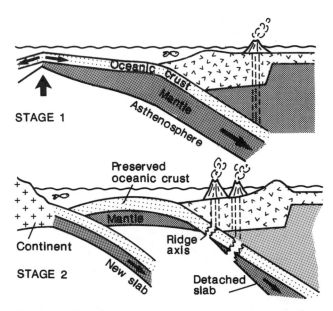

Figure 8.4. Normally, the old oceanic crust is recycled into the mantle rather than being preserved on the continents, but special circumstances may permit its survival. In stage 1 a mid-ocean ridge is about to be subducted under an island arc. Subduction, driven by the weight of the cold hanging slab, ceases when this happens, and the slab is detached and sinks. A new subduction zone forms farther out at sea, preserving an intervening stretch of oceanic crust. Eventually (stage 2) the continents collide, and the preserved oceanic crust is thrust up onto one or the other continent as an ophiolite complex.

dies, while the cold, heavy slab detaches and continues to sink. The continental plate continues to advance, and a new subduction zone forms somewhere on the far flank of the former ridge. A strip of oceanic crust remains attached to the continental plate that will be compressed and folded into the fore-arc of the new subduction zone, and eventually incorporated in the collision of the two continents, testimony to a vanished ocean basin.

Trenches with double arcs resulting from the collision of oceanic plates are common in Indonesia, the southwestern Pacific, and the Philippines, and parts of the Lesser Antilles are of this type too. Continent–ocean collisions occur on the west side of the Americas from Chile to Mexico, less vigorously along the coasts of Oregon and Washington, and strongly again from southern Alaska along the Aleutian chain to Kamchatka and Japan. This is the "ring of fire" that marks the world's largest ocean as it is slowly being reduced in size. The Pacific is dying.

8.2 OROGENY AND GEOSYNCLINES: CLOSING THE THIRD ATLANTIC

Orogeny is the geological term for the folding and thrusting, metamorphism, intrusion of igneous rocks and uplift that build most mountain ranges (orogens). Orogens contain large volumes of sediment, once laid down in great thickness in elongated, narrow, subsiding troughs. Such troughs have long been called geosynclines and scholars have debated for over a century whether trenches are the early stages of geosynclines, before they are deformed and made into mountains. Unhappily the definition of the term geosyncline includes its ultimate conversion to a folded mountain range, and so we cannot know whether our modern trenches are geosynclines until they become mountain ranges. The term is outdated now, but it is a classical example of how one can paint oneself into a corner in a matter of terminology.

In plate-tectonic terms, a geosyncline is a trench together with its sediment fill, fore-arc wedge and volcanic arc. Subsidence, sedimentation and deformation occur simultaneously rather than, as in the classic geosyncline, sequentially.

During the assembling of Pangaea, North America was joined successively to several then still separate pieces of Europe, thereby closing a Paleozoic Atlantic also known as the "Iapetus Ocean" (Figure 8.5). The collisions raised mountains from Texas to northeastern North America, and all over western and northern Europe. It was for the Appalachians that, in the late 1960s, Jack Bird and John Dewey, then respectively at the State University of New York in Albany, and at Cambridge, pioneered the application of plate tectonics to major orogens. The task has turned out to be far more complicated and arduous than their early publications assumed.

The earliest evidence for a subduction zone in eastern North America goes back about 500 my, and a phase of mountain building, the Taconic orogeny, followed soon after. It was accompanied by volcanic activity, by considerable folding, and by the uplift of a land area to the east of the present Atlantic seaboard. Prior to the Taconic orogeny, the sediments that filled the subduction trench, forming there the deposits of the so-called Appalachian geosyncline, had come from the west, but afterwards a huge delta spread from the new eastern lands, laying down a thick wedge of redbeds in a subtropical climate.

The Taconic orogeny was only the beginning, and it closed no ocean. Volcanic activity and sedimentation continued, and so did subduction,

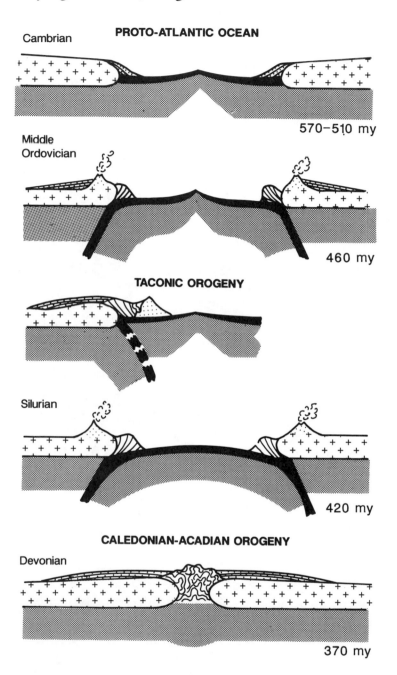

Figure 8.5. In the Paleozoic the Iapetus Ocean or proto-Atlantic separated North America from Europe and Gondwana. At first, it had passive margins, but subduction began in the Cambrian. The Ordovician witnessed a phase of intense mountain building, the Taconic orogeny, possibly caused by a collision with a hypothetical, small continental block offshore (not shown). After more uneventful subduction, northeastern North America collided in the Devonian with the little continent Baltica (now northwestern Europe). This Caledonian–Acadian orogeny raised a vast mountain range that shed its sediments across two continents.

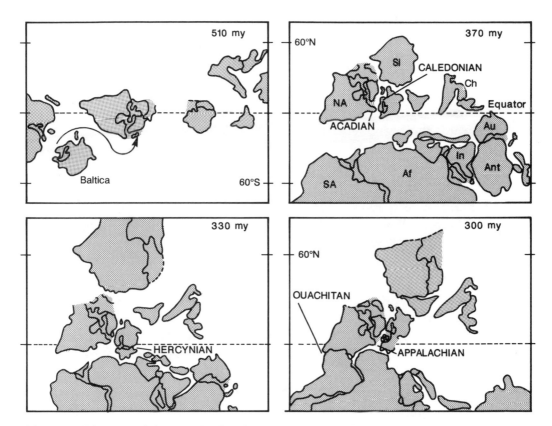

Figure 8.6. These maps help us to visualize the complicated series of events that accompanied the construction of Pangaea's northern part. First Baltica, following a curved path, collided with North America (NA) during the Caledonian–Acadian orogeny. Then central North Africa (Af) pushed into Baltica, while Gondwana, starting with its South American corner, became attached to North America during the Ouachitan and Appalachian orogenies. Ch: China; In: India; Si: Siberia; Au: Australia; Ant: Antarctica; SA: South America.

until about 370 my ago the northeastern tip of North America collided with the small continent Baltica, now northern Europe (Figure 8.6). This Acadian–Caledonian orogeny raised a huge mountain chain, of which the remains are found in New England, eastern Canada, Scotland and Norway. The Acadian and Caledonian mountains shed sediments to both sides: westward across eastern North America to form the Catskill delta, and southeastward over England and Europe as the Old Red Sandstone.

The final closure of the Paleozoic Iapetus came during a third collision, the Appalachian orogeny. It began when the South American corner of Gondwana ran into Texas and Oklahoma and raised the Ouachita mountains. Next, the southern Appalachians were thrust up by a collision with northwestern Africa. An African limestone plateau attached itself to

North America; it is now the State of Florida. Finally, a number of smaller pieces squashed between Baltica and Africa formed the present core of central and western Europe. The time was 300 my ago.

It is a fascinating story, but there are problems with it. Consider the southern Appalachians. They experienced the Taconic and Acadian orogenies during a long period of continent–ocean subduction, but afterwards there was still an ocean to the east and the true clash of continent against continent was yet to come. The great volcanism, widespread intrusion of granite, and deformation that accompanied both orogenies seem excessive if what happened was merely intensified subduction or collision with an island arc. More surprisingly, the two orogenies raised land areas large enough to feed vast deltas from the east. Island arcs do not usually support big rivers nor do they produce adequately copious amounts of sediment.

Seeking the more substantial bang that seems to be required, the most favored solution is a collision with a continent assumed to have been lounging conveniently offshore, and blessed with the lovely name of Avalonia. Intensive orogenies providing a prolific sediment source are more plausible during continent-to-continent encounters, even if one of them was small and had more or less to be pulled out of a hat. The hypothesis explains the facts, but the need to draw on such creativity to keep things straight is inelegant and somewhat perturbing.

Since the birth of these ideas in the 1970s, other concepts have come to the fore that suggest that, perhaps, Avalonia was not a real continent, but a raft of fragments of oceanic (and perhaps continental) crust, gathered together by chance and subsequently welded onto the North American continent. There are, for example, large geological and paleontological differences between components of the Caledonian and Acadian orogens that suggest that their origins were far apart in time and space.

8.3   FLOTSAM AND JETSAM

The classical case of such geological driftwood is found in western North America. Once upon a time, in the Precambrian, this was a rifted margin, its opposite continent long gone, to become welded to Asia. Now located in eastern Nevada, Idaho and British Columbia, this margin remained passive for hundreds of millions of years. A subduction zone may have lingered offshore, but not until the late Cambrian is there any evidence for it on the continent. Since then, the active Pacific margin of North America has steadily grown westward by accretion, overriding the Pacific plate in several steps, two in the Paleozoic, others in the Jurassic and Cretaceous. Each step was accompanied by an orogeny.

The Pacific Ocean is the descendant of Panthalassa, but the entire Paleozoic crust of the superocean has been subducted and eventually most of its Mesozoic crust too. At the same time, the North American continent increased greatly in size. The process is illustrated by the history of coastal California. In the early Mesozoic, a major island arc was driven against the continental margin, leaving a suture that can still be seen in the foothills of the Sierra Nevada. Steady subduction followed, adding fore-arc wedges to the continent, while the fore-arc basin occupied the Great Valley and on the continent behind rose andesitic volcanoes above large intrusions of granite deep in the crust. Today, after prolonged erosion, these granites form the backbone of the Sierra Nevada. In the early Cenozoic, a final, intense phase of orogeny caused the fore-arc to emerge and form the present Coast Ranges.

Shortly afterward a reorganization of plate movements replaced the subduction zone with a transform fault, the much-feared San Andreas which connects the subduction zone off Central America with a mid-ocean ridge off Oregon. Along this fault, a coastal sliver of California bearing the cities of San Diego, San Francisco and Los Angeles, is drifting inexorably north toward the Gulf of Alaska where, in a few tens of millions of years, it will be a thin continent out at sea, somewhat like New Zealand.

Again we encounter the long-lasting, apparently continuous subduction punctuated by separate orogenic phases, a problem that bedevils the application of plate tectonics to the Appalachians also. Moreover, the major deformation and large intrusions of granite involved in those events are not easily reconciled with subduction, fore-arc tectonics and volcanism.

Various explanations have been offered, each meritorious, but none adequate. One of the best suggestions, today probably the ruling hypothesis, rests on the fact that throughout the Pacific Ocean, especially in its older parts, we find crustal sections that rise too high and are too thick for normal oceanic crust. Some may be old island arcs, some huge undersea volcanoes, others fracture zone ridges or hotspot plateaus. A few may even be tiny continental blocks. What happens when such pieces arrive at the trench? Imagine an attempt to subduct Iceland or the Hawaiian Islands; it would thoroughly disturb the smooth course of events. In fact, it might be easier for the subduction zone to jump seaward beyond the obstacle and start anew, leaving the indigestible morsel attached to the margin of the overriding plate (Figure 8.7). In the few

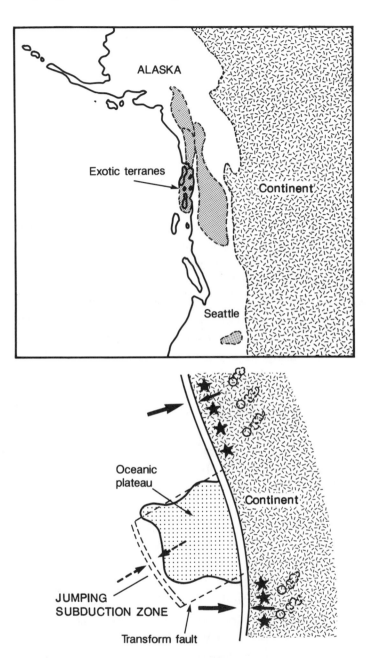

Figure 8.7. The broad strip of crust added on to western North America by subduction and accretion contains crustal blocks that are very unlike the surrounding rocks (top). These "exotic terranes" consist mainly of island arcs or oceanic plateaus, both overthickened pieces of oceanic crust. They traveled far, even from south of the equator, before they collided with western North America. The collisions caused spasms of mountain building, because the terranes were too thick to be swallowed, and added greatly to the continental mass. When such a terrane reaches the plate boundary, the subduction zone usually jumps seaward, also initiating two transform faults (bottom).

162

places where today an uncommonly massive oceanic crust impinges on a subduction zone, as on the west coast of Ecuador, we find a break in the earthquake and volcano belt behind the trench showing that subduction has been deactivated there.

Do we encounter such alien chunks of crust in the subduction complexes of the past? The answer is yes. In British Columbia and southeastern Alaska we find rock formations strikingly different from those that surround them. By the nature of their rocks and fossils these formations appear to have come from elsewhere, from beyond the subduction zone (Figure 8.7). The earliest of these exotic terranes arrived in northwestern Nevada during the Permo-Triassic orogeny. Another one, a long, narrow block that now forms the coastal islands of British Columbia, was born between 90 and 50 my ago far down in the southern hemisphere, judging by its paleomagnetic properties. It traveled to its present home at the astonishing speed of at least 15 and more likely 28 cm/year. The fragments of another terrane, named Wrangellia, lie scattered from Idaho to Alaska. Paleomagnetic data indicate that Wrangellia began its existence at 15° latitude, but whether north or south we cannot say.

To a surprising degree, geologists are romantics (remember Avalonia?) and this wreckage, gathered in subduction zones like driftwood on beaches, has been attributed to a lost continent of Pacifica, once located in southwestern Panthalassa, its fragments now scattered around the western and northern Pacific. The notion, alas, does not suit; few, if any of the exotic terranes and anomalous plateaus consist of continental crust, and most are pieces of ordinary island arcs, torn from their roots. Continental fragments adrift are not impossible, however; the Seychelles islands in the Indian Ocean are one example, New Zealand is another. Eventually, California too will run into the Alaskan subduction zone and be attached to another continent as a truly exotic terrane. Perhaps some future geologist will name it Avalonia then.

The idea of exotic terranes has caught the fancy of many a geologist concerned with the margins of the Pacific, and examples have multiplied like rabbits, in a manner reminiscent of the proliferation of hotspots when those first became fashionable. It is not clear just how exotic many proposed terranes really are, nor can it be confirmed that all came from far away. This frenzy to find exotic terranes appears a bit overdone and not likely to speed up our understanding of coastal orogenies, but the concept is useful and in a sizable number of cases well documented.

If this seems untidy to an observer interested in the history of the earth, it is nothing compared with the confusion surrounding the Tethys, the great eastern embayment of Pangaea of which today only the Mediterranean remains (Figure 7.1). A swarm of microcontinents broke away early from Gondwana's eastern and northern shores, drifted north and westward along various routes, and eventually participated in the construction of Spain, Italy, Greece, Anatolia, Arabia and Iran. Other pieces were torn from southern Asia and traveled west to end up in the Balkans. Along the way these fragments twirled, collided with each other, broke apart again, or were separated by short-lived mid-ocean ridges. Ultimately they produced the tangle of mountain ranges extending from the Pyrenees by way of the Alps to the Balkans and the Caucasus and, in doing so, founded a profitable tourist industry.

This traffic jam continues. Arabia now pushes against Iran, a rift is opening from the Gulf of Aden to the Dead Sea, Turkey is forcing its way west into the Aegean, and the seafloor north of Egypt and Libya is being subducted under Italy, western Greece, Crete, south Turkey and Cyprus. The geological evolution of the Mediterranean has obviously been no less turbulent than its human history, and it does not surprise us that the various accounts of this sequence of events are so far much at odds with each other.

The exotic terrane concept implies that a considerable amount of oceanic lithosphere has accreted over time against the ancient Precambrian cores of the continents, the cratons. David Howell of the United States Geological Survey, a pioneer of the exotic terrane concept, has estimated that the process plasters on the average as much as one cubic kilometer per year against the continents, a non-trivial amount over the duration of earth's history. In the beginning the rate may even have been faster.

8.4   THE FATE OF THE SUBDUCTED SLAB

The amount cited above is for accretion to the continents only. If we estimate the average volume of crust subducted in the mantle, a better word would be recycled since that is where it came from in the first place, we obtain even larger figures, up to 4 km^3/year. Over the past 150 my a strip of ocean floor 8,000 km long vanished under the Andes, the North Pacific shrunk by 13,000 km, and the Mediterranean engulfed 1,000 km-worth of African plate. The rate has varied with time: the Cretaceous was especially voracious whereas at present a lot less oceanic lithosphere is being consumed, as was also true in the Jurassic. The variations in the rate of crustal consumption have an impact on sea level changes as we shall see in Section 9.7.

What happens to the subducted slab, where does it go, and what does it do when it gets there? Much information regarding this question has been obtained from the travel paths of earthquake waves through the crust and mantle (see Figure 13.1), aided by a new method called seismic tomography which uses earthquake waves in a manner analogous to the medical X-ray CAT scans that reveal our insides in three dimensions.

The descending slabs are colder than the surrounding mantle and therefore denser, and sink, most to between 200 and 500 km, but some to a depth of 670 km where they come to a stop at the top of the lower mantle. The boundary between upper and lower mantle is due either to a chemical difference, as is the case for the Moho discontinuity, or to a sharp change in the density, compressibility, or rigidity caused by the increasing pressure. Such a transition in the state of a rock not accompanied by a change in composition is called a phase change. In either case, the impact of a cold, heavy slab should dent the boundary, but if the boundary was chemical the dents would be 100–200 km deep, whereas a phase change would merely dimple on a scale of 30 km or thereabouts. Seismic tomography has shown that the boundary at 670 km is indeed bumpy, but only on a small scale, so confirming that we have to do with a phase change. When the slab meets this barrier it flattens and warms up, eventually melts, and is then recycled into the upper mantle.

Plate-tectonic reconstructions, supported by the findings of seismic tomography, show that the places where subducted slabs have gone (and still go) to their final rest form a broad belt from Antarctica across Australia into southeast Asia where it curves to the northeast. From northern India a branch reaches westward into the Mediterranean, but the main zone crosses the Americas from north to south, then joins its own Antarctic tail. There is a growing suspicion that this pattern, even though it is so very different from that of plate motions, may reflect mantle convection. To understand what that means we must wait for more seismic tomography.

8.5 CONTINENTS COLLIDING

If less crust is generated in an ocean basin than is subducted, continents on opposite sides of the basin must collide in due time. Australia will some day push into southeast Asia, but at the moment India's encounter with Asia is the prime live example of a continental collision. It began 40 my ago when India touched the underbelly of Asia and the subduction of ocean crust came to an end (Figure 8.8). The suture of

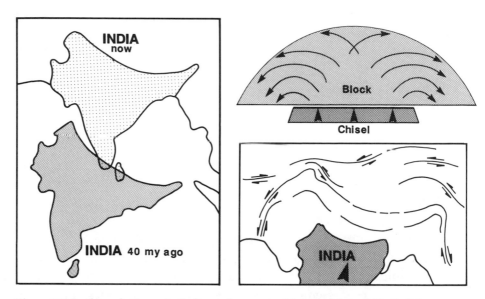

Figure 8.8. In the early Cenozoic, India made contact with central Asia (left). It did not come to a halt, however, and still continues to penetrate deeper into the other continent, accommodated by strike-slip faults similar to the cracks that form if a rigid block is hit with a chisel. The many disastrous earthquakes from Iran to China are thus the result of India's inability to stop and of attempts of the Asian crust to get out of the way.

this collision is not, as one might think, the Himalayan range, but lies well north of those lofty mountains which are themselves the product of later events.

Since that first encounter, India has moved 2,000 km farther north. Half of that distance can be accounted for by horizontal compression and by the compression that created the Himalayas, but about 1,000 km is the measure of India's penetration into Asia. This process, which still continues at about 5 cm/year, is akin to pushing a chisel into a block of metal (Figure 8.8). The chisel drives a triangle of metal inward, and forces aside parts of the block along shear zones that behave like the horizontal strike-slip faults of Figure 6.6 (top). The lithosphere is too inhomogeneous for a perfect copy of this experiment, but with India the chisel and the Tibetan plateau the triangle, we can spot the strike-slip faults along which the crust of central Asia is getting out of the way. The calamitous earthquakes that destroy villages and cities as far away as central China and northwestern Iran are the direct consequences of the collision between India and Tibet.

Two thousand kilometers is an impressive amount of movement, but it may be an underestimate. The late Cretaceous and early Cenozoic faunas associated with the hotspot lavas of the Deccan Traps (Section

7.5, Figure 7.14) suggest that the continent-to-continent collision began earlier, some 65 my ago, and that the shortening may have been twice the 2,000 km usually assumed. Here, as with Pangaea (Section 7.2), we see how useful it is to correct and complement plate-tectonic hypotheses with geological data.

Recently, tectonicists like Dan McKenzie of Cambridge University, Peter Molnar of the Massachusetts Institute of Technology, and Philip England at Oxford have shown that the model of Figure 8.8 (bottom right) is oversimplified. Instead, they believe that sets of strike-slip faults of opposite sense enable large blocks of exotic terrane or continent to twist out of the way of the chisel by rotating on a vertical axis. Pieces of Greece, for example, have rotated a full quarter turn in five million years in response to the stress applied by the northward movement of the African plate and the westward push of Anatolia.

Continental collisions begin with the consumption of all oceanic crust, followed by abandonment of the subduction zone. The shallow sea that remains from the former ocean fills with marine and later river deposits as mountain ranges rise on one or the other side. This process was recognized early in the Alps where the marine deposits received their traditional name of flysch while the river sequence is called molasse. The Persian Gulf is a good active example of this stage. Eventually flysch and molasse too are caught up in the folding, and granites in profusion are intruded into them. In addition, the crust often increases greatly in thickness to form a kind of root. The formation of a root and the heat source for the granites are not required by the first principles of plate tectonics and demand an explanation.

An ingenious suggestion by Peter Bird of the University of California, Los Angeles, goes as follows. The lithosphere below the Moho is relatively cool and heavy and, if not kept afloat by the light continental crust, would sink in the asthenosphere. When the subducted slab ceases to function and becomes detached (Figure 8.9), it pulls at the lower lithosphere, causing part of it to tear away and sink. This opens a path for the asthenosphere to rise and heat the crust, creating granitic magmas that push their way up. Also, a space is provided where, as the collision continues, the continental crust of India can slide under that of Asia, forming a thick root.

In a more conventional mood, Peter Molnar believes that one does not need these special processes. He proposes a huge north-dipping fault, formed when the Indian plate was thrust under the Asian continent. Along this fault a large sliver of the Indian continental crust was shoved on top of the edge of Asia rather than under it as Peter

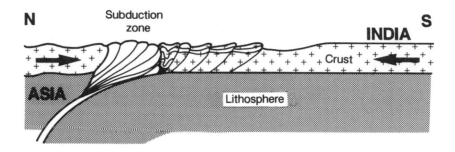

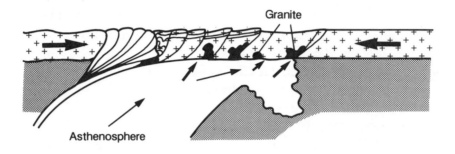

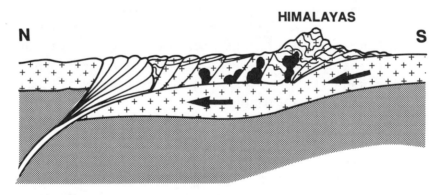

Figure 8.9. When the last remnant of oceanic crust between Asia and India had been subducted, the compressed subduction zone became the suture between the two continents. According to Peter Bird, the drag of the now useless subducted slap tore away the lower part of the lithosphere, thereby providing access for enough heat to produce large granite intrusions. The southern segment of the lithosphere became wedged under the now thinned northern part, more than doubling the lithosphere in thickness, and raising the Himalayas as much by isostatic compensation as by compression.

Bird suggested. The result is a thickened crust. Both models involve a sliver of crust, but the slivers have opposite origins and move in opposite directions. Both fail to explain some of the features and neither can really be shown to be true at this time. This has inspired other geologists to propose a variety of models that combine elements of both.

At first sight these new views on the grand deformations of continents appear to destroy further the simple elegance of plate tectonics. In a few years, however, when these ideas have matured, the tectonic map of Eurasia, the only continent under major stress, will no doubt provide its response with a graceful logic of its own.

PERSPECTIVE

Like so many of its kind, the geological revolution was slow in coming and not so very novel in its most enduring achievement, the acceptance of continental drift. Still, although it now seems odd that we managed so contentedly and for so long with a world of fixed continents, the consequences of this acceptance are far-reaching and we have not yet fully accommodated them.

Recently our thoughts have turned toward restoring some of the fundamental differences between continents and oceans that perhaps had been too casually abolished in the first flush of enthusiasm, but other problems remain. Episodicity, for example, is evident in mountain building, but it is not demanded by the steady subduction of oceanic crust, is not observed on divergent mid-ocean ridges, and would not be expected, given the enormous momentum of the plates.

Slow vertical movements relative to each other of large regions or whole continents, known as epeirogeny, dominate the history of the continents. Such movements lack major folding and faulting and fit poorly into plate tectonics. Because of them geologists living far from active plate boundaries, in particular in the former USSR, long failed to see the need for the new paradigm. Now we begin to see that the key to epeirogeny is the behavior of the mantle underneath the plates where an uneven distribution of temperature and therefore density, combined with convective flow, is capable of changing the level of the surface over which the plates move. This causes the plates to rise or fall by hundreds of meters over tens and hundreds of millions of years. This inquiry, to which we shall return in Section 9.7, is in its infancy and rests so far mainly on theoretical models underpinned by seismic tomography. Still, it is already clear that matching these models against the geological history of continents and oceans holds great promise.

With that in mind, we turn in the following chapters to that part of the earth where the impact of plate tectonics, continental drift and a whole host of new geological methods has not yet been fully discounted, the history of the earth's continents and oceans. A quiet revolution is in progress here, as Nick Shackleton of Cambridge University has argued.

FOR FURTHER READING

This time we are in luck; the geological revolution has inspired a vast literature, scientific as well as popular. I must choose and my choices are not likely to satisfy everyone, but I trust that the following will provide a reasonable overview. On the history of the geological revolution see: Hallam, A. (1979). *A Revolution in the Earth Sciences* (New York: Oxford University Press); and Oreskes, N. (1988). The rejection of continental drift, *Historical Studies in Physical Sciences*, **18**, 312–48. A keen eye witness of the revolution was Walter Sullivan (1974). *Continents in Motion, the New Earth Debate* (New York: McGraw-Hill).

The following books elaborate on plate tectonics: Cox, A. V. (ed.) (1973). *Plate Tectonics and Geomagnetic Reversals* (San Francisco: W. H. Freeman; a collection of classic papers); Cox, A. V. & Hart, R. B. (1986). *Plate Tectonics: How it Works* (Oxford: Blackwell; a do-it-yourself plate tectonics); and Uyeda, S. (1973). *The New View of the Earth* (San Francisco: W. H. Freeman; an overview with a personal touch).

SPECIAL TOPICS

Anderson, D. L. & Dziewonski, A. (1984). Seismic tomography, *Scientific American*, **251**, 60–68.

Bloxham, J. & Gubbins, D. (1989). The evolution of the earth's magnetic field, *Scientific American*, **261**, 30–37.

Bond, G. C. & Kominz, M. A. (1988). Evolution of thought on passive continental margins from the origin of geosynclinal theory ($\approx$1860) to the present, *Bulletin of the Geological Society of America*, **100**, 1909–33.

Boucot, A. J. & Gray, J. (1983). A Paleozoic Pangaea, *Science*, **222**, 571–81.

Burke, K. & Wilson, J. T. (1976). Hotspots on the earth's surface, *Scientific American*, **235**, 45–57.

Cook, F. A., Brown, L. D. & Oliver, J. E. (1980). The southern Appalachians and the growth of continents, *Scientific American*, **243**, 156–68.

Decker, R. & Decker, B. (1981). The eruptions of Mount St. Helens, *Scientific American*, **244**, 68–80.

Francis, P. & Self, S. (1983). The eruption of Krakatau, *Scientific American*, **249**, 172–87.

Frohlich, C. (1989). Deep earthquakes, *Scientific American*, **260**, 32–39.

Gass, I. G. (1982). Ophiolites, *Scientific American*, **247**, 122–31.

Gordon, R. G. & Stein, S. (1992). Global tectonics and space geodesy, *Science*, **256**, 333–42.

Hoffmann, K. A. (1988). Ancient magnetic reversals: clues to the geodynamo, *Scientific American*, **258**, 50–59.

Howell, D. G. (1985). Terranes, *Scientific American*, **253**, 116–25.

Hudnuth, K. W. (1992). Geodesy tracks plate motion, *Nature*, **355**, 681–82.

Jones, D. L., Cox, A. V., Coney. P. & Beck, M. (1982). The growth of western North America, *Scientific American*, **247**, 70–85.

Jordan, T. H. & Minster, J. B. (1988). Measuring crustal deformation in the American West, *Scientific American*, **259**, 32–40.

Kutzbach, J. E. & Gallimore, R. G. (1989). Pangaean climates: megamonsoons of the megacontinent, *Journal of Geophysical Research*, **94**, 3341–57.

Kutzbach, J. E. & Guetter, P. J. (1990). Simulated circulation of an idealized ocean for Pangaean time, *Palaeoceanography*, **5**, 299–317.

Lay, T. (1992). Wrinkles on the inside, *Nature*, **355**, 768–69.

Drifting continents, rising mountains

Minster, J. B. (1990). New plates, rates, and dates, *Nature*, **346**, 218–19.

Molnar, P. (1986). Structure of mountain ranges, *Scientific American*, **255**, 70–79.

Molnar, P. & Tapponnier, P. (1977). The collision between India and Eurasia, *Scientific American*, **236**, 30–41.

Moody, J. D. (1975). Distribution and geologic characteristics of giant oil fields, in *Petroleum and Global Tectonics*, A. G. Fischer & S. Judson (eds.), pp. 307–20 (Princeton: Princeton University Press).

Mutter, J. C. (1986). Seismic images of plate boundaries, *Scientific American*, **254**, 66–76.

Parrish, J. T. (1993). Climate of the supercontinent Pangaea, *Journal of Geology*, **101**, 215–35.

Richards, M. A. & Engebretson, D. C. (1992). Large-scale mantle convection and the history of subduction, *Nature*, **355**, 437–40.

Scotese, C. R. & Sager, W. W. (eds.) (1988). *Mesozoic and Cenozoic Plate Reconstructions* (in *Tectonophysics*, **155**, 1–399).

Smith, A. G. & Livermore, R. A. (1991). Pangaea in Permian to Jurassic time, *Tectonophysics*, **187**, 135–79.

White, R. S. & McKenzie, D. P. (1989). Volcanism at rifts, *Scientific American*, **261**, 44–55.

Changing oceans, changing climates

If we are to believe Wegener's hypothesis, we must forget everything that has been learned in the past 70 years and start all over again.
R. T. Chamberlin at the 1926 meeting of the American Association of Petroleum Geologists

It is the clarity of the radically new and absolutely simple idea which catches us as if it were an intuition.
Henri Bergson, *The Creative Mind*

CONTINENTAL DRIFT AND
ANCIENT ENVIRONMENTS

Attempts to describe the landscapes and seascapes of the past are as old as geology itself. Paleoclimatology, one component of this arcane art, has grown into a separate science; another, related one, paleoceanography, took on a life of its own some two decades ago. None other than Alfred Wegener wrote, with his father-in-law Wilhelm Köppen, an early book on the climates of the past. Those that followed, however, skillfully erected climate histories based on fixed continents that satisfied everyone for decades, something that is difficult to understand today. The successes of paleoclimatology and paleoceanography in the new context of plate tectonics and continental drift are the subject of the next four chapters.

Reconstructions of the environmental history of the earth rest on sediments and the fossils they contain, interpreted with liberal handfuls of climatic and oceanographic theory. Because the data cannot truly be regarded as what physical scientists call "robust" evidence and the applicability of the theory is not always certain, the product is seasoned with courage and creative imagination. Fossils, especially, are prone to lead us on the treacherous path of circular reasoning. For the recent past we infer their preferred environments by comparing fossil species with close living relatives, so gaining confidence in the interpretation. But what about a group that has been extinct 100 my or so? How shall we deduce its environmental tastes, except in broadest terms: marine or terrestrial, shallow or deep, forest or desert? Not surprisingly, the facies deduced from the presence of certain fossils has, unknowingly or even knowingly, occasionally been used to explain why those organisms lived there. This dubious procedure is doubly undesirable when we inquire after the influence of a changing environment on the evolution of life.

Obviously, I have put this a little sharply to make my point. Nonetheless, it is the beauty of plate-tectonic theory, and an inspiration to earth historians, that it enables them to reconstruct the geography of the past without immediate recourse to sediments or fossils. This, combined with the theory of oceanography and climatology, allows us to infer many things about the physical environment with a rigor that was hitherto not possible. We can then test the results with the aid of inferences drawn from sediments and fossils, with greater confidence and a diminished concern about circular reasoning.

All this is still in its infancy. Paleoclimatologists and paleoceanographers have only just begun to hone their tools, to formulate their questions and to make the first attempts at solving them. Still, even these hesitant steps illustrate the enormous impact of plate tectonics on our understanding of the earth in all its aspects. In tandem with the availability of large computers, this has led to a proliferation of computer-based models of the climates and ocean conditions of the past, models that rest mainly on an as yet uncertain data base and on ideas about the behavior of atmosphere and ocean derived from the present and still of dubious applicability to the remote past. In a recent symposium on sea level and climate of the past more than half the contributions were purely armchair models; what little geology they contained was window dressing. Because computers produce impressive maps with great ease, whereas the search for geological confirmation or rejection is tedious and its outcome often ambiguous, a totally unfounded conviction has developed that we understand much more than we actually do.

9

The sea comes in,
the sea goes out

Twice daily on an open coast the edge of sea moves out, then in; the land grows slightly, then shrinks again. During an ice age, several times each 100,000 years, the silver line between land and sea travels much farther. Over the eons, the shore has rarely stood still, the area of land increasing and decreasing with the tides of time. Much geology is about this eternal movement of the coast; much of the rock record exists only because of it.

For many years geologists have contemplated this spectacle and wondered whether they saw the land rising or the level of the sea falling (or the other way around), a local problem. Now, believing that we understand the local part, we wonder why the relation between land and sea has been forever so unsteady.

Geological maps depicting ancient shorelines provide direct information; from them we may obtain the proportions of land and sea as they have varied with time. Many such estimates exist, each a little different from the others, all marred by lack of data, by the ravages of time, and by defects of the time scale. Nevertheless, on the grandest scale a coherent picture has emerged from two centuries of effort. Over the last 400 my, the continents have emerged from the flooded condition of the early Paleozoic only to submerge again in the Mesozoic, until the Cenozoic brought the deepest fall of the sea ever. This broad pattern is adorned with many flourishes on a smaller scale.

9.1 TRANSGRESSIONS AND REGRESSIONS

The sea covers and uncovers the land either because the water level itself changes, inevitably a worldwide event, or because the land sinks or rises. No reference point exists to calibrate the vertical motions of past seas. They must be inferred from the shifting sedimentation patterns when

177

rise or fall of land or sea moved the shore: a transgression as the sea encroaches, and a regression as it withdraws. On a nearly level coastal plain, a small rise of the sea may move the beach far, but on a steep coast a large one will hardly displace it at all. Unless we know the slope of the seafloor and the adjacent land, we cannot estimate a vertical change of sea level from the distance over which the shore migrated.

Intuitively one equates transgression with a rising sea or sinking land, and regression with the reverse, but even when the sea is stationary, erosion may cause the coast to recede, while the deposition of a delta moves it seaward. The point is not trivial: sedimentation is a powerful force which must be taken into account when we consider whether a given change in the level of land or sea will cause a transgression or a regression. Stated formally:

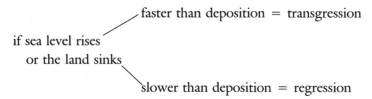

faster than deposition = transgression

if sea level rises

   or the land sinks

slower than deposition = regression

For erosion the inverse is true, but erosion is not often important, because waves do not cut deep and soon make a shallow platform that breaks their strength and protects the coast. Even though houses carelessly placed above the beach do get swept into the sea, loss of land by erosion tends to be modest.

Local transgressions and regressions are tediously common in the geological record, but those that are global raise important questions. We have already encountered the many sea level changes of the ice age, but they are not the only ones. Superimposed on the trend of Figure 9.1 are numerous other fluctuations of the boundaries between land and sea. Some seem to be synchronous across the world and so must be eustatic, although they are not obviously connected with glacials and interglacials.

9.2 VAST AND SHALLOW SEAS

In the present emerged state of the world, we tend to consider the shelf seas as mere transitions between continent and ocean, but in the early Paleozoic and later Mesozoic vast shallow seas without modern counterparts ranked equal in importance to land and deep-sea.

It was in those shallow seas that the abundant life evolved that makes most Paleozoic deposits so different from anything we find today. Life's

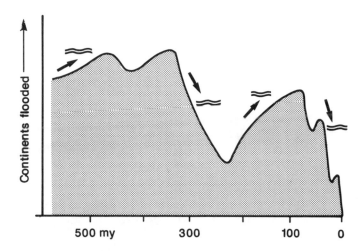

Figure 9.1. Until recently, curves depicting changes of sea level over time were based on the area of land covered by marine sediments of various ages. During most of the Paleozoic the sea stood high, albeit with fluctuations of shorter duration. About 400 my ago, the sea began to fall, reaching a low point in the early Mesozoic, then rose again to a Cretaceous high stand. The Cenozoic has been mainly a time of falling sea level, and the present stand seems to be the deepest ever. Other, more modern and detailed curves are presented in Figure 9.3.

sudden, explosive diversification at the start of the Paleozoic, and its abundance and good preservation, afford us useful environmental and temporal distinctions to correlate events that happened in distant places. Regarding the early Paleozoic in this bright light, we find it a wet world (Figure 9.2), its continents inundated far more than they have ever been since then, and the rise of the sea continuing. Before this rise ended, very little land remained above water.

Of the successive transgressions that accomplished this in North America, each began with widespread sandstones laid down on beaches and in shallow nearshore waters. Offshore, on the clear and sun-lit shoals that were common in the submerged parts of the continent, algae and sponge-like organisms formed extensive limestones. As the sea advanced farther and the lands shrank, limestone began to dominate the rock record, products of a sea filled with the largest reef complexes the world has ever seen. Needing sunlight and plenty of food, the reefs of the North American continent, then located just north of the equator, grew best on the windward edges of banks and shoals. There wave turbulence furnished nutrients for a rich growth of plankton, which fed a complex community of reef organisms. Altogether, the reefs, lime shoals and channels formed a seascape not unlike that of the present Bahamas, on a huge scale and in the middle of a drowned continent instead of in the middle of an ocean.

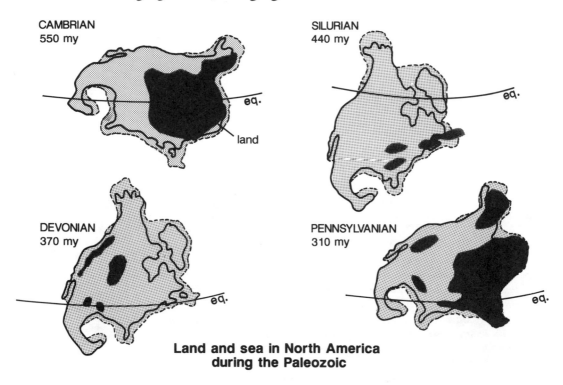

CAMBRIAN
550 my

land

SILURIAN
440 my

eq.

DEVONIAN
370 my

eq.

PENNSYLVANIAN
310 my

eq.

**Land and sea in North America
during the Paleozoic**

Figure 9.2. During most of the Paleozoic the sea stood high, widely flooding the continents, as these maps of North America, properly oriented with respect to the equator of the time, show. Land is dark, shallow seas are lightly shaded. The present coast of the continent is given for guidance, but does not reflect the geography of Paleozoic North America.

If we look closer, we notice that the sea rose and fell in several phases, now known as cycles (Figure 9.3). The first four cycles roughly coincide with the first four major subdivisions of the Paleozoic, not much of a surprise really, because each low stand of the sea caused a huge uncon-formity. A fifth cycle spans the later Carboniferous (Pennsylvanian) and Permian together, but it is anomalous; starting with a sharp drop of the sea, it was in reality a long slow regression that ended near the start of the Mesozoic with one of the lowest stands of the sea ever.

The Paleozoic cycles are of very unequal duration, ranging from the brief Silurian one (30 my) to the long Pennsylvanian–Permian cycle of 110 my, but there is general agreement that they were worldwide in nature and hence eustatic in origin.

Seen in more detail, the five major cycles are composed of shorter second and higher order oscillations (most of them not visible in Figure 9.3). These range in length from a few million years for the second-order Silurian and Devonian cycles to numerous higher order Pennsylvanian

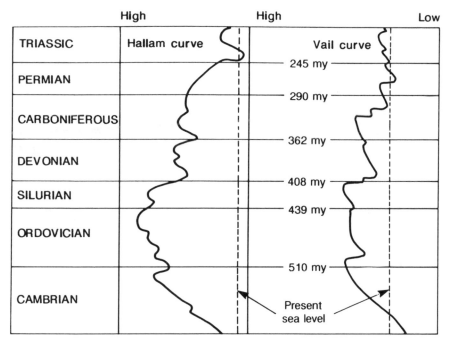

Figure 9.3. Two views of global sea level history derived from the available data by two expert geologists, each with his own perspective (compare with Figure 9.1). Superimposed on the broad first-order rise and subsequent fall of the Paleozoic sea are five second-order rises followed by falls. The main regressions coincide quite well, but that is not surprising; they mark the boundaries of the basic stratigraphic units of the Paleozoic. In detail, the differences are large, and especially in the Carboniferous the sharp changes postulated by Vail are regarded as much more gentle by Hallam.

cycles in central North America that are measured in tens to hundreds of thousand of years.

About 300 my ago the first age of reefs came to a close as the sea began a slow withdrawal that in the late Permian led to a nearly complete emergence of the new-born supercontinent Pangaea. Already in the Pennsylvanian, much of eastern and central North America (Figure 9.4) had emerged to form a vast plain across which many rivers wound their way, bringing debris to the sea from the Appalachian mountains towering in the east. The low coast resembled that of the Amazon delta today: vast swamps, low, muddy shores, and sluggish streams that built deltas in a warm, turbid sea. A rich land flora had evolved since the Silurian (Section 18.5), and for the first time forests covered the plains, laying down the coal now mined in eastern North America, Europe and Siberia.

The Pennsylvanian strata of the central United States show many curious repetitions of the same sediment types. Each cycle begins with

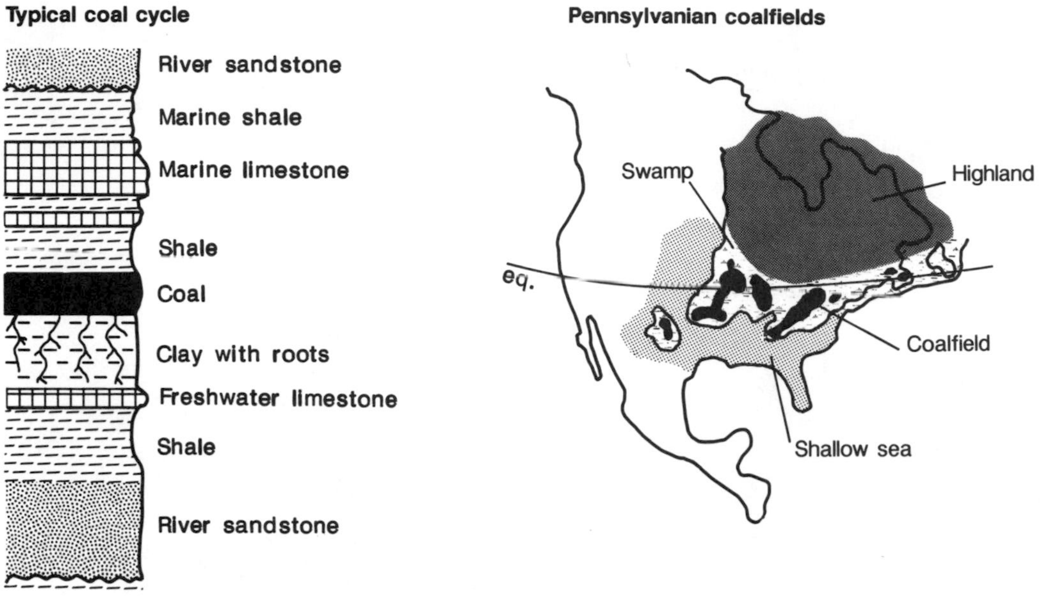

Changing oceans, changing climates

Figure 9.4. In the shallow Pennsylvanian sea of central and southern North America, sediments were deposited in many cycles, usually called cyclothems. Each cyclothem begins with a river facies, evolves by way of a coastal swamp to very shallow marine deposits, and ends with an unconformity and a regression. The cyclothems, possibly related to glacial eustatic sea level changes, include major coal deposits.

river and coastal sandstones resting on an eroded land surface (Figure 9.4 left). Clay with tree roots follows and then coal. Next come swamp deposits that spell the beginning of an inundation, and then a shallow marine facies with shales and algal limestones. Another unconformity announces the next cycle.

At least 50 such cycles, ponderously known as cyclothems, can be recognized in these Pennsylvanian strata, each a regression followed by a transgression. Such a rhythm comes naturally to large deltas when the river, having built seaward as far as possible in one place, seeks a shorter path to the sea to one side or the other. As a new delta is begun, the old one subsides and is taken over by the sea. The Mississippi delta south of New Orleans, set between abandoned and now eroding subdeltas, is a type example of this behavior. The Pennsylvanian cycles are thin, they seem brief, and in this way fit the pattern. But they are also extensive, far too extensive for a single delta, no matter how big the river might have been. Moreover, other cyclothems, deposited at the same time in Kansas and elsewhere, record similar sea level changes but on an open coast a bit like the western Gulf of Mexico today. Similar brief sea level oscillations occur in the Silurian.

Brevity is relative in the distant past, when isotopic dates have uncertainties measured in millions of years, but, dividing the number of minor and major cyclothems into the time available, we find maximum durations of 40,000–100,000 years for minor and 240,000–400,000 years for major ones. That is much too long for a migrating delta which wanders about on a time scale of centuries.

Already in 1936, Harold Wanless and Francis Shepard wondered if the waxing and waning icecaps in distant Gondwanaland might have produced the many sea level changes recorded in the Silurian and Pennsylvanian cyclothems of North America. Long-distance correlations and better estimates of their duration have now persuaded many that this is an idea whose time has come. An estimate, admittedly a courageous one, suggests that, when the Pennsylvanian icecap accumulated on Gondwana, the sea fell about 60 m. That is less than in a Quaternary glacial, but the Gondwana caps were smaller. The range seems about right for the sea level changes recorded by the cyclothems, but cannot yet be confirmed.

So do we see here the distant echo of Paleozoic ice ages, with durations of 40,000 to 100,000 years, quite similar to the Milankovitch cycles? Perhaps, but estimates of brief intervals so far in the past are insecure. Caution is advisable, but my own guess is we are indeed dealing with glacial/interglacial cycles.

9.3   A MAJOR TRANSGRESSION

During the Cretaceous, the inundation of the continents reached an extent not seen since the early Paleozoic, and in the end little more than half of the present land area remained above the waves. In North America, successive transgressions created a vast inland sea from the Gulf of Mexico to the Canadian Arctic (Figure 9.5). In the west this sea lapped onto a mountain range bordering a Pacific subduction zone and studded with volcanoes, while in the east it wetted the shores of a lower, gentler land. Sediments dumped in from each side lined the shores with deltas, lagoons, swamps and beaches, very much like the Gulf Coast today, and the haunt of the last of the dinosaurs. The coastal wedges bear the imprint of many transgressions and regressions of the Cretaceous flood, five of which may have been roughly simultaneous across the world. Europe and Africa were similarly covered with shallow seas, but Australia, curiously, rose high and dry, eroding slowly to the featureless plains that are the dominant landscape now.

It is not easy to say exactly how high above its present level the

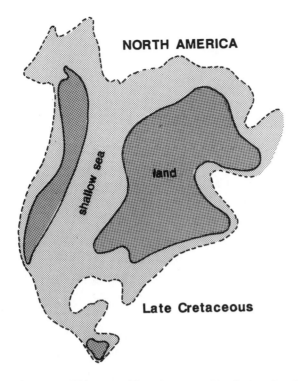

NORTH AMERICA

shallow sea

land

Late Cretaceous

Figure 9.5. This map of late Cretaceous North America shows how large a difference even a sea level change of a few hundred meters made in the geography of the continent that had been eroded deeply during the Mesozoic. A two hundred meter (600 ft) rise of the sea today would have far less effect.

Cretaceous sea may have risen. The deposits of many Cretaceous shores crop out on land, but their elevations need correction for compaction, for the isostatic compensation for loading by later deposits, and for tectonic movements. In the mid-Cretaceous, before the transgression peaked, the sea was more than 200 m above its present level, that much is certain. Its maximum was probably just below 300 m. Today, a similar rise would inundate a smaller area, because the continents stand so high, whereas Mesozoic lands were low and flat after eons of erosion.

What do the many ups and downs of the Mesozoic sea mean? It helps to remember that the Mesozoic began when a newly completed super-continent rose high above the ocean, and ended with Pangaea fragmented and the embryonic Indian and Atlantic oceans coming into their own. It is not so far-fetched to speculate that those momentous events affected the capacity of the ocean basins and the elevation of new continental margins, and therefore the level of the sea.

9.4 THE CONTINENTAL MARGIN AS A SEA LEVEL GAUGE

Changes of the level of the sea over time can be inferred from ancient shorelines displayed in the rocks, but even at its best this record is woefully incomplete. It would be nice to study a complete sequence of transgressions and regressions without having to hunt far and wide for the pieces of the puzzle (and failing to find many of them). Fortunately, what we seek exists on the passive continental margins. There, as the shelves slowly subsided because the crust cooled and an ever thicker load of sediment pressed them down, a record was kept of transgressions and regressions that is far more complete than the shore deposits now stranded inland.

We can read this record because of a technique based on the old method of echo-sounding at sea. If we send a sound signal to bounce off the ocean floor and measure the time until its return, we can calculate the water depth, provided we know the velocity of sound in seawater. If we use a more powerful source of energy with a lower frequency, part of the energy will penetrate the bottom and be reflected by boundaries between sediment layers. On this principle rests the method of continuous marine seismic profiling (Figure 9.6), originally developed by oceanographers and later expensively perfected by the oil industry. A seismic profile resembles a geological cross-section but it is not, because the horizontal scale represents time as the ship travels, not distance, and the vertical scale measures echo time, not depth. We can obtain the distance if we keep track of the ship's position, but to convert echo time into depth we must know the sound velocity in sediments and that varies with kind and depth. Fortunately, because the velocity is difficult and expensive to measure, we may with caution use estimates instead. And if we wish to know the nature and age of the reflectors, we must, of course, drill, core and date the samples, usually with fossils.

Most echoes return from surfaces that once were the bottom of the sea. Because usually each reflector represents a single instant in time, a time line, the course of past transgressions and regressions can be inferred from seismic profiles (Figure 9.7). Sometimes the character of the reflector also tells us something about the sediments themselves: on a seismic profile coastal deposits differ from others laid down in deep water. In short, the use of seismic reflection profiles has evolved into an important mode of investigation now known as sequence analysis, meaning the study of sequences of transgressions and regressions.

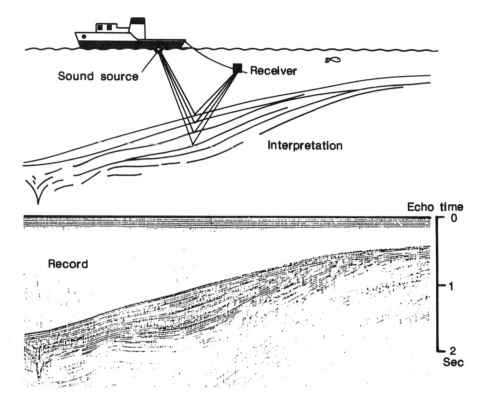

Figure 9.6. The deposits of the sea bottom can be examined by seismic profiling. A survey ship sends out sound pulses that are reflected by the bottom and by various internal surfaces in the underlying strata. The record shows the return times of the sound plotted against the travel time of the ship. With some caution it can be read as if it were a geological cross-section.

9.5 SEQUENCE STRATIGRAPHY AND THE VAIL SEA LEVEL CURVE

Sequence analysis of seismic reflection profiles has shown that they often record cycles of transgressions and regressions stacked on top of and against each other. If we knew their ages, we would know the history of local sea level changes. Correlation of a large, globally well-distributed set of local histories would then tell us which cycles, if any, were synchronous, world-wide, and hence by definition eustatic. Such information is not easily come by. Oil companies possess by far the best seismic reflection data and, through drilling, have acquired much age information as well, but for sound, and sometimes not so sound, economic reasons, they tend to be reluctant to release the fruits of their labors. The first and so far only global compilation of sea level changes from this source is by the Exxon Corporation and is today known as the

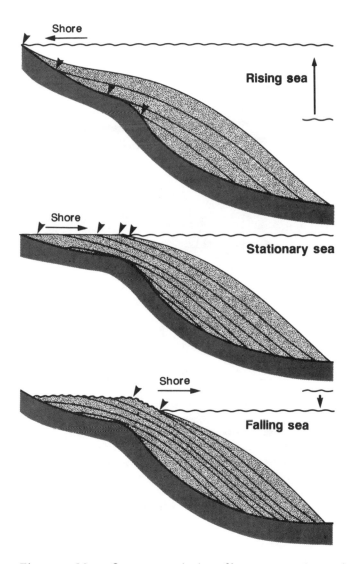

Figure 9.7. Most reflectors on a seismic profile represent ancient seafloors. The highest point of each (arrow) will thus mark the highest sea level when it was formed. As the sea rose, another termination formed a little above its predecessor; when the sea was stationary, the reflectors were displaced seaward but their tops stayed at a constant level. A falling sea level caused each termination to be lower than the previous one. Reflection records from continental margins so enable us to determine sea level changes. The method is known as sequence analysis and is widely used in the exploration for oil and gas and in marine research.

Vail curve, after Peter Vail, the geologist who led the study while working for Exxon.

The primary trend of the Vail curve is much like other sea level curves (Figure 9.8, left) with high sea levels early in the Paleozoic and again during the Cretaceous. Superimposed on this trend are "supercycles,"

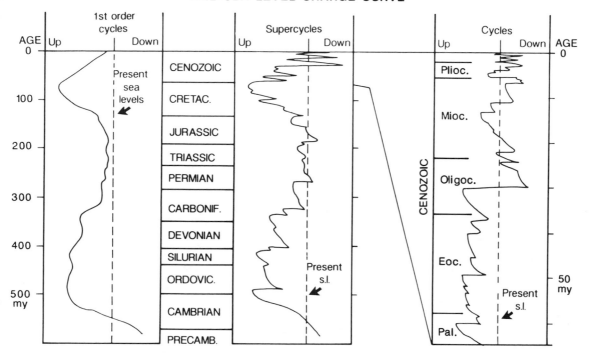

Figure 9.8. Sea level changes based on sequence analysis of seismic profiles with a global distribution show change on three scales. The first-order changes or megacycles resemble Figure 9.1, although with a less extreme range. Superimposed are variations of similar amplitude but lasting 4 to 15 my (20–50 my in the Paleozoic), called supercycles. Supercycles are composed of cycles, again of similar amplitude but of much shorter duration. Higher order cycles, such as the Quaternary glacial/interglacials exhibit, exist but are not shown here.

large transgressions and regressions which exhibit a remarkable episodicity, an earth rhythm. Like the teeth of an abused saw-blade, each supercycle rises rapidly, slows to a standstill, then drops precipitously, in sharp contrast to the symmetric behavior many experienced geologists believe they have seen in Jurassic and Cretaceous outcrops on land.

The supercycles are themselves composed of cycles, which are shorter but otherwise similar in shape and amplitude. Higher (fourth and fifth) order cycles also exist; some are our old acquaintances, the glacial–interglacial sea level changes of the Quaternary, but others, possibly more local in extent, do occur.

The parents of the Vail curve have steadfastly maintained that their supercycles and cycles represent globally synchronous sea level changes, and are therefore eustatic. This has generated a lot of controversy, because for long the stingy release of data regarding the spatial distribution of the evidence and the nature of the dates by Exxon prevented the verification of the claim. Geologists using means other than sequence analysis found many discrepancies, and by no means all seismic records available to academics support the Vail curve. Responses to the critics have been slow in coming, but the latest version, a few years old now, finally dealt with the biostratigraphic base in some detail, and included sequences on land. The global and temporal distributions of the evidence, however, still remain inadequately known, and much non-Exxon data fits poorly into the Vail curve. This, I note, is not gratuitous carping; it is of the essence of science that others can verify the evidence for any claim.

With the biostratigraphic base came a considerable revision of the definitions, numbers and dates of supercycles and cycles. Now 27 supercycles mark the Mesozoic and Cenozoic, lasting from 5 to 15 my each, with about 10 my the most common number. One to seven cycles enliven each supercycle, more than a hundred in all; they vary in duration from 1 to 5 my.

Sequence analysis and Vail curve have caused an outpouring of long, wordy papers and a bulky body of theory, generated by a smallish group of authors in varying combinations. Characteristic are a stubborn adherence to the belief that most sea level cycles are eustatic, a tendency to cite only supportive literature by insiders, and complex theoretical models that ignore modern work on depositional processes and geomorphology. This gives sequence analysis, useful and stimulating as it is, somewhat the flavor of a cult. If that seems a bit startling for science, I add that it is not unheard of. We shall encounter it again.

9.6   CAUSES OF EUSTATIC SEA LEVEL CHANGES

Two decades after its birth, sequence analysis is widely in use, but the Vail curve, in its emphasis on global validity, remains an intriguing proposal rather than a demonstrated reality. Local and regional sea level changes can be attributed to tectonics, but why should the level of the sea have risen and fallen eustatically throughout geological history? What can we learn from the time scale of the cycles, from the slow rate of rise of no more than 1–3 cm/1,000 years, from the almost instantaneous falls, or from the remarkably uniform vertical range of a few hundred meters for all cycles?

Like tea in a cup, the global level of the sea depends on the volume of ocean water and the size of the containers, the ocean basins. Each may change in several ways, some capable of causing a large change in sea level, others trivial in their consequences.

Most water on earth was released early from the interior (Section

14.1), but volcanoes and hotsprings keep adding a little more every day, although most of that is recycled groundwater. The addition is far too small to raise sea level by even a few centimeters per 1,000 years. Moreover, we can add, but we cannot subtract; there is no known way to dispose of excess water when the time comes for the sea to fall once more. The only way to effect changes in both directions is to remove water from the ocean, store it elsewhere for a while, then release it again, but the total capacity of streams, lakes and groundwater is much too small to have the impact we need. Only icecaps are large enough, but they change too fast, by meters per 1,000 years, and besides, much of the Paleozoic and the whole Mesozoic had no icecaps.

Other processes, a change in the moisture content of the atmosphere or the effect of warming and cooling on the volume of ocean water for example, are capable of altering sea level, but not nearly enough. Sea level changes due to increases and decreases in the volume of water in the ocean are, except for ice ages, ruled out.

What about the container? Can we change the capacity of the ocean basins to produce the right effect? Some proposals are easily dismissed. Closing and evaporating a small ocean such as the Mediterranean would raise global sea level by no more than 15 m. Might we dump into the oceans enough continental sediment, of which there is an abundant supply, to do the job? Sea level would rise, but too slowly and not enough, and we would have to invoke episodic accelerations of subduction to dispose of the sediment layer each time the sea was to fall again.

Still, plate tectonics may be involved. Because of the mid-ocean ridges, the ocean floor is not concave upward like a dish but convex like the bottom of a wine bottle. The young, hot crust at the crest is shallow, but as it is trundled away it ages, cools and subsides in proportion to its age (Figure 9.9). When a ridge spreads rapidly, young, hot, and hence shallow ocean crust is found farther from the ridge axis than when spreading is slow. Therefore, when spreading is accelerated, the volume of the ridge increases, the capacity of the ocean basin is reduced, and the sea rises. When spreading slows down, the level of the ocean falls. Obviously, the same effect is achieved by increasing or decreasing the number or length of the mid-ocean ridges.

Will this be adequate? Take a ridge 10,000 km long, about one-seventh of the present system, and triple the spreading rate smartly from 2 cm to 6 cm per year, both perfectly reasonable values. A new equilibrium profile will be complete in 70 my (Figure 9.9), no sooner, no later, and the sea rises, rapidly at first, then more slowly, to reach a new height of 120 m above its old level. If we decelerate to 2 cm per year, the sea will

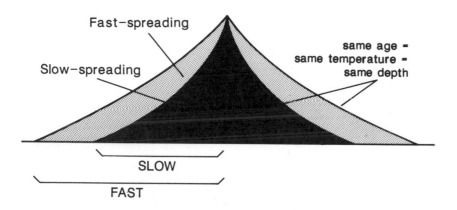

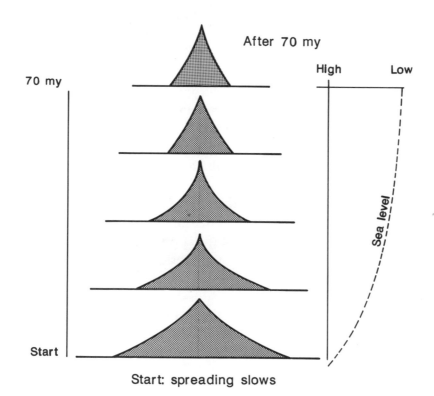

Figure 9.9. The cross-section and volume of a mid-ocean ridge depend on its spreading rate. A fast ridge has a larger volume and displaces more water than a slow one. If the spreading rate is cut from 6 to 3 cm/year, the ridge profile becomes narrower, taking 70 my to reach equilibrium. This increases the volume of the ocean basins and causes sea level to fall. An increase in the spreading rate has the opposite effect. In this way spreading rate changes cause sea level changes of long duration, but many supercycles and all cycles are too short to be explained in this manner.

recede to its former level in another 70 my. To match the 200 or 300 m observed in the Mesozoic, all we need do is to change the spreading rate more or lengthen the ridges. But it will always take many tens of millions of years to achieve an effect of the right magnitude.

Are such modest changes in the rate of spreading or in the number or length of mid-ocean ridges in accord with observations? The answer is yes: since the Cretaceous the mean spreading rate has decreased enough to explain the fall of the sea during the Cenozoic, even if we take into account all possible errors.

So spreading rate changes can explain first-order sea level changes and perhaps the longest supercycles as well, but higher order cycles are not so easily accounted for. For one thing, they tend to be brief, and, for another, there is no sign of the hundred or more global spreading rate changes needed during the last 200 my.

Before we reject the hypothesis out of hand, we may consider other things. The shoreline forms where the rate of sedimentation equals the rate at which the sea rises. If the sea begins to rise faster, sedimentation fails to keep pace and the shore migrates landward. If, on the other hand, the rise slows down, too much sediment arrives at the water's edge, and the coast will build seaward, starting a regression even when sea level is not actually falling. In this way, small changes in the rate at which sea level moves produce regressions and transgressions like those of the Vail curve without actual reversals of the rise or fall of the sea. Changes in the rate of subsidence of continental margins have the same effect.

These processes produce curves of the appropriate short duration, but are hardly likely to do so all over the world at the same time. If we believe that Vail cycles are eustatic, this explanation is no help. Different, and differently changing rates of subsidence and sedimentation in different places could be cleverly combined to obtain eustatic cycles of fall and rise, but the plausibility of the argument decreases as its complexity increases. All it does is to make it difficult to prove that the cycles are not eustatic.

9.7   A VIEW FROM THE CRATON

So far we have considered the sea level problem from the sea, a plausible perspective for me, a geological oceanographer, but not the only one. What do we see if we regard it from the continents, standing perhaps on their ancient cores, the cratons? We still see the sea rise and fall relative to our perch, but now we note, as others have done before, that many major regressions coincide at least approximately with times of mountain

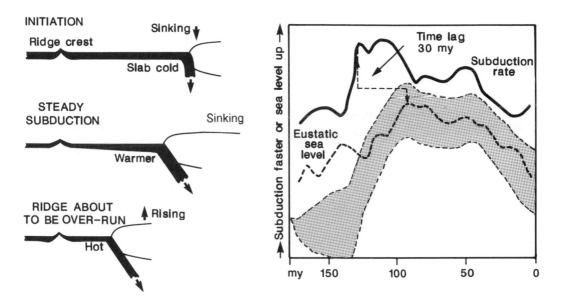

Figure 9.10. Plate collisions also affect sea level (left diagram). The sinking slab of a new subduction zone drags the adjacent plate edge down, causing a transgression which moves inland as the subduction matures. If the plate later overrides a mid-ocean ridge, subduction comes to an end, and the buoyant hot oceanic crust raises the overriding plate, bringing about a regression. It is more difficult in this case than at mid-ocean ridges to calculate the consequences, but over a time span of 50 my or more the effect is significant (right). The shaded band, wide because of the uncertainties of the calculation, depicts the response of sea level to changes in the subduction rate (heavy curve). Note the time lag between cause (subduction) and effect (estimated sea level change). Observed sea level changes are shown with a dashed line.

building, with orogenies. Lest we get carried away, however, I note that correlating mountain building with sea level changes has never produced much else than controversy, because the durations of the events and the uncertainties of their ages are often of the same order as their spacing in time. Anyone possessed of firm resolve can usually find a correlation that suits, but persuading others is a different matter.

The explanation that attributes eustatic changes to changing rates of seafloor-spreading ignores subduction. This is wrong, as Michael Gurnis of the University of Michigan has pointed out. When a subduction zone is created (Figure 9.10), the old, cold, and therefore heavy oceanic lithosphere at first sinks more or less straight down, dragging the edge of the overriding plate a few hundred meters down with it, even if it is a continent. As time passes, the sinking slab warms, assumes a gentler slope, and the zone of continental subsidence moves landward, accompanied by a regional transgression. Subduction comes to an end when a mid-ocean ridge approaches (Figure 8.4), causing ever-younger crust to be subducted. This crust is hotter and lighter, and its increasing buoyancy pushes up the continental margin to form a plateau. The result is a regression.

The power of subduction to cause sea level to change is thus clear, and over time the state and extent of subduction zones do change also. Today, active margins constitute a mere 30 percent of the total, but during the late stage of its formation Pangaea was surrounded by active margins that may have had a considerable impact on sea level.

But let us proceed; what else can we observe when looking at the sea from the craton? In addition to local subsidence and uplift, are there other, more widespread tectonic effects that we need to consider? There are: continents, in part or in their entirety, rise and fall because of an ill-defined process called epeirogeny. It should not be ignored, as most plate tectonicists have done, and adherents of eustatic sea level changes still do. On many cratons vast marine deposits in shallow basins preserve histories of submergence and emergence of considerable length. Eustatic sea level changes do not suffice to account for these or are improperly timed. Instead, extensive vertical motions of the continental crust of a few hundred meters, the same range as Vail sea level changes, are needed.

Epeirogenic movements often affect more than one continent at a time. From 125 to 115 my ago, all continents, including Australia, were becoming widely inundated by a rising sea. Then, while others submerged further, Australia began to rise and while the transgression reached its maximum elsewhere, Australia fell dry, implying an uplift of hundreds of meters to compensate for the rise of the sea.

Vertical motions that affect regions thousands of kilometers across and that happen without relation to plate boundaries must have their cause below the lithosphere. Imagine an asthenosphere with large hot and cold masses dispersed throughout. Hotspots, mantle plumes, detached subducted slabs, and convection currents are all possibilities. The colder masses are dense and by gravity will pull down the surface of the asthenosphere, whereas warmer, lighter ones will raise it. A relief of highs and lows thus forms at the base of the lithosphere on a horizontal scale of thousands of kilometers. Seismic tomography (Section 8.4) has revealed the existence of such hot and cold masses, allowing us to make this more than a thought experiment. How high might the relief be? The best answer is a few hundred meters, again a number similar to the amplitude of cycles and supercycles. With as yet sparse data we cannot go very far in this reasoning, but it would seem quite possible for continents to travel across this undulating surface, suffering epeirogenic rise and fall as they go, accompanied by transgressions and regressions on a time scale of 10 to 100 my.

Add that eustatic changes due to changing spreading rates (or to glaciations) may occur at the same time, and local tectonics too, and it will be clear that tracking all possible causes of a change in sea level is tricky. It also does require the opposite of what the proponents of the Vail curve have done. Instead of constructing a global sea level curve from dispersed pieces of information and then looking for the cause, we need separately determined causes for sea level histories from all over the world.

Be all this as it may, it is clear that the simple, very old story of the ups and downs of the sea leads to weighty questions about how the earth works. Not only are the questions far from being answered, there is not even unanimity about which questions are the right ones.

10

Other times and other oceans

A quarter century ago, few geologists gave the oceans of the past much if any thought. The emphasis on the ocean basins by the plate-tectonic revolution has changed that and we have come to realize how important ancient oceans are for our understanding of the earth as a whole. As yet we cannot discuss with confidence any oceans older than those of the late Cenozoic, and next to nothing is known about those of the Precambrian. Still, in two decades a rich harvest of new information has been reaped, due above all to a major program of ocean drilling that began in 1968. New concepts are worth testing, and as regards chronology we have advanced considerably. If what follows is hedged with doubts and cautions, another decade or two should change that.

10.1   HOW THE OCEAN WORKS

The great ocean rivers, the surface currents, are driven by the planetary winds, by the trade-winds on either side of the equator and by the westerlies at 45–60° N and S. A simple ocean bordered by land would have two equatorial currents, pushed westward by the trades (Figure 10.1). At the western barrier, some of the water would be reflected back along the equator, while the remainder would form large gyres in the northern and southern hemispheres, eventually returning east under the influence of the westerlies. Another gyre, flowing counterclockwise, would exist in each subpolar zone. The surface currents of real oceans are a bit different, because our continents are not so simply shaped and arranged, but the basic patterns are the same (Figure 10.2).

Like the atmosphere the ocean carries heat from low to high latitudes, but it does so more efficiently, because water holds much more heat than air and releases it more slowly. The equatorial currents are warmed by the sun and their waters are warmest where the return flow, having tra-

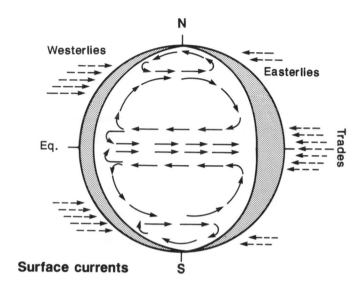

Surface currents

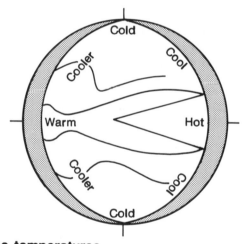

Surface temperatures

Figure 10.1. The surface currents of the ocean are driven by the planetary winds. In this simple ocean enclosed by land, some of the water of the two equatorial currents, blown westward by the trade-winds, returns as an equatorial countercurrent after encountering the western barrier. The rest is deflected north and south and forms two large gyres in the northern and southern hemispheres. The temperature distribution at the surface mirrors the current pattern, with the warmest waters at the end of the equatorial countercurrent because they traveled the equatorial distance twice.

versed the equatorial ocean twice, meets the east coast. The two gyres bring back to the equatorial zone water that has been cooled during its voyage in high latitudes, making the surface temperature distribution at mid-latitudes asymmetric (Figure 10.1). The salinity distribution is also

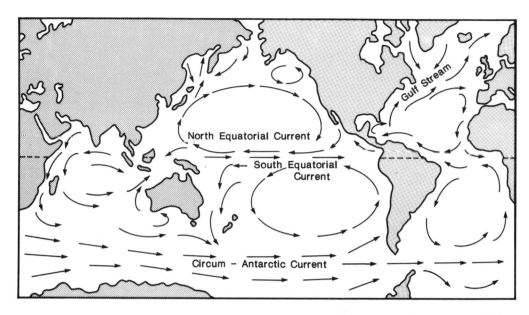

Figure 10.2. The current patterns of the geographically more complex real oceans are variations on the theme of Figure 10.1 that reflect the realities of the shapes and locations of the continents. The map also brings out the importance of gateways that connect oceans with each other, such as the straits through the maze of the Indonesian Archipelago or the Drake Passage between South America and Antarctica.

asymmetric, low in the eastern equatorial region because of tropical rains, but increasing westward as the water warms and evaporates. It is highest inside the subtropical gyres, where rain rarely falls.

Surface currents affect only the upper 100 to 200 m of the oceans. The present ocean is a large bowl filled with two fluids: a thick layer of dense water at the bottom and a thin, lighter one on top. Both are seawater; the density difference is due to a sharp temperature change at shallow depth called the thermocline (Figure 10.3). High salinity also raises the density of seawater, but plays a minor part in most places.

Above the thermocline the water is stirred by currents and waves, and the temperature is quite uniform. Underneath it is cold, cooling only a little more toward the bottom. The density difference across the thermocline is small, less than 0.05 g/cm³, but that suffices to render it difficult to raise deep water to the surface or to blend one water mass into the other. The stratification of the present ocean is thus very stable, a condition that severely limits its fertility.

The biological productivity of the ocean, often regarded as virtually limitless and a guarantee of a vast and mostly untapped supply of food, is actually just marginally greater than that of a desert, and already seriously over-exploited. Only in a few areas, along many coasts and in

198

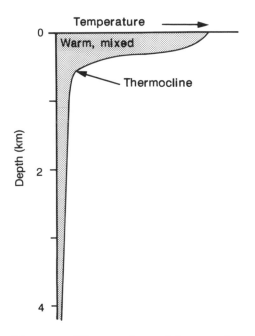

Figure 10.3. The upper layer of the ocean is warmed by the sun and mixed by waves and currents. A sharp drop in temperature, the thermocline, sets the shallow zone off from the cold abyss. The water above the thermocline is less dense than that below, creating a stable stratification that impedes the exchange of water between the surface and the deep.

major ocean currents, is the ocean really rich, and less than one-tenth of its ocean surface produces an amazing 90 percent of all organic matter. Microscopic floating plants living above the thermocline, the phytoplankton, use sunlight, water and carbon dioxide, all generously available, to manufacture organic matter by photosynthesis. They are the ones responsible for the primary productivity on which all other oceanic life depends. To do so, they require nutrients such as nitrogen, potassium and phosphorus. These essential elements are dissolved in seawater, but in such small amounts that a single healthy plankton bloom can exhaust the supply in hours or days.

Once dead, the organisms sink, decay, and well below the thermocline release their nutrients and their carbon dioxide. The deep ocean is therefore an enormous reservoir of nutrients, but unless the water is recycled across the thermocline, those are lost to the cycle of life at shallow depths.

Turbulent currents can mix some of the deeper water upward, but the process most important in ocean fertilization is called upwelling (Figure 10.4). Like the winds, the ocean currents are affected by the Coriolis force (Section 3.3) which pushes them to the right in the northern hemisphere, so that a current flowing south along a west coast, for example

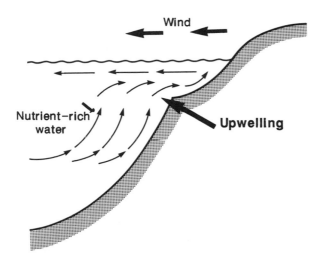

Figure 10.4. The stable stratification of the ocean limits the supply of nutrients to the surface. Although currents stir the sea to limited depth and so fertilize the surface zone, the main agent in bringing up nutrients from the deep is upwelling. A seaward wind, or a current along the coast deflected seaward by the Coriolis force, drives the surface water away from the shore. It is replaced by water that wells up from between a few hundred and a thousand meters down, far below the thermocline, and is therefore rich in nutrients.

the California Current, is forced away from the coast. The gap is filled by water that rises from several hundred meters down, bringing nutrients with it. On the southern hemisphere the situation is reversed. The north-flowing Humboldt Current off western South America is forced to the left, and fertile waters rise to the surface near the coast. Upwelling is responsible for the fertility that supports most major fisheries of the world, those off California, for example, and off Peru and Ecuador, West Africa, Japan and Newfoundland.

By far the largest volume of ocean water lies below the wind-driven surface circulation and the thermocline. This deep water is not stagnant; it flows in orderly patterns, driven not by the wind but by density differences. Dense water masses sink and lighter ones rise and take their place. In today's ocean, the density differences are usually due to temperature contrasts which have a wider range and a larger effect than salinity. Freshwater reaches its maximum density at 4 °C, but seawater is densest at its freezing point which depends on the salinity but is normally between 1 and 2 °C below zero.

Around Antarctica, the Circum-Antarctic Current carries very cold water that has spent many years in high latitudes (Figure 10.2). In the southern winter, sea ice forms along the Antarctic ice edge; as it freezes,

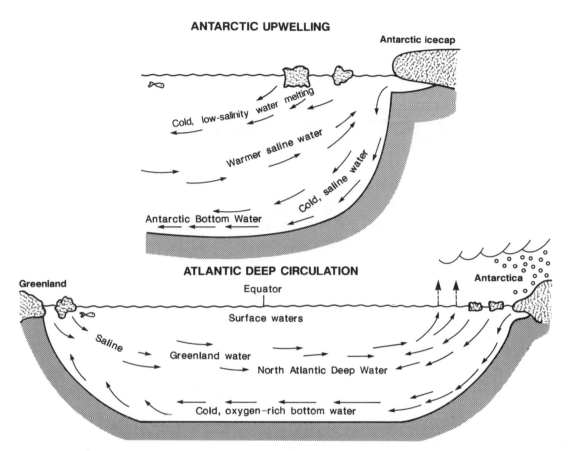

Figure 10.5. The deep circulation of the oceans is driven by density contrasts that are caused by differences in temperature and salinity and is therefore called a thermohaline circulation. When the sea freezes, very cold, saline water forms around Antarctica, then sinks due to its high density. This water travels north in all oceans as an abyssal current of Antarctic Bottom Water. Farther offshore, icebergs melt and another water mass is formed, also cold but less saline. It travels at much shallower depth and not so far. Near Greenland, North Atlantic Deep Water is formed by the same process and by addition of cold waters from the Arctic. Traveling south at intermediate depth, this water surfaces near Antarctica. Being warmer, it evaporates, bringing moisture to the air and snow to the southern continent.

it extrudes most (*c.* 70 percent) of the salt. The salt increases the density of the very cold water below the ice, and it sinks. Driven by the pull of gravity, this Antarctic Bottom Water travels north along the bottom, forming abyssal currents in all oceans (Figure 10.5). To replace it, slightly warmer water of normal salinity rises from intermediate depth in a vigorous upwelling that makes the circum-Antarctic region one of the most fertile places in the world. In spring and summer, ice floes and icebergs break off the icecap, float north and melt along the way. The meltwater is only brackish, but so cold that it too sinks and travels north at an intermediate depth.

In the North Atlantic, cold water is formed also, but on a smaller scale. Cold, saline water is produced off Labrador and eastern Greenland by the freezing of sea ice, an influx of cold Arctic water and evaporation, and sinks (Figure 10.5). This North Atlantic Deep Water we have already met under the name "salt conveyor" (Section 5.4; see Figure 5.6). It is neither as plentiful nor as dense as the Antarctic Bottom Water, but is nevertheless able to traverse the Atlantic at a depth of 2–3 km. Close to Antarctica it rises, supplies nutrients and, because it is relatively warm, evaporates and provides snow for the Antarctic icecap. A much smaller cold-water source occurs in the northwestern Pacific.

Surface water temperatures range from about 27 °C on the equator to −1.5 °C near Antarctica, but because all abyssal water comes from the same Antarctic source it begins with a temperature of just below −1 °C. Taking about four centuries for its voyage north through the Atlantic and 1,500 years in the Pacific, the Antarctic Bottom Water is slightly warmed by the heat flowing out of the ocean floor and by mixing from above with warmer water. When it arrives at its northern destination, it is light enough to rise and begin a return trip.

10.2 CONTINENTAL DRIFT AND OCEAN CIRCULATION

Deep, narrow passages, such as the channels east of the Falkland Islands in the southern Atlantic and near Samoa in the western Pacific, are the only outlets that lead from the Antarctic abyss to the northern basins. These deep "gateways" control the pattern and direction of abyssal flow. At the surface, the Drake Passage, which allows the Circum-Antarctic Current to pass between South America and Antarctica, and the Straits of Gibraltar between Spain and Africa illustrate the key role of gateways for the ocean circulation. Continental drift controls most of them, but even minor tectonic events may open or close some gates and have an impact on the ocean circulation as large as the stately drift of the continents themselves, and often more abruptly.

To appreciate the role of gateways, let us return for a moment to the ocean circulation model of Figure 10.1 and arrange the four continents so that the equatorial currents flow freely around the world (Figure 10.6). The equatorial current is heated more than before, because much of it circum-navigates the earth more than once, and the north- and south-

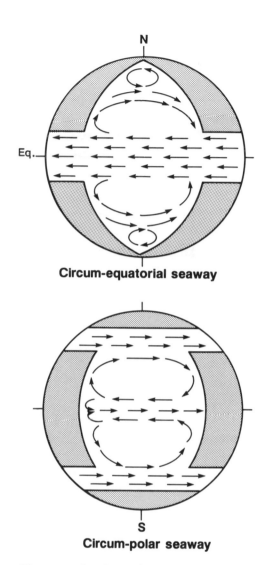

Circum-equatorial seaway

Circum-polar seaway

Figure 10.6. Barriers and gateways play an important role in the circulation of the oceans. If the equatorial current can pass around the earth one or more times before being deflected north and south, a more even heat distribution across the latitudes is the result. If the equatorial current is restricted, but the way is open for circum-polar flow, the polar continents are insulated from the warm water at low latitudes, and the temperature gradient from equator to poles steepens.

flowing currents that diverge from it are warmer too. Therefore the ocean is warmer overall and yields more moisture to the atmosphere, making it a warmer and wetter earth.

Conversely, we may block the equatorial flow and set up gateways at high latitude to allow circum-polar currents to pass. These currents insulate the polar continents from warmer seas and cause the polar temperature to drop. If an adequate supply of moisture were available, an icecap might form. The return flow to the equator is colder too, the whole earth is cooler, and there is a larger temperature contrast across the latitudes.

The supercontinent Pangaea (Figure 7.1) and its breakup in the Mesozoic provides examples. At the equator, Panthalassa, the world ocean, had an equatorial current system spanning almost 80 percent of the circumference of the earth, and one main gyre in each hemisphere. Circum-polar flow existed only in the far north. The ocean was warmer than the present Pacific, the temperature difference between equator and poles was smaller, and without icecaps to generate cold, dense water, the deep circulation was more sluggish than it is now.

About 200 my ago, a major gateway broke through to the west from the Tethys embayment (Figure 10.7) and, assisted by the opening of the southern North Atlantic, wedged the northern and southern continents apart. The now circum-global equatorial flow warmed up, north- and south-flowing currents took this warm equatorial water to high latitudes, and in the absence of cold circum-polar currents the temperature gradient from equator to poles was further reduced.

Starting with the separation of Africa from Antarctica about 125 my ago, the breakup of Pangaea rendered the ocean circulation ever more complicated (Figure 10.7). But late in the Cretaceous, the circum-equatorial current still ran uninterruptedly, the high latitude ocean was quite warm and, as regards its climate, even the early Cenozoic earth resembled its Mesozoic ancestor more than its modern descendant.

10.3   MESOZOIC WORLD

The images of Figure 10.7 are paleoceanographic reconstructions based on the outlines of the continental blocks and on common sense. They are as good as we can make them, but each represents an interval much longer than the whole Quaternary, during which much may have changed. Also, because they have only the outlines of blocks of continental crust, they cannot take account of real shorelines and even of epicontinental seas. Because of the many and large sea level changes of the Mesozoic (Figure 9.8), that is a serious defect. Living in a time when epicontinental seas are rare and much smaller than those of the Mesozoic, we find it difficult to understand the oceanography of such a world.

Can we confirm the deductions in some detail? Picking a time around 100 my ago (mid-Cretaceous), compilations of sedimentary and paleontological data indicate an earth that was much warmer than today at

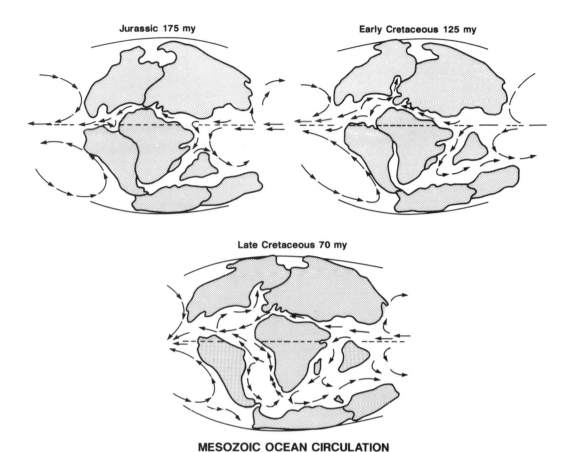

Jurassic 175 my

Early Cretaceous 125 my

Late Cretaceous 70 my

MESOZOIC OCEAN CIRCULATION

Figure 10.7. During the Mesozoic, the surface circulation of the oceans evolved from a simple pattern in a single ocean with a single continent to a more complex situation in the new oceans of the Cretaceous. Throughout this time, the open circum-equatorial path and the absence of circum-polar currents resulted in a temperature distribution that was more even than today.

middle and even at high latitudes (Figure 10.8). Warm-water faunas mark the shallow seas between 45° N and 45° S, and coral reefs reached much farther north than they do now.

Closer to the poles the predictions are also supported by the geological record. Forests occurred in Antarctica and Canada up to 85° latitude. Coal beds found beyond 60° N and S reveal not only the expected high rainfall, but also show that large deciduous trees of cool- and even of warm-temperate forests survived far beyond the latitude where they grew at any time during the later Cenozoic. At such high latitudes, the winters were long and dark as they are now, and the average winter temperature probably below freezing, although not nearly as cold as today. Experiments have shown that modern tree species related to those of the high-latitude Cretaceous survive without damage the long nights, provided

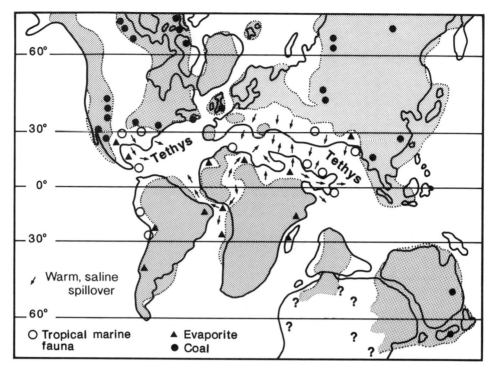

Figure 10.8. The world of the middle Cretaceous, about 100 my ago, was a watery one. The area of land (shaded) was sharply reduced and shallow seas were widespread, especially at low latitude. Coal beds, formed in warm or temperate swamps (black dots) are found above 60° N and S, and tropical fossil faunas and tropical and subtropical salt deposits extended north and south far beyond where they occur today. The latitudes are·those of the middle Cretaceous. The arrows show where dense, warm, saline waters from shallow seas might have spilled into deep basins, thus forming warm rather than cold abyssal waters.

the temperature is distinctly on the cool side. Besides, the disadvantage of long winters may have been at least partly compensated by endless summer daylight.

There is no evidence for polar icecaps, but coarse sediment ranging from sand to boulders probably brought there by drifting icebergs occurs here and there in marine mudstones. Adherents of the Vail curve see in this the confirmation of eustatic sea level changes, but the evidence is at best spotty, tide-water glaciers do not necessarily imply icecaps, and the fossil forests argue strongly otherwise.

The north–south temperature contrast was thus very different from the one we know today. This condition is borne out by oxygen isotope temperature measurements on fossils from surface waters of the Mesozoic oceans. They are more sparse than one would like, but consistently indicate water temperatures of 25–30 °C in the equatorial region, declining to 12–15 °C at 60° to 70° N and S. Measurements on shells of bottom-dwelling (benthonic) organisms that lived at depths near 2000 m give values of around 15 °C, much warmer than the 1–4 °C of today. The source of deep water is the densest surface water on earth, and that usually occurs in polar regions. With polar surface water temperatures of 10–15 °C, no icecaps could exist.

Taken together, the evidence for a world with warm polar regions, a green-house world, is consistent and impressive. One can argue whether there were many glaciers in high latitudes or few, or whether the water at −2000 m was really the coldest water on earth. Until something more solid comes along, however, that rather seems like quibbling.

The unusual warmth of the Cretaceous deep water has other implications. The density contrast between surface water of 25 °C and deep water of 15 °C is so small that salinity also becomes an important factor in the density of the water. It is therefore possible that the Cretaceous deep circulation was driven partly or even entirely by salinity differences rather than only by temperature as today.

In the Mediterranean, the evaporation in summer exceeds the total influx of freshwater. This results in a salinity of 3.8 percent which, when in winter the surface water cools to 15 °C, makes it as dense as abyssal water of 2 °C and a salinity of 3.6 percent. It sinks, flows out through the Straits of Gibraltar and, being only slightly less dense than the much colder but less saline deep Atlantic water, spreads out across the central North Atlantic just above the bottom (Figure 10.9).

It is conceivable that, during the Cretaceous, supersaline water formed by evaporation in epicontinental seas or in new, shallow oceans was a source of deep water. The thick evaporites of mid-Mesozoic age that line the young South and North Atlantic oceans on both sides (Figure 10.8) confirm that such salty waters did exist. One can imagine the young South Atlantic rift spilling dense salty waters into adjacent ocean basins where they sank to the bottom. From there they might have flowed north and south to rise in the polar regions, in a reversal of the present sense of flow. Fed in this manner by waters of more than one origin, the Cretaceous deep-sea would have lacked the orderly circulation pattern of the deep oceans of the later Cenozoic. There is indeed isotopic evidence for the existence of surface waters of high salinity in mid-Cretaceous shallow seas.

10.4 THE BLACK SHALES OF THE CRETACEOUS

In a world with such small temperature differences, the driving force for the abyssal ocean circulation must have been small, and the flow therefore

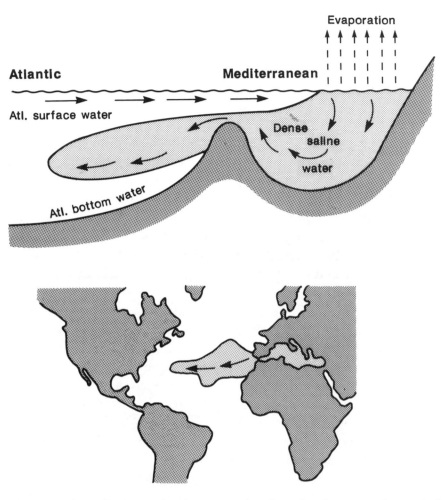

Figure 10.9. The Mediterranean Sea demonstrates how dense abyssal water can form in shallow equatorial seas rather than in cold, high-latitude regions. Evaporation in the Mediterranean in summer exceeds the influx of freshwater, and a saline brine forms. When the brine cools in winter, it sinks, and dense 15 °C water flows across the sill at the Straits of Gibraltar and spreads across the central Atlantic, just above the much colder but less saline Antarctic abyssal water.

more sluggish than today. Sometimes and in some places, the Cretaceous abyss may even have been stagnant, as the Black Sea is now, a dark, putrid pool.

We have abundant evidence for unusual deep-ocean conditions in the form of black shales rich in organic matter (up to 30 percent sometimes). The deep-sea sediment record of the Cretaceous, especially between 115 and 125 my ago and again around 90 my, is full of them (Figure 10.10). Black shales were also common in marginal Cretaceous seas, where they became the source of some of the world's most prolific oil fields, for

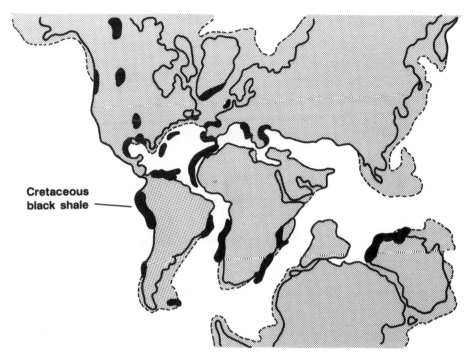

Cretaceous
black shale ———

Figure 10.10. Black shales rich in organic matter formed in abundance at various times during the Cretaceous. They were deposited in the oceans to depths as great as 3,000 m and also in coastal seas, and became important source beds for oil.

example those of Venezuela. The high organic carbon content shows that there was too little oxygen in the water to decompose all of the organic matter produced at the surface before it sank to the bottom and was buried there.

Another factor contributing to the accumulation of organic matter in Cretaceous deep-sea sediments was the warmth of the deep water, because much less oxygen dissolves in seawater of 15 °C than at 2 °C. Therefore, the Cretaceous abyssal waters right from the start contained less oxygen than those of today.

In the long run, a negative feedback protects the abyss from turning permanently anoxic (free of oxygen). When the oxygen supply runs out, less organic matter is decomposed and fewer nutrients are recycled. This lowers the fertility of the surface waters, the biological productivity slows down, and less organic matter sinks to the bottom. With less demand, the oxygen content of the deep water recovers, and more nutrients reach the surface. For this reason, black shales should be deposited intermittently; very black layers should alternate with others not so rich in organic matter and, indeed, we do find this kind of lamination in Cretaceous black shales. The presence of the lamination is itself evidence that the oxygen content was low enough to deny organisms the energy to burrow for food and destroy it.

Modern equivalents of black shale accumulate in basins that are separated from the ocean by a sill that prevents oxygen-rich seawater from entering the deeper parts. The Santa Barbara basin off the coast of southern California receives oxygenated water only once in a while, and the deep Black Sea is permanently oxygen-free; both have sediments very rich in organic matter.

Moreover, even in the oxygen-rich oceans of today the biological productivity can be so high that the rain of dead organic matter consumes virtually all oxygen in the water column before it even reaches the bottom. This creates an oxygen-minimum zone in mid-depth, and organic-rich deposits accumulate where the zone intersects the continental slope, in the Gulf of California, for example, and on the continental margins of Peru and Ecuador. Below the oxygen minimum zone, deep water currents supply enough oxygen to support a benthonic fauna that removes most of the remaining organic matter, and the deepest sediments are almost free of organic matter. Most Cretaceous black shales seem to have been deposited on continental slopes with an oxygen minimum zone rather than in an oxygen-free, Black Sea-like abyss.

The burial of so much organic matter means that equally large amounts of nutrients were removed from circulation and so were no longer available at the surface. Paradoxically, the Cretaceous ocean, producer of many rich oil source beds, was not especially fertile, much less fertile than the present ocean.

When carbon dioxide is converted to organic matter and taken out of circulation by burial, less oxygen is consumed and the atmosphere becomes enriched in this biologically important gas. If we can estimate how much organic matter was buried at various times, we would obtain valuable insight into the variation with time of the oxygen content of the atmosphere.

10.5 CARBON BURIAL, CARBON ISOTOPES, AND THE ATMOSPHERE

To estimate changes in atmospheric oxygen due to the burial of organic matter, we turn again to stable isotopes, this time to those of carbon. In addition to ^{14}C, the unstable carbon isotope used mainly to date archaeological finds (Section 2.2), there are two stable ones, ^{12}C and ^{13}C, which have proved to be of great value. As with oxygen isotopes and usually together with them, they are determined by mass spectro-

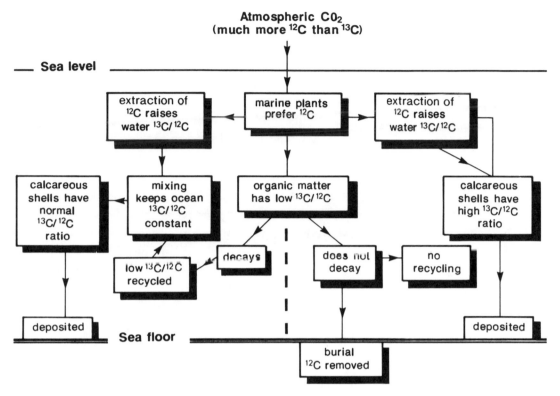

Figure 10.11. The carbon isotope system resembles that of oxygen isotopes and is treated the same way. It provides insight in the sources of organic material and the ocean–atmosphere cycle of carbon dioxide and oxygen, rather than evidence on temperature and icecaps. See the text for a discussion.

metry, and we use $^{13}C/^{12}C$ ratios (as well as $\delta^{13}C$) in exactly the same way as $^{18}O/^{16}O$ ratios (Section 4.1).

Plants making organic matter by photosynthesis prefer the light carbon isotope. Hence, the carbon isotope ratio of organic matter, whether vegetal or animal (animals obtain their organic matter from plants), is enriched in ^{12}C relative to the carbon dioxide (CO_2) of the atmosphere and the ocean and has a smaller $^{13}C/^{12}C$ ratio. Therefore, the CO_2 in the water is depleted in ^{12}C and has a higher $^{13}C/^{12}C$ ratio. As with oxygen isotopes, there is a temperature effect, but it is small and can be ignored.

Planktonic plants living near the surface make tissues with a lower $^{13}C/^{12}C$ ratio. When they die, the organic matter sinks to the deep-sea floor and decays by oxidation to CO_2 (Figure 10.11, left) which dissolves in the water. This brings the $^{13}C/^{12}C$ ratio of the dissolved CO_2 back to its original value.

If the organic matter is not oxidized but buried instead, the ^{12}C buried with it is removed from the system (Figure 10.11, right). The CO_2 dissolved in the water retains its higher ^{13}C content and larger $^{13}C/^{12}C$ ratio. Calcareous fossils deposited with the organics, having derived their $CaCO_3$ from isotopically heavier ocean water, are heavier themselves with a larger $^{13}C/^{12}C$ ratio.

This sounds a bit complicated, but it means that, even if we cannot study the organic matter itself, changes with time of the carbon isotope ratio of calcareous fossils tell a story of burial and recycling of the organic matter in the oceans. If the $^{13}C/^{12}C$ ratio rises, carbon was buried and ocean and atmosphere contained less CO_2 than before. Less oxygen was consumed and this gas (O_2) was enriched in water and air. Nutrients used in the production of organic matter were buried with it and the oceanic fertility was reduced. We can estimate the amount of buried carbon from the $^{13}C/^{12}C$ ratio of calcareous fossils and from this number compute the amount of oxygen that was not consumed and therefore retained in the ocean and the atmosphere.

Conversely, a decrease in the carbon isotope ratio of fossil carbonate says that buried carbon was exhumed by erosion of the seafloor deposits, oxidized and put back in circulation as carbon dioxide. Because this process consumes oxygen, a decrease in the oxygen content of the atmosphere was the result. In this way, the fossil shells of oceanic organisms contain a record of changes in the amount of atmospheric oxygen. A large set of carbon isotope data from oceanic sediments extending into the latest Cretaceous (Figure 10.12) shows major changes, both broad overall trends and sharp, sudden changes.

By late Eocene time about 100 trillion tons (10^{20} grams) of carbon in excess of the amount stored today had been buried in Mesozoic and early Cenozoic marine sediments. As a result, the atmosphere was much enriched in oxygen, perhaps so much that it became a fire hazard. Concentrations of tiny charcoal grains in Paleocene ocean sediments do indeed suggest great forest fires. Then, in the Middle Miocene, some 15 my ago, a major decrease of the $^{13}C/^{12}C$ ratio shows that buried carbon was being exhumed and oxidized, releasing CO_2. From the decrease Nicholas Shackleton has estimated that enough carbon had been converted into CO_2 to take up an amount of oxygen equivalent to 20–25 percent of the present atmospheric content. The late Cenozoic atmosphere, much poorer in oxygen, also became about twice as rich in CO_2 as before.

The erosion of the carbon-bearing sediments was due to a strong flow of abyssal water that began 15 my ago. It removed much sediment in all oceans, and produced large, widespread unconformities. The oxida-

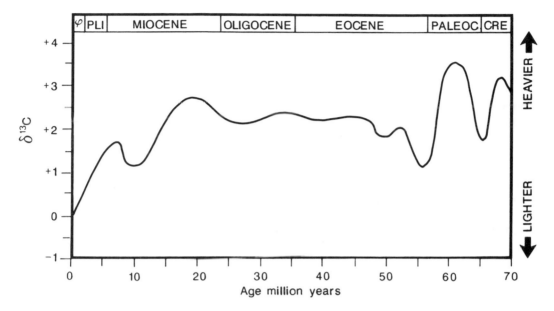

Figure 10.12. Carbon isotope ratios of pelagic carbonate, here represented by the deviation $\delta^{13}C$ from the standard, varied greatly throughout the Cenozoic. The early Cenozoic microfossils were isotopically heavy, meaning that light ^{12}C was being buried, but this ceased in the Eocene. Little change occurred until the middle Miocene, when the carbon ratio of the calcareous oozes began to shift rapidly toward lighter values, indicating that a large amount of buried carbon was being exhumed and oxidized. As a result the late Cenozoic atmosphere became enriched in carbon dioxide, but lost about one-fifth of its previous oxygen.

tion was enhanced by a 10 °C cooling of the abyssal water that rendered it capable of holding about 20 percent more oxygen than before.

As with the oxygen isotopes, some changes in the carbon isotope record (Figure 10.12) are quite sharp, although one must remember that as much as a hundred thousand years may have been involved in each case (Section 2.3). There was a sharp change to lighter values at the end of the Cretaceous, and another one at the Paleocene/Eocene boundary. Then, for a while, little carbon was added or subtracted from the ocean reservoir, but declines around 15 and 6 my ago, although not quite so abrupt, sent the system firmly into the carbon oxidation mode. Each change can be correlated approximately with a large drop of the temperature of the deep water, suggesting that an ever more vigorous abyssal circulation was involved in the recycling of buried carbon.

10.6   A TALE OF TWO OCEANS

The Cretaceous ocean circulation definitely promoted the transfer of heat to high latitudes, but was it sufficient to explain those warm poles? Other

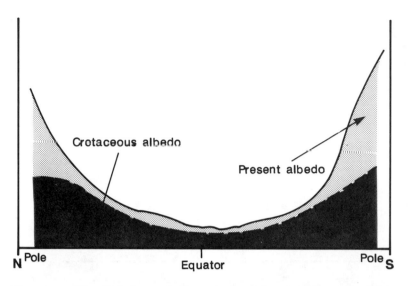

Figure 10.13. The albedo is a major factor in the surface temperature distribution of the earth (Section 3.4). Because the Cretaceous lacked icecaps and had a different configuration of land and sea with respect to latitude, the average albedo was much lower than it is at present. A warmer earth and a smaller temperature gradient from equator to poles was the result, but how much warmer is not clear.

factors need to be considered, such as possibly large changes in the earth's albedo and their effect on the planetary heat budget. Because deserts have a much higher albedo than water, transgressions and regressions in the dry subtropical latitudes are accompanied by large albedo changes. Subtropical lands reflect much solar heat that an equivalent area in the tropical or temperate zones will retain because of the low albedo of their forests. Replace the sea with land in the temperate zone, and the winter snow will greatly increase the albedo. And replacement of sea by land anywhere in the world will emphasize seasonal differences, because land loses more heat in winter than water does, and heats up faster in the summer.

As the continents drifted, the area of land in the southern temperate zone diminished, while in the subtropics deserts grew greatly in size, as did somewhat later snow-covered land in the high north. With the help of many simplifying assumptions, the albedo changes that were the result can be estimated (Figure 10.13). The obvious effect could well have played a role in the climate change of the last 100 my.

Albedo changes are only one of the many factors that should be considered as we search for past analogues of the green-house of our future. The role of sinking saline water masses in the heat transport of the oceans, the green-house effect of a very moist atmosphere, the function of vast

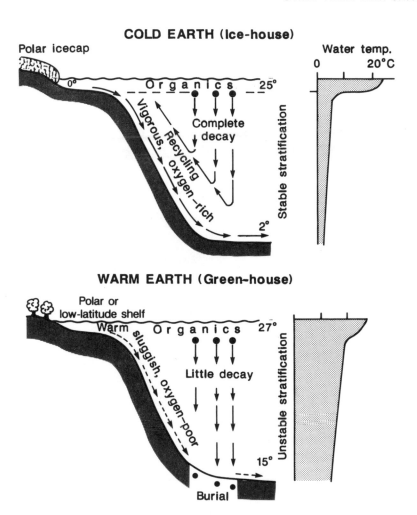

Figure 10.14. The ice-house and green-house states of the earth envisaged by Alfred Fischer are distinguished by drastic differences in ocean temperature, in the oxygen content of the deep water, and in the recycling of organic matter and nutrients.

shallow seas as heat reservoirs, and biological effects such as changes in the abundance of oceanic phytoplankton may all be important. For the Mesozoic an atmosphere rich in CO_2, a Mesozoic green-house, is also a real possibility. Many, perhaps all of those processes do interact, and that does not make it any easier to understand what they did and might do.

The Cretaceous and Quaternary oceans represent two extreme states (Figure 10.14; Table 10.1). The cold modern ocean is characterized by steep temperature gradients with latitude and depth and therefore possesses a vigorous circulation and a stable stratification (Figure 10.14, top). The nearly complete decay of organic matter recycles all nutrients and insures a high but localized fertility, notwithstanding the difficulty

Table 10.1. *The two extreme states of the oceans*

Ice-house state (Example: present earth)
Ocean highly stratified;
Very stable: recycling of nutrient-rich deep water is difficult;
Surface water temperatures range from <2 °C (circum-polar) to >25 °C (equatorial);
Bottom water ranges from *c.* +2 °C (interglacial) to +1 °C (glacial);
Vigorous flow of bottom water, rich in oxygen, hence strong oxidation, little storage of organic matter;
Environments diverse; high productivity in areas of upwelling.

Green-house state (Example: Cretaceous oceans)
Much less stably stratified than icehouse state;
Surface water temperatures not much higher than ice-house state at equator, but 12–15 °C in high latitudes;
Deep water temperatures from 15 °C (equatorial) to *c.* 10 °C (circum-polar);
Low-density surface water leads to slow bottom water flow; water at 15 °C holds half the oxygen of 2 °C water, hence little oxidative power;
Therefore little recycling of organic matter, much buried in sediment; nutrient recycling much reduced;
Low productivity, but good oil source beds.

of transporting nutrients across the barrier of a stable density stratification. In contrast, the oceans of the Cretaceous were weakly stratified, lacked a vigorous deep circulation, and constituted such effective sinks for organic matter and nutrients that their overall fertility was low, although the oil potential was high (Figure 10.14, bottom).

Alfred Fischer, then at Princeton University, once compared the coming and going of warm and cold ocean states with other events that vary with time, and pointed out some vague but rather curious correlations. Warm oceans correspond to transgressive seas (first-order cycles and large supercycles), to the aggressive dissolution of the calcareous shells of oceanic fossils, and to the deposition of black shale. Cold oceans coincide with regressions, with copious deposition of calcareous oozes, and with full recycling of organic matter. Warm oceans contain a very diverse life in complex biological communities, but the number of individuals in each ecological niche is small. Cold oceans possess abundant life, but it is less specialized; the number of species is smaller and the community structure simpler, but the individuals are without count. Extinctions accompany transitions from a warm state to a cold one, and evolutionary explosions are common during the reverse. A rhythm is vaguely discernible: long intervals of warmer oceans are punctuated about every 30–35 my by brief colder states (Figure 10.15). Until we know much,

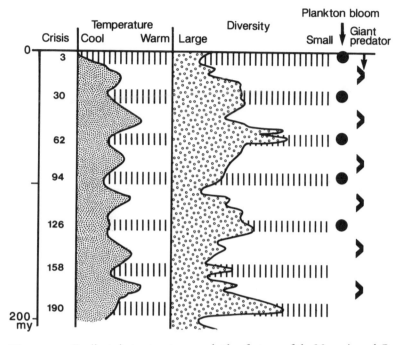

'CYCLES' IN THE OCEAN

Figure 10.15. Fossils, paleotemperatures and other features of the Mesozoic and Cenozoic suggest that the ocean may oscillate between long warm and brief cool states. During a warm interval, faunas and floras are diverse, and giant predators, such as huge sharks and marine reptiles, flourish. During colder intervals, shown here with vertical hatching, life is more abundant but less diverse, and plankton blooms, often consisting of a single species, are common. A faint periodicity of c. 30 my is implied.

much more, this is just food for cautious thought, but big thinking, if done for the pleasure of it, can be very stimulating.

Not everyone is happy with an earth that dwells for long in a warm state. Some, having a stake in eustatic sea level changes, are quite passionately opposed to the thought, and wrestle with the evidence to demonstrate how faulty it is and how, if it were better, all would see that a warm earth is a figment of our imagination. Many climate modelers have joined the doubters, because, being unable to create models that generate warm poles, they feel that something must be wrong with the evidence. I suppose there is a bit of the uniformitarian in all of us, but it would be well to rely on common sense here. The geological data are far from ideal, that is true, and have been misinterpreted at times. But if the Mesozoic world was indeed so unlike the present one, models derived from one state are unlikely to illuminate the other, until we understand the dynamics of atmosphere, earth and ocean far better than we do.

Changing oceans, changing climates

For the time being it seems wise to put one's trust mainly in the evidence, and that leaves us with the tale of two oceans or two earths, the warm and the cold, the green-house and the ice-house, as Alfred Fischer has called them.

II

Onward to the Ice Age

Continental drift influences the oceans and hence the atmosphere in many ways. It shifts land masses across latitudes and climate zones and opens and closes gateways, controlling the circulation of surface and deep water. It raises mountains that cast rain shadows and make deserts. Erosion grinds the mountains down and again the climate changes. Plates move and hotspots rise, and the resulting sea level changes alter the climate as they increase or decrease the size of shallow seas. Volcanoes exhale carbon dioxide and enhance the green-house effect, but weathering consumes carbon dioxide and the earth grows cooler.

To address the question left open at the end of Chapter 5, "Why do ice ages occur?", we must examine the 100 million years of history that have led to the present ice-house world. It is too early to do this with finality, because we are continually becoming aware of other forces that drive the climate, fully understand only a few, and are a long way from being able to say whether one is more important than another or how they interact. In this regard, the past ten years have brought a descent from confidence into humility, but that is not rare in the course of scientific progress. I start with the oceans in the naive belief that we understand them best.

II.I GOING THERE BY SEA

When the Cenozoic began, there were small Atlantic and Indian oceans and a huge remnant of the superocean Panthalassa, now the Pacific, but the arrangement of the continents and therefore the pattern of ocean currents were vastly different from those of today. Circum-global tropical currents dominated the surface flow, but the drifting of the continents had begun to produce modern-looking mid-latitude and subpolar gyres except in the North Atlantic (Figure 11.1).

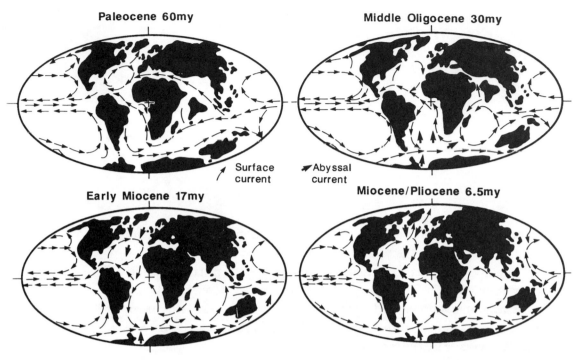

Figure 11.1. The Cenozoic history of the ocean circulation is dominated by two events: the opening of the Antarctic circum-polar seaway 25–30 my ago, and the closure of the circum-equatorial seaway that was completed in the Pliocene when the Isthmus of Panama emerged. Whether the resulting changes made a major contribution to the late Cenozoic ice age is still an open question. The surface current patterns shown here are derived mainly from the changing geography.

As the southern continents continued to push away from Antarctica, the Tethys began to close. First the Indian continent got in the way when it approached Asia, although a high sea level allowed water to flow westward between India and Asia until the early Oligocene. In the western Pacific, islands and submarine ridges were shoved together by the Australia–New Guinea plate on its way toward southeast Asia, blocking the flow from the Pacific to the Indian Ocean. In the Near and Middle East, collisions between microplates reduced the space between Africa and Europe, and when Spain rotated toward Morocco in the Miocene, the Mediterranean became an inland sea. The last equatorial gate closed about 3 my ago when the Isthmus of Panama linked North and South America.

In the southern hemisphere, matters went the opposite way. There Australia's northward progress had cleared a deep gateway south of Tasmania late in the Cretaceous, and the opening of the Drake Passage south of Chile some time between 30 and 25 my ago completed the circum-polar ocean.

As a result, the surface circulation of the oceans changed a great deal. The equatorial flow weakened and the waters turning away from it to the north and south became less warm. Cool water returning to the equator along the west coasts of the continents reduced evaporation and hence rainfall there. As a result, the deserts of western Chile and Peru probably developed in the mid-Cenozoic. When in the early Pliocene the two Americas joined, the equatorial Atlantic waters were diverted northward to form the Gulf Stream which warms the coasts of western and northern Europe and creates a comfortable climate at a latitude where Greenland and Labrador are barely fit for living.

The impression we get from the changing circulation of the oceans is that of a gentle slide toward colder polar climates and a less warm equatorial ocean. We are in a far better position to check such intuitions for the Cenozoic than for the Mesozoic, because of the paleontological and isotopic data generated by more than 900 core-holes drilled in the ocean floor by the research drill-ship *Glomar Challenger* and its successor.

The paleontologist's conclusions about past oceanographic conditions can be checked against temperature information from oxygen isotope ratios of the calcareous shells of planktonic and benthonic microfossils (Section 4.1). For the early Cenozoic we need no corrections for water locked up in icecaps, because there is no evidence for them. Although Antarctica had occupied a polar position for millions of years, evidence for glaciers does not turn up until the late Eocene. Instead, the southern continent was covered with a temperate broadleaf forest not unlike that of southern Chile today, and so was the Canadian Arctic. On Svalbard, already then north of 60° N, broadleaf woods deposited coal beds that are much exploited now.

Estimates based on oxygen isotopes from deep-water benthonic microfossils also indicate oceans almost as warm as those of the Cretaceous, and at middle and high latitudes on land the early Eocene may have been even warmer (Figure 11.2). It began to cool late in the Eocene, but the oceans remained warmer than today, and bottom waters everywhere were at most 5–10 °C cooler than those at the surface.

11.2   A COOLING STORY

The real cooling did not come until the Eocene/Oligocene boundary when in the far south the surface water suddenly got colder, producing a sharp temperature drop of 4–5 °C in abyssal depths. It was the first of

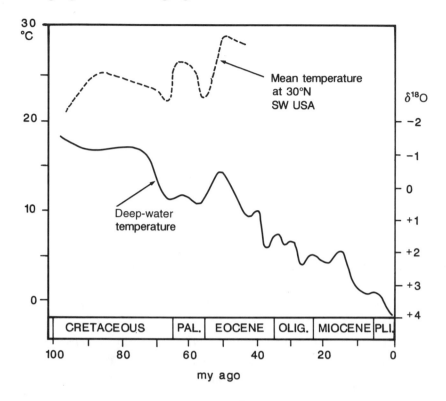

Figure 11.2. The mean temperature on land in the later Cretaceous and early Cenozoic, as inferred from plant remains, reached a maximum in the Eocene after brief cold spells at the Cretaceous/Paleocene boundary and in the middle Eocene. In the deep-sea the water temperature, recorded by oxygen isotope ratios of micro-fossils and displayed here as $\delta^{18}O$, fell almost 20 °C over the last 100 my. It reached another maximum early in the Eocene, but then the decrease became marked and several times took the form of sharp steps, for example late in the Eocene, in the middle-late Oligocene and the middle Miocene.

several steps down toward the current value of about 2 °C, each almost instantaneous even on our time scale (Figure 11.3). The deep ocean basins swiftly filled with cold water, with dire consequences for a bottom-dwelling community accustomed to more comfortable conditions. Many members became extinct, and the benthonic fauna remained impoverished for a long time. Cold water also covered the Antarctic shelf and penetrated into embayments where sea ice formed, while glaciers grew on land. Cores obtained by deep-sea drilling show that in late Eocene and early Oligocene times here and there on the Antarctic shelf deposits were laid down by a glacier complex and by icebergs.

In middle and lower latitudes, surface water temperatures remained within the present range (Figure 11.3), and the northern hemisphere does not seem to have been greatly affected by the events in the far south.

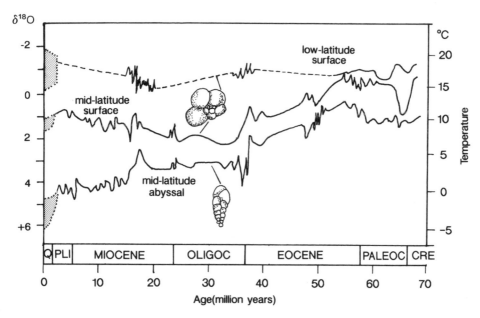

Figure 11.3. In the course of the Cenozoic, the abyssal ocean temperature declined, usually slowly, but sudden major cooling occurred at the Eocene/Oligocene boundary and in the middle Miocene. In low latitudes, surface temperatures changed little, but in the far south the Eocene saw a drastic drop from the Paleocene high of c. 15 °C. Thereafter surface temperatures remained roughly constant. The temperature scale on the right holds true only until about 15 my ago when the ice storage effect becomes dominant (Section 4.1). If a significant Antarctic icecap existed during the Oligocene, the true temperatures would have been slightly higher than shown for the same reason. The shaded areas reflect the rapid, large temperature fluctuations of the last two or three million years. The curves are based on oxygen isotope ratios of planktonic and benthonic Foraminifera, displayed here as δ¹⁸O.

The warmth-loving trees that grew in northern Canada in the Eocene, however, made way for more temperate types.

During the following Oligocene, the supply of warm water to high latitudes diminished further as the Tethys gradually closed. The Drake Passage opened, widened to about its present size in a few million years, and the Circum-Antarctic Current was born. Its encircling waters insulated the polar continent from the warmer seas to the north, so increasing the production of cold, dense water. More glaciers and the formation of an icecap seem to have been the consequences, although the cooling of the surface waters must have reduced precipitation. It is fitting that at this time, some 30 my ago, we finally have proof of Antarctic glaciers at sea level, because cobbles appear in deep-sea sediments offshore that can only have been brought there by ice. Still, Antarctica was far from fully covered with ice, and stunted forests grew here and there.

In the far north the Arctic Ocean, originally wide open to the Pacific,

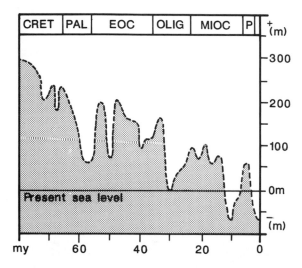

Figure II.4. From the late Cretaceous onward, the level of the sea dropped, with many ups-and-downs, toward its low position in the Quaternary. The final position shown is that of the last glacial maximum. The long decline is marked by a number of sharp falls, usually followed by a more gradual rise. Especially striking are the regressions at the Eocene/Oligocene boundary and in the later Miocene. Both coincide with changes in the oxygen isotope ratio (Figure II.3) that confirm that major increases of the Antarctic icecap were taking place at the same time.

had been fenced in by the northward drift of North America and Eurasia, but in the Eocene a new gateway, this time to the Atlantic, opened between Norway and Greenland. This set off an exchange of surface water between the Arctic basin and the North Atlantic, but major southward flow of deep water was still blocked by an undersea barrier, the Greenland–Iceland–Faeroes Ridge. In general, the north does not seem to have yet been much affected by the steady drive towards a southern ice age, but in the Oligocene the first evidence for sea ice in the Arctic basin appears and the southward flow to the Atlantic gets a bit colder.

A striking feature at the start of the Oligocene is a sharp rise of the $\delta^{18}O$ of benthonic microfossils (Figure II.3). This was almost certainly due to a sizable increase in the Antarctic ice cover, because without the water-stored-as-ice correction the oxygen isotope ratio would imply a bottom water temperature as low as or lower than that of today and that does seem unlikely. An ice volume of about one-third of the present one would make for a more reasonable bottom water temperature in line with known surface temperatures at the time. Also, it would cause sea level to drop about 40 m, and that did indeed happen (Figure II.4).

From then the oxygen isotopes increasingly reflect changes in water volume due to glaciations and deglaciations, as well as temperature changes. The two effects work in the same direction, but can be resolved if we consider the benthonic and planktonic records simultaneously. If the $^{18}O/^{16}O$ ratios increase together, storage of water as ice is the cause, because both draw their oxygen isotopes from water that is then higher in ^{18}O. If they vary independently, probably only temperature is involved. And just recently, it has been shown that oxygen in the opal (SiO_2) of diatom shells faithfully preserves the surface water temperatures and can be used to calibrate the oxygen record.

It is obvious that the Oligocene was an important time in the transition of the earth from a green-house to an ice-house state, and that this happened approximately when the shift from a circum-equatorial to a circum-polar circulation took place. At the same time, the earth changed from a high sea level to a low one or, if you will, from a world of submerged to one of emerged continents (Figure 11.4). Emergence and the drift of continents to high northern latitudes increased seasonal changes, and the climate at high latitudes became more severe.

About 20 my ago, early in the Miocene, the ocean basins and surface currents resembled those of today, and cool Antarctic water flowed through the deep ocean basins, but the main climatic change was still to come. The Miocene began rather cool, then warmed for a while, but about 16 my ago the Antarctic icecap grew suddenly to nearly its present size. This caused a large drop of sea level (Figure 11.4), and much ^{12}O was stored in the icecap. As a result the oxygen isotope ratios upon which Figure 11.3 is based from then on primarily reflected ice volumes rather than ocean temperatures. The growth of the icecap also accelerated the production of very cold water, and the full-blown circulation of Antarctic Bottom Water (Section 10.1) dates from that moment. Elsewhere, the world was not yet particularly cold, but once the great Antarctic icecap had formed, further global cooling became inevitable.

Why was the formation of the Antarctic icecap delayed 10 or 15 my past the onset of the Circum-Antarctic Current, and why did it occur during a time that was not especially cold? The answer may be found in the far North Atlantic, where the Greenland–Iceland–Faeroes Ridge continued to sink until Arctic water could travel south at intermediate depths. When this North Atlantic Deep Water began to well up around Antarctica, its temperature, warm for the chilly local conditions, increased evaporation and augmented the snowfall on the continent. Thus, the delay in the arrival of the Antarctic icecap may have been more a matter of insufficient precipitation than of temperature.

From that moment on, the world was in an ice age. The Miocene, in the accustomed manner, went through two more brief episodes of climate amelioration before it ended with a drastic cooling about 6 my ago. A drop in sea level accompanied this event, presumably due to another increase of the ice volume on Antarctica, because there is no evidence yet for icecaps in the north.

The low stand of the sea also isolated the Mediterranean which evaporated swiftly to dryness, leaving a thick deposit of salt in the deepest basins. This removed enough salt to lower the salinity of the oceans by about 6 percent, thereby raising the freezing point of seawater a little. As a result, sea ice could form at slightly lower latitudes than before, increasing the albedo and so causing a drop in temperature. Still, the effect was surely not sufficient to explain the Miocene cooling.

The Miocene cooling had other consequences in addition to an increased temperature contrast between equator and poles. One of those was the reduction in average global rainfall due to the colder oceans. This may have been the reason that in East Africa and southern Asia the tropical forests were extensively replaced by savanna. This lured some forest-dwelling primates out of their trees into the open, starting the evolution to *Homo sapiens* (Section 17.5).

Once again the world briefly warmed in the Pliocene, but icecaps finally began to develop on the northern hemisphere some 3 my ago, their birth revealed by another abrupt shift in the oxygen isotope ratios, again reflecting water storage in the icecaps (Figure 11.3), by glacial deposits on the northern continents, and by ice-rafted cobbles in the far northern Pacific and Atlantic oceans. Frequent sea level changes began also as the northern icecaps, in contrast to the more permanent Antarctic ice, waxed and waned.

The onset of the northern hemispheric ice age coincided with the closure of the Panamanian Isthmus, suggesting a linkage. The closure may have caused the Gulf Stream to carry warm Caribbean water farther north than before, thereby bringing more moisture to northeastern North America and northwestern Europe, and so facilitating the formation of the icecaps.

11.3   A GOOD AND SUFFICIENT EXPLANATION?

Two major oceanographic changes, both due to continental drift, mark the Cenozoic course toward the Ice Age: the segmentation of the circumglobal equatorial current system and the initiation of Circum-Antarctic flow. A third, the development of the modern deep circulation, may have been their joint consequence. Do these three events furnish an adequate tectonic explanation for the global cooling at high latitude, for the increased north–south temperature gradient, and for the eventual forma-

Table 11.1. *Summary of the evolution of the Quaternary ice age*

Time (my)	Events
c. 100	Gateway between Australia and Antarctica opens;
>50	Free circum-equatorial ocean flow. Warm climate and ocean even at high latitudes. Deep water much warmer than now; circulates slowly. No ice on Antarctica.
c. 55	Warm period even in high latitudes.
48–40	Gradual cooling. Gateway between Antarctica and Australia fully open. Gateway between Greenland and Norway opens; Arctic surface water flows into the Atlantic.
36–34	Cooling of surface water in the south and of deep water everywhere. Antarctic now source of cold deep water; abyssal circulation speeds up. Major deep-sea benthonic faunal extinction followed by development of a cold-adapted one. Extensive sea ice and large glaciers in Antarctica.
35–30	Northward advance of India and barriers in western Pacific restrict circum-equatorial ocean flow. Drake Passage starts Circum-Antarctic circulation some time after 30 my ago. Development of Antarctic glaciation reaches sea level; start of an icecap? Sea ice in Arctic basin. Marked drop in global sea level around 30 my ago.
25	Drake Passage fully open. Circum-Antarctic Current established. Antarctica much colder, but no complete icecap yet; some forest remains.
15–10	Following brief warm interval, sharp cooling occurs. Greenland–Iceland–Faeroes ridge sinks; North Atlantic water wells up around Antarctica, provides moisture; increased snowfall and growth of icecap to present size, accompanied by drop in sea level. Present abyssal circulation fully established.
6	Sharp cooling; Mediterranean salinity crisis.
5–3	Isthmus of Panama closes. Gulf Stream intensified; more precipitation on northeastern North America and western Europe.
3–2	Icecaps in northern hemisphere; ice age is under way.

tion of icecaps? The story summarized in Table 11.1 seems persuasive, but doubts remain on two main fronts. The hypothesis fails to account for some anomalous major climatic and oceanic events, and it ignores other forces that might have contributed to the change in the state of the earth's surface.

To consider the anomalous events first, many changes in sea level and oxygen isotope ratios were surprisingly abrupt, taking their course in a geologically very short time (Figures 11.3, 11.4). This abruptness, followed by long intervals of little change, poorly fits the slow, steady drift of the continents. Gateways do not open or close with a snap, and the pace at which continents march across the climate zones is unlikely to bring sudden climate changes. We may invoke positive and negative feedbacks of the kind we found so useful before in explaining sudden changes (Section 5.3), but no suitable tectonic processes come to mind.

These dissonants do not stand alone. Equally anomalous is a strong warming of the Weddell Sea off Antarctica in the last half million years of the Cretaceous, followed by a sharp cooling just before the Paleocene. Another is the warm early Eocene mentioned before (Figure 11.2) but ignored in the continental-drift model. The warm surface water and large increase in abyssal temperatures may tell us that the oceans shifted from a condition where deep water formed at cool high latitudes to one where high salinity water sank in subtropical seas, the opposite of what the drift of continents would lead us to expect.

Critical for the question whether plate tectonics played a major role in the onset of the Ice Age is the date of the first Antarctic icecap. Not long ago it was taken for granted that we need not make an allowance for large icecaps when interpreting the oxygen isotope record until about 15 my ago. Now evidence is mounting that Antarctica had a significant icecap a lot earlier, perhaps as early as 35 my ago and certainly at 30 my. This is disturbing, because one of the main arguments in favor of the tectonic hypothesis is that the encirclement of Antarctica by the Circum-Antarctic Current was a necessary precondition for major build-up of ice. It now seems that a sizable icecap formed well before this event took place, although admittedly paleogeographic reconstructions of gateways cannot be very precisely timed.

But that is not all. If we believe that the sea level record (Figure 11.4) is an ice volume gauge, the Oligocene icecap did not last long and Antarctica, if not wholly ice-free, carried much less ice between 25 and 16 my ago. This reversal of the trend toward an ice age, if true, need not kill tectonic control outright, but would definitely reduce its influence.

The second group of doubts rests on the uncertain role of other possibly major forces in the climatic deterioration of the Cenozoic. We have already noted (Section 10.6) that the albedo must have changed as the continents drifted north across the earth's surface. Seen originally as possibly a large element in long-term climatic change, recent model studies suggest that its contribution may not have been so very important, but the case remains open.

We know even less about other factors, such as changes in the atmospheric carbon dioxide content (Section 14.6) or of volcanic emanations in the atmosphere. Farsighted people, among them Eric Barron of Penn State University, a pioneer of the study of Mesozoic and Cenozoic climates, felt a decade ago that variations in atmospheric CO_2 could be the explanation for part or all of the alternation between green-house and ice-house earths. At that time one could do little more than speculate about what the CO_2 and oxygen contents of ancient atmospheres might have been, but there has been some progress since then, although nothing conclusive yet.

The gradual fall of global sea level that reduced the area of shallow seas to about one-fourth of what it was in the late Cretaceous also had climatic consequences. We know that large bodies of shallow water, almost non-existent at present, have an impact on the surface circulation of the oceans, on ocean temperatures, and probably also on the formation and movement of deep water. Having no good modern examples to study, however, the climatic role of large shallow seas remains ill-defined.

There are other, less obvious examples of potentially major climatic forces. Grass was a successful and important innovation of the Cenozoic, and the spread of prairie, savanna and steppe, not seen in the world before the Oligocene, not only altered the albedo, but also patterns of evaporation and precipitation, although we can but guess at the magnitude of this biological factor. It is clearly wise to be prepared for other, as yet unexpected changes in our understanding of the workings of the oceans and climates of the past, present and future, because almost everything said in this chapter was unknown 25 years ago.

II.4   RISING MOUNTAINS AND THE ICE AGE

There is another aspect of tectonics that has recently attracted attention as a possible climatic force, namely orogenesis. Unlike the oceans, land presents to the wind a rough surface of plains, mountains and high plateaus. When crossing a high mountain range the air rises and, as it expands, becomes less capable of holding on to its moisture. Rain falls on the upwind flank, and a rain shadow develops on the downwind side. In South America, the Andes intercept the moisture brought by the trades. Dense forests cover their eastern flank, but the ocean side is a desert. In this way, even minor mountain ranges may affect the regional albedo.

The influence of very high plateaus and mountains reaches far up into the stratosphere where they may deflect the course of planetary winds or intensify the monsoon. The later Cenozoic was a time of major mountain building, especially in western North America, along the Pacific margin of South America and in the rift system of East Africa. Largest among those recent uplifts, with an area half that of the United States and an average height of 5,000 m, as high as the highest peaks in North America,

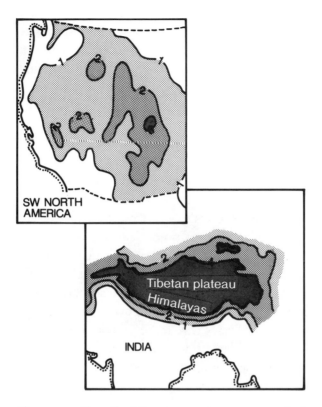

Figure 11.5. Since the Miocene, mountain ranges and plateaus have risen to the sky in southwestern North America and central Asia. The elevation contours are labeled in kilometers. The appearance of these giant barriers across the planetary circulation of the northern hemisphere may have given the final push to the northern hemispheric glaciation.

are the Tibetan plateau and Himalayas raised by the collision between Asia and India (Figure 11.5).

The late Cenozoic also brought large changes in the plant cover of the northern hemisphere. Beginning late in the Miocene, warm–wet forests disappeared from large areas of northern and eastern Africa and west-central Asia, and savannas took their place. This event has usually been attributed to the mid-Miocene cooling of the oceans and resultant reduction in average global rainfall by ocean-oriented paleoclimatologists who have dominated Cenozoic paleoclimatology for two decades, but they may have missed its true significance.

The idea that the vigorous mountain building of the Cenozoic might have contributed to the change toward an ice age climate is an old one, but it was ignored until it was resuscitated recently by Maureen Raymo and Bill Ruddiman, then at Columbia University. Looking for reasons why the glaciation of the northern hemisphere arrived so much later than the Antarctic one, they considered the possibility that its onset was triggered by late Cenozoic uplift of the Tibetan plateau in central Asia and of the Sierra Nevada, Cascade Ranges, Rocky Mountains and Colorado Plateau in western North America. The uplifts began in the late Miocene, but about half of the present elevation might have been achieved in the last 5 my, because that is the time of a late orogenic phase in Tibet and the end of subduction under southwestern North America.

Using a compilation of geological data, Raymo and Ruddiman, with John Kutzbach of the University of Wisconsin in Madison, compared models of the atmospheric circulation before and after the two kilometers of uplift of the Tibetan plateau. The models showed that raising major mountain barriers in the way of the mid-latitude airflow on the northern hemisphere would not only strengthen the monsoon, but have an impact over the entire northern hemisphere. The uplift would convert the equable, moist Miocene climate in East Africa, India, and parts of eastern Asia to one of much greater seasonal contrast between warm and cold, and wet and dry conditions. It would greatly increase regional differences, and could account for the observed changes in plant cover since the early Miocene. Its consequences, for instance in the form of wind-borne dust from newly dry central Asia, can be seen in late Cenozoic marine sediments from the Mediterranean to the North Pacific.

The argument goes as follows. At mid-latitude on a large continent the atmospheric circulation is monsoonal, as seasonally reversing winds, driven by temperature differences between land and sea, blow toward the heart of the continent in summer, and toward the sea in winter (Section 3.3). A high plateau brings a stronger monsoon than a low, flat continent, because the thin air increases the contrast with the adjacent lowland and strengthens the monsoon winds. As the air over the plateau rises, it spreads mushroom-like at high altitude, reducing the already low pressure over the plateau while increasing it in the periphery where it, now cooler, sinks. The peripheral air flows inward but, diverted to the right by the Coriolis force, it swirls counterclockwise around and onto the plateau. In winter, the situation is the opposite; because of the elevation, the plateau air is cold and heavy, a high-pressure zone forms, and the air flows out toward the periphery at a low level, this time in a clockwise direction.

The model also suggests that, as the Colorado Plateau and the mountains of the western United States rose during the last 10 my, the summer along the coast turned dry, causing the deciduous woodlands of the mid-Cenozoic (which needed summer rain) to vanish (Figure 11.6). In the

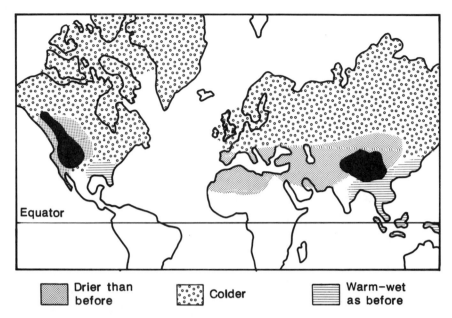

Figure 11.6. *The climate change brought about by the uplift of the western North American and Tibetan highlands greatly altered the plant over. The change in climate, as predicted by computer models, is shown here; it corresponds to the present situation and is sharply different from the conditions in mid- and low latitudes during the Miocene and earlier. Whether this change in climate was sufficient to give the final push to the glaciations of the northern hemisphere, however, remains open to doubt.*

great plains to the east of the ranges, the winds shifted in winter, when the jetstream is strong, from mainly westerlies to today's dominantly northerly winds, bringing polar air and cold, dry winters that favor prairie over woodland. In Asia, the rise of the Tibetan plateau induced even greater changes (Figure 11.6), except in eastern India and southeast Asia where the warm, wet climate was not affected.

The influence of a large, high plateau is not limited to its immediate surroundings, because a mid-latitudinal plateau of great height and size deflects the jetstream northward (compare Figure 3.5). The air rising above the Tibetan plateau, for example, travels as far as the Mediterranean and east-central Asia before it sinks. This air is dry because it comes from a high level in a dry mid-continental region, and as it sinks and is compressed it becomes drier still. This is normal for the subtropics, but the Tibetan plateau spreads the condition far beyond them. In winter, the influence of the cold air mass sinking down and away from the plateau is felt as far as distant oceans in subpolar regions.

In short, while Antarctica, driven mainly by changes in the oceans, switched to a full glacial state more than 15 my ago, the northern continents would have waited for an extra push to flip them into the same state, a push that was provided by the late Cenozoic uplift of vast highlands.

This is not, unfortunately, the final answer to the question. The precise time of uplift of mountains is difficult to establish, and many critics feel no more can be reasonably said than that the plateaus in question reached their present heights some time during the last 15–20 my. Also, the models failed to generate a high-latitude drop of the summer temperature large enough to cause snow to persist throughout the summer and so give the northern icecaps a solid start (Section 5.2).

But atmospheric circulation, temperature, rainfall and vegetation are not all that is affected when mountains are raised very high. Faster runoff encourages erosion and lays down thick deposits in the river valleys and lowlands surrounding the raised region. Furthermore, steep slopes, less vegetation and the cold of high altitude alter the patterns and products of weathering. Weathering is a large consumer of the atmospheric carbon dioxide that makes the carbonic acid essential for dissolving rocks. As a result, the amount of this green-house gas in the atmosphere is, over the very long term, controlled mainly by the balance between the input of new CO_2 by volcanoes and its consumption by rock weathering. Volcanoes do not seem to have been unusually active in the late Cenozoic, and so the consumption of CO_2 by weathering of rising mountain ranges could have drawn down the supply, and so brought about the cooling trend that initiated the ice age.

This proposition involves a very complex geochemical system and at the present time we do not have the data to evaluate it properly. One needs to know how much CO_2 was there to begin with, and how much was lost by burial of organic matter and in the form of calcareous sediments. How much was added to the atmosphere by erosion of sediments containing organic matter and by dissolution of limestones? Also, the estimate of how much the consumption of carbon dioxide by weathering was increased by the uplift must be quite precise. For the time being, we are mired in a swamp of inadequate data, and cannot hope to attack the problem firmly by this route.

Even so, the rising-mountain hypothesis is a useful addition to the set of possible causes of ice ages, and it furnishes an excellent example of the use of climatic modelling in geology, in contrast to the more premature applications cited in Chapter 10.

12

A matter of rhythm

For two centuries geologists have been content to re-create the history of the earth as a series of snapshots that brought to life the "age of coal forests," the "time of the dinosaurs", or "icecaps on the world." Gradually, the focus has sharpened, the color improved, and more detail has been added, but the result was still a series of vignettes, strung together to show the way the world was rather than the way it became so. Now the plate tectonics revolution has taught us that we can and should study change itself, a major shift in our focus, from product to process, from rocks to what makes rocks.

Unfortunately, except for the Quaternary, the traditional approach is likely to be insufficient for this purpose, as the two previous chapters have made clear. Either we must greatly improve our ability to tell time or we shall find ourselves severely limited in the kinds of processes we can profitably study. What we would like to know is clear; it is everything. What we can know depends on how well we can tell time.

12.1 THE DESIRABLE AND THE POSSIBLE

Events on the surface of the earth cover a range of time scales. Continents move and mountains rise taking tens to hundreds of millions of years. By comparison a climate change, the building of the Mississippi delta, or the growth and decay of icecaps are almost instantaneous (Table 12.1). Each component of the global system responds to the forces that drive it on a different time scale; the atmosphere almost daily, the oceans yearly, tectonics in millions of years. The geological time scale suits events of long duration such as orogenies, but it is useless to distinguish Paleozoic interglacials from glacials.

For most of the Cenozoic we can date events to the nearest 100,000, occasionally even 20,000 or 10,000 years, but in the late Cretaceous the

234

Table 12.1. *Response time to change of various processes operating at the earth's surface*

Process or system	Response to change in
Atmosphere	days, months, few years
Ocean surface waters	months to a few years
Deep waters of the ocean	decades to centuries
Forest vegetation	centuries
Ice sheet buildup or melting	centuries to millennia
Glacio-eustatic sea level changes	millennia
Other sea level changes	millions of years
Weathering and sediment input in oceans	millions of years
Plate tectonics	tens of millions of years
Mid-ocean ridge swelling/subsidence	tens to hundreds of millions of years

resolving power declines to half a million years and in the Jurassic to a few million, equal to the entire Pleistocene. To misjudge the duration of an event is bad, but to be able to say no more than that two events happened at some instant during an interval much longer than each, spells defeat. Therefore we ought to spend much effort on perfecting our ability to determine time and to correlate events from place to place.

The task is daunting. The magnetic polarity reversal time scale, so useful in tying sediments to isotopically dated igneous rocks, is limited by the frequency of reversals to a resolution of hundreds of thousands of years, dropping to millions in the magnetically quiet zones of the Mesozoic. Biostratigraphic zones are similarly limited by the rates of evolution. As regards the isotopic dates, we have noted already that their uncertainty, an inevitable consequence of the process, increases with the age of the rocks. So what else should we do?

12.2 REFINING THE PAST

Many events that leave a mark on the geological record act and disappear swiftly, within 100,000 years or less. Some, such as glacials and inter-glacials, are roughly predictable, while others, great earthquakes for example, appear at random intervals. Most events are local, but some have regional and a very few even global consequences. If we use the traces left by regional and global events as markers between the golden spikes implanted by "hard" dates, we could refine our stratigraphy to a level where short-term phenomena would become accessible to our curiosity.

The present world is very active, mountains build and many volcanoes erupt, its land masses stand high and its oceans are low, our shallow seas are few and small, and our climate varies rapidly in time and in space. Its geological record is packed with brief local events, but it does not offer us many regional or global ones to subdivide time finely and over large regions. From this world we have fashioned the uniformitarian key to the past, a key that does not encourage us to hope for an event stratigraphy that is usable over wide areas.

Throughout most of the Phanerozoic, however, the world had no icecaps, the climate was more equable, its continents had been leveled by erosion, and the ocean often stood high. Therefore, we may reasonably expect that in the Phanerozoic brief regional or global events were reflected more often in the geological record, especially in the strata of marine intracontinental basins far away from high mountains, sediment sources, and plate boundaries. Under the cumbersome name of high-resolution event stratigraphy, this expectation was given form a few decades ago, detailed quantitative observations were wedded to statistical techniques to strengthen the base, and several expensive books were written on the subject. Since then it has successfully filled many a gap between the smallest stratigraphic units of the geological time scale, below the million-year or even hundred-thousand-year level. Because we rely on brief events, our observations must be spaced more closely than is customary; we examine the rocks on a scale of meters or even millimeters.

The traces left by such events are manifold (Figure 12.1). They may be physical: volcanic ash deposits, great scours from hurricane waves, gravel fans produced by big floods (or the sudden uplift of a mountain range), changes in the incidence of storm-flood deposits, or brief regional unconformities. There are chemical indicators of short-term events too, sharp variations in oxygen and carbon isotope ratios, the brief deposition of black shales, shell beds, or extensive evaporites. Mass mortalities or the rapid colonization of fresh erosion and deposition surfaces add biological markers.

The principle is not new. The sea level cycles of the Vail curve are an event stratigraphy of sorts, and so is the oxygen isotope stratigraphy now widely used for the later Cenozoic. The magnetic polarity reversal time scale is an event stratigraphy, and most major boundaries of the geological time scale are erosion events expressed by great unconformities. It is not the principle but the scale that distinguishes the high-resolution event stratigraphy from any other.

The deposits of the shallow seas that during the Cretaceous covered the interior of North America from Canada to the Gulf of Mexico and from the Appalachians to the Rocky Mountains document the power of the method (Figure 12.1). Numerous physical, chemical, and biological

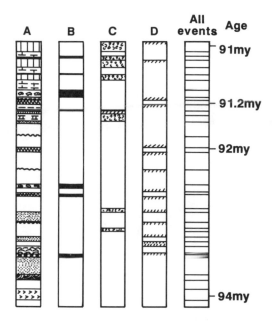

Figure 12.1. The high-resolution stratigraphy for the mid-Cretaceous inland sea of North America is based on several kinds of events. Column A displays the sequence of sedimentary rocks (the types are not important here). Chemical deposits, mainly soils and farther down evaporites, are in column B. Column C lists biostratigraphic markers, and column D special events such as storm floods or mass mortalities. At right all event markers are combined; they divide the 3-my-long interval into no fewer than 35 units.

events subdivide an interval of 3 my into units ranging in length from 50 to 100,000 years. Not all of them were equally widespread, but all were more than local.

12.3   GOOD OLD MILANKOVITCH

Many regional and global events recruited for stratigraphic purposes generate repetitive depositional sequences known as sedimentary cycles. A famous example are the cyclothems of the Pennsylvanian of North America (Figure 9.4). Because sedimentary cycles reflect time on short scales from thousands to hundreds of thousands of years, they have great stratigraphic potential, provided they are of regional extent.

Cycles that recur irregularly are episodic, whereas the term periodic is reserved for events that recur at fixed intervals. Confusion comes easy here: a 100-year flood is not a flood that returns every 100 years; it is not a periodic event. The term merely says that it will, on the average, occur no more often than once per century, but it may happen at any time.

237

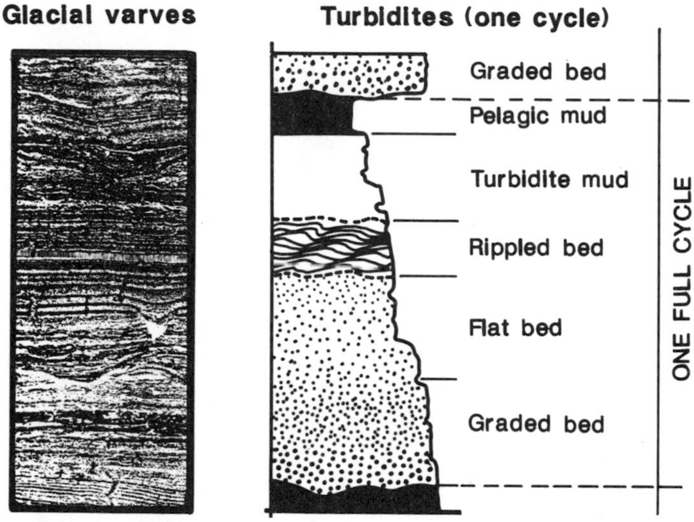

Glacial varves   **Turbidites (one cycle)**

Graded bed

Pelagic mud

Turbidite mud

Rippled bed

Flat bed

Graded bed

ONE FULL CYCLE

Figure 12.2. Two common cyclic deposits are glacial varves (left) and turbidites (right). Varves are alternating light and dark layers, each pair representing one year. Turbidite cycles are more complex, ranging from graded coarse basal deposits to fine-grained pelagic mud. The turbidite cycles repeat themselves, but not at regular intervals. Varves are periodic events, turbidites are episodic.

Mount St Helens volcano on the Pacific coast has erupted many times in the past and will do so again, but the dormant periods between eruptions have been of different lengths; its eruptions are episodic. Winter and summer predictably recur every six months and are therefore periodic. Glacials and interglacials are also periodic, but because they obey three forces with three separate periodicities, their own periodicity is not immediately obvious. Periodicity has also been claimed for sea level changes, for earthquakes, for the arrival and departure of ice ages, and for the impact of extraterrestrial bodies. At the moment, the chance seems slim that those claims are all true.

A perfect example of predictable recurrence are the varves that form annually in glacial lakes as the sediment input varies with the seasons (Figure 12.2). Other varves, annually deposited in the ocean by seasonal plankton blooms, are periodic too. Far more common are the cyclic deposits of bottom-following turbidity currents that along many continental margins constitute the bulk of the deep-sea sediments. They start high on the continental slope and, because they carry large amounts of suspended sediment, they attain high speeds. They deposit gravel and sand that grade upward into silt as the flow decelerates. Turbidity currents recur on a scale of decades to centuries and produce strikingly repetitive

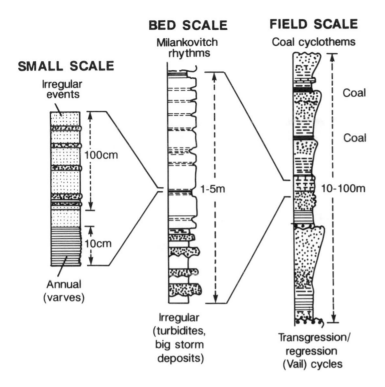

Figure 12.3. Sedimentary cycles occur on many time scales. On the smallest scale we have periodic varves, or the episodic sand deposits that wash over river banks during great floods that recur from years to centuries. Milankovitch rhythms observed in pelagic sediments and episodic turbidites occupy an intermediate range of thousands to hundreds of thousands of years. Coal cyclothems and Vail cycles range from a hundred thousand to a few million years; both are episodic and unpredictable.

sequences (Figure 12.2), but they are not predictable and therefore not periodic.

Pelagic oozes also often show a cyclic variation of their carbonate content on time scales ranging from tens of thousands to half a million years, variations that relate to interglacial-to-glacial climate changes. Less well understood is a form of cyclic bedding that is common in shallow water limestones of Jurassic and Cretaceous age in Europe.

Sediment cyclicity can thus be observed on many scales from readily visible in the field to the microscopic, and some deposits display more than one order of cycles (Figure 12.3). Recognizing cycles is not usually difficult, but how do we decide whether they are episodic or periodic? The point is an important one, as the Pleistocene time scale which is based on Milankovitch cycles in oceanic sediments demonstrates (Section 5.2).

Whether a recurrent phenomenon is periodic and what its period or

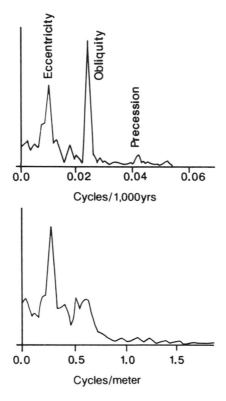

Figure 12.4. Two examples of the result of a time series analysis of cyclic deposits. The vertical axis represents the relative importance of the various frequencies. The upper one is a Pleistocene deep-sea sediment core that clearly displays the three orbital frequencies. The other, also a core, is undated and the frequencies of the events had to be measured in cycles per meter. We can still see the eccentricity peak clearly, but the obliquity is blurred, because the thickness scale is not a consistent measure of time.

periods are, can be determined with a straightforward computation called time series analysis (Figure 12.4). We need not concern ourselves here with the mathematics; they are standard and many geologists do simply crank their data through computer programs without bothering to remember what the mathematics are.

As the name implies, the analysis requires that we have a series of time-dependent measurements of a geological variable, for example oxygen isotope ratios. For the orbital parameters of the earth we know the time exactly, but in the geological record we are often forced to use sediment thickness as a substitute to interpolate between the real dates which tend to be widely and unevenly spaced.

The use of sediment thickness as a proxy for time is fraught with pitfalls, because it depends on estimates of sedimentation rates. Sedimentation rates are controlled by numerous factors of which many, such as 100-year river floods, major storms, slumps due to earthquakes or turbidity currents, operate briefly, then stop for a while, and are replaced by a different, much slower process of sedimentation. Hiatuses, especially brief ones, are another problem. In the hunt for cycles, geological sequences that contain such gaps should be avoided, but they are not always obvious. Another problem is the compression of sediments under the weight of the overlying deposits, their compaction. The finer and more porous a deposit is to begin with, the greater the compaction. Sand loses about one-fourth of its volume when fully compacted, calcareous pelagic ooze one-half, and some siliceous oozes three-fourths. Start with a meter and one ends up with 25 centimeters.

Whatever the cause, there is a fair chance that the relation between sediment thickness and time has been distorted to some extent. This means that we can sometimes prove that a sequence is cyclic, but not usually that it is not. Even if we succeed, we have a periodicity labeled in centimeters or meters, and what might that be in years?

Knowing the length of the period in a periodic sequence is essential if we are interested in its cause, and especially if we suspect that orbital cycles are involved. So far, the orbitally controlled climatic cycles are the only ones we know of that are capable of affecting sedimentation on a global scale, but to prove in a given case that Milankovitch was there is not simple.

The difficulties do not end there. The Milankovitch periods, brief and closely spaced, range from 20,000 to 400,000 years. The last one is another eccentricity period that was not important in the Pleistocene, but shows up at other times. Those periods are uncomfortably close to or just beyond the limit of our ability to determine time. We can get around that if we check whether the Milankovitch cycle ratios of 400:100:41:21, or roughly 8:5:2:1, are present in the cycles we extracted from a thickness-based record. Unfortunately, some sedimentary processes that are not controlled by climate also yield a narrow range of thicknesses with periods that mimic these proportions. The daily, lunar and seasonal periodicities of tidal-flat cycles (Chapter 1, Preamble) are a warning that Milankovitch is not the only one who cycles.

Notwithstanding the difficulties, orbital control has been claimed for many past cyclic events, although some claims demand a cheerful optimism for acceptance. The thick oceanic sediments of Triassic to Miocene age in the central Apennines of Italy, for example, include sequences said to fit the Milankovitch model.

The extension of Milankovitch dating to early Pleistocene and uppermost Pliocene deep-sea drill cores rests on a solid footing, and cyclic

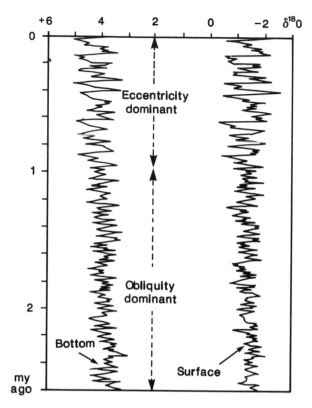

Figure 12.5. The relative importance of the three orbital cycles – eccentricity, obliquity and precession – has not been constant in the past. The oxygen isotope ($\delta^{18}O$) curves from a pelagic sediment core spanning the last 2.6 my show that the 41,000-year obliquity cycle dominated the early part, with the result that glacials and interglacials were of roughly equal length. For the last 800,000 years, the 100,000-year eccentricity cycle has dominated, and the duration of the glacials became five or ten times that of the interglacials. The reason is not clear (but see the Perspective at the end of this chapter).

deposits from southern Sicily have helped to carry it into the early Pliocene. The Milankovitch periods can be traced throughout this interval, but with increasing age their relative importance shifts. During the late Pleistocene, the 100,000-year eccentricity peak dominates (Figure 12.5), quite why is not yet clear. Beyond one million years or so, the 100,000-year eccentricity peak declines sharply, and the curve, now dominated instead by the 41,000-year obliquity period, shows a fine-scale complexity. In the Pliocene the 400,000-year eccentricity cycle is the most conspicuous, while the obliquity, having persisted for a while, gives way to the 21,000-year precession cycle throughout the late Miocene (Figure 12.6).

Precession cycles are suspected also in the lower Jurassic of Switzerland

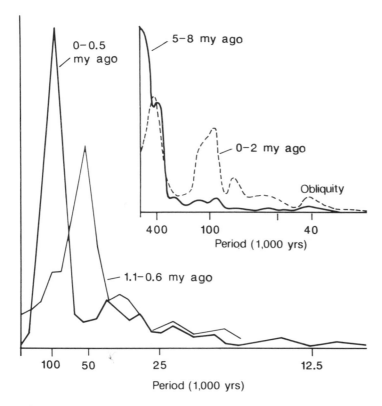

Figure 12.6. During the late Miocene, the second component of the eccentricity, which has a 400,000-year period, was much larger than it is now, and little is to be seen of the 100,000-year eccentricity and 41,000-year obliquity cycles. This is in sharp contrast with the recent past shown below.

and in Cretaceous deposits of the Apennines where they combine in sets of five to produce the 100,000 years of a pseudo-eccentricity cycle. The obliquity cycle, on the other hand, has so far rarely been found beyond the Pleistocene.

Why the orbital parameters trade off their positions over time is unclear, and it makes the slow advance of our research into the remote past a voyage of adventure. Is it, in fact, even reasonable to expect that the orbital periods themselves would have remained constant over time? After all, the orbital behavior of the earth is affected by changes in its spin rate that result from tidal friction, from variations in the orbits of the inner planets, or from the changing distance between earth and moon. The decrease in the number of solar days (Figure 1.1) and lunar months per year since the early Phanerozoic suggests that the orbital periods of the earth may have changed as well.

Professor A. L. Berger of the University of Louvain, Belgium, a major

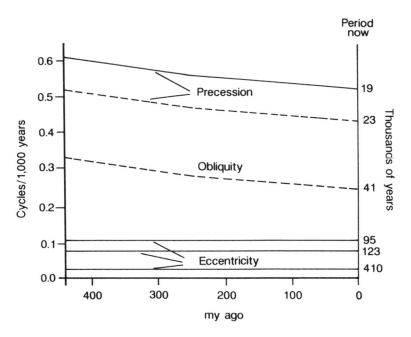

Figure 12.7. There is no particular reason to assume that the Milankovitch orbital periods would have been constant through geological time, and indeed they have not. Influenced by many celestial forces, some have changed and others have not. The three eccentricity periods are stable, but the two precession periods and the obliquity have slowly lengthened. Current periods are shown on the right.

contributor to the study of astronomic aspects of the Milankovitch model, has shown that since the early Phanerozoic the eccentricity has remained constant, but that the obliquity and precession periods have shortened considerably (Figure 12.7). We have so far failed to spot this in the geological record, but that may be due to a poor conversion from sediment thickness to time. There is a clever way around this, however, if we use the constant eccentricity cycle as a clock to time the periods of precession and obliquity periods. One such attempt found that 100 my ago the obliquity period was 39,000 rather than 41,000 years, in reasonable accord with Berger's calculations.

Some Milankovitch periodicities from the pre-Miocene record look quite solid and are beginning to build a foundation for a new high-resolution chronology, but for others a healthy skepticism is still warranted.

12.4 THE PULSE OF THE EARTH

Long ago, in the difficult years of 1939 to 1945 and partly while imprisoned by the German Gestapo, the great Dutch geologist J. H. F.

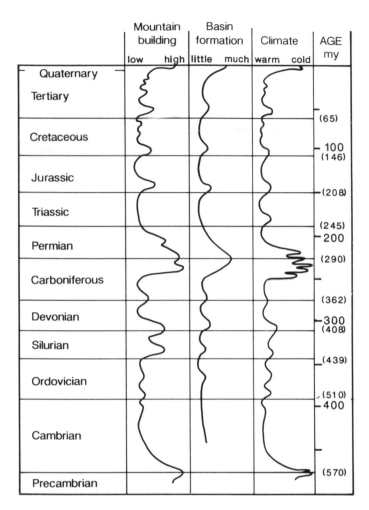

Figure 12.8. Many people have hoped for rhythms in the history of the earth. One of the earliest attempts to see whether there actually is a "pulse of the earth" was made by J. H. F. Umbgrove. Part of his vast compilation is given here and it does show surprisingly close correlations in time between mountain building, basin formation and climate. The correlations have not held up, however, because the dating available to Umbgrove was still quite poor, and his diagram is now obsolete. The figures in parentheses are the stratigraphic boundary ages as used nowadays and in this book (Figure 2.4).

Umbgrove wrote, under the title I have borrowed above, a seminal book about rhythms he saw in the history of the earth (Figure 12.8). His modest opening sentence, "Problems of current interest related to the earth's physical history will be discussed in this volume," does no justice to its contents, which started a debate about the earth that has not yet ceased.

The wish to see orderly patterns, not just random wiggles superimposed upon a long-term trend of cooling and evolving life, in the history of the earth, or in human history too for that matter, comes naturally to us. Attempts to define this order fall in two classes; periodicists look for repetition at predictable intervals, while cyclomaniacs are happy to live with irregular intervals if the recurrent events are dramatic enough.

Several elegant major cyclicities of the earth have been proposed that appeal to our need for meaning in the universe. A cyclic earth history is a simple, sensible earth history, whereas a random sequence of events is difficult to establish, difficult to understand, and impossible to like. Great cycles also help us to think through our current knowledge on the grandest scale. So far they have provoked more critique than applause, but that is a mark of health; without proposal and rejection no progress.

In Section 10.6 we encountered Alfred Fischer's tentative suggestion that the normal warm state of a green-house earth is interrupted periodically by an ice-house condition. Others believe that major extinction events follow each other with monotonous regularity every 26 my (Figure 19.11). Some speak of a plate-tectonic period called the Wilson cycle after the father of the transform fault, during which the oceanic crust goes from creation to subduction in 200 my.

The weak point of all proposals is that the dates assigned to the events are so uncertain that it is impossible to prove that they do or do not return at fixed intervals. Geologically, it is not unreasonable to call an event that returns every 35 my, give or take five, a periodic event, but it is not possible to prove rigorously that it actually was periodic. As long as we do not care about the cause, that matters little, but when specific connections with demonstrably periodic processes are at stake, it becomes critical. Imagine that we could only say that day will break every 24 hours, give or take an hour or two; we would be unable to relate the event to the rotation of the earth.

Truly grand is the concept that the earth oscillates between a supercontinental and a fully dispersed state in a "megacycle" of 300–400 my. It has been around ever since Wegener proposed continental drift, and was revived in various forms shortly after the plate-tectonics revolution. Again we owe a typical example to Fischer. The best-documented part of his megacycle began with a major ice age just before the Phanerozoic (Figure 12.9), followed by a warm early Paleozoic occupied by continental dispersal. Late Paleozoic ice ages happened while Pangaea was being assembled, then came a warm Mesozoic to early Cenozoic, another green-house, and another continental breakup with an ice age. Superimposed on these roughly 300-my-long megacycles are the (also roughly) 30 my oceanic cycles Fischer had proposed earlier (Section 10.6), upon which the high-order relief of orbital perturbations has been finely etched.

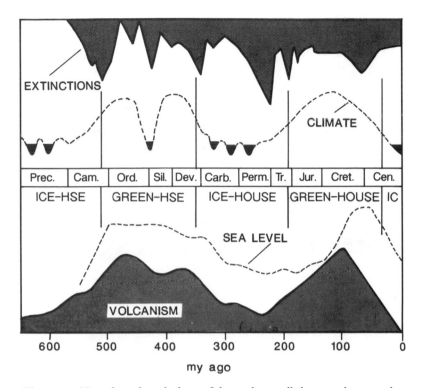

Figure 12.9. Nowadays, the pulse beats of the earth are called megacycles, a much more impressive term. Shown here are Alfred Fischer's megacycles. Based on several variables of which four appear here, he divided the history of the earth into ice-house and green-house phases.

Possible cyclic behavior of mountain building and volcanism also has never been far from the minds of geologists since it was put forth in the 1930s by the German geologist Hans Stille. It has reappeared many times since and we shall encounter it again (Section 13.4) disguised as episodic growth of continental crust.

We can embellish Fischer's megacycle. Supercontinents may be such good insulators (Section 7.5) that large hotspots develop underneath them and cause continental breakup and drift. The heat forms vast domes in the top of the asthenosphere that cause major sea level changes as the continents slide on and off the bulges. This ties sea level changes to the cycle of continental assembly, breakup, and recombination accompanied by subduction, collision, volcanism, and mountain building. The volcanism contributes CO_2 to the atmosphere which generates a greenhouse effect, but is consumed by weathering when, during the supercontinent stage, the high mountains and plateaus are eroded down to plains. Also, continental aggregation and dispersal have a significant, though not necessarily dominant influence on the circulation of ocean and

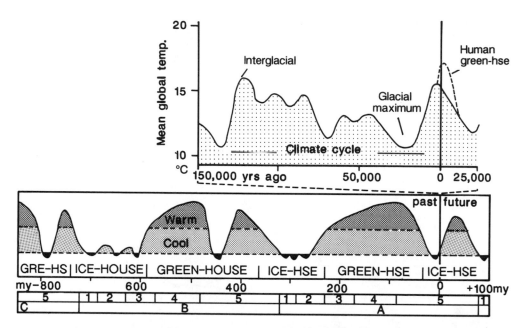

Figure 12.10. Improved megacycles have sprung up like mushrooms in recent years. John Veever's model is not so very different from Fischer's, but it goes much farther back in the past and, for good measure, a little into the future as well. Because this concerns us directly, an enlargement of the more recent past and immediate future is given at the top.

atmosphere. Putting all those eggs in one basket, as John Veevers of Macquarie University has done, we have a lovely, improved version of Alfred Fischer's megacycles (Figure 12.10).

These schemes rest on compilations of major events plotted against time (Figure 12.8). When contemplating such works of art, we should always remember that the beginning, climax and end of major events are not sharp and clear as are the death of kings or the elections of presidents. They are malleable and suited to a bit of subjective interpretation, seasoned at times with just a little wishful thinking. Megacycles are great devices to lure the mind away from the trees and focus it on the forest, but they are not yet the precursors of a grand synthesis.

There may, of course, be no pattern at all, because nothing says that nature should be patterned, and, for the time being, I regard the design of grand schemes as equivalent to the attempts of a baby to walk; it is lovely to watch but not yet very secure.

PERSPECTIVE

Does this mean that I believe that the history of the earth (and therefore probably of life also), is merely seductively chaotic; that nothing at all is predictable? That conclusion, it seems to me, would be premature.

One hears these days much of chaos theory, best remembered from the story that nature, being so complex, is so unpredictable that a butterfly fluttering through an embryonic hurricane could prevent it from being born. To expound chaos theory here would take an additional chapter, but its essence is simple. We tend to think of cause and effect as related in a linear fashion: double the force and the effect will be twice as great. That is untrue for almost all environmental processes we have drawn upon; they are non-linear, often to a high degree. Besides, many of them are so complexly (and non-linearly) related that, as they interact, more than one outcome is possible. Although superficially they may seem to repeat themselves, close examination indicates two important things: they do not in fact do so exactly, and it is not possible to predict which one of the many possible outcomes will turn up. Quasi-periodicity, not true periodicity, is what we may expect. The shift from a dominant 41,000-year obliquity cycle in the early Pleistocene to the 100,000-year eccentricity cycle later is very likely merely a matter of chaotic behavior.

Therefore, even if we should fully and quantitatively understand all processes that combined over millions of years to make the Grand Canyon, we shall not be able to describe its history in a mathematical equation with one and only one outcome. Does it matter that we may never be able to explain the past nor forecast the future quantitatively? I think not, because I do not believe that we need that level of knowledge.

Geologists have long been aware of this. They have insisted that geology is a historical science in its own right rather than a mere application of the laws of physics and chemistry to the past. And so they have studied that past, not listening to those who suggested that they should just wait a while until all the basic physical and chemical equations had been worked out whose application would tell us all we need to know. Chaos theory and the realization how non-linear the universe really is have shown us that we were right all along: predicting the past is not so simple a matter.

This brings to a fair end our exploration of the impact of plate tectonics on our understanding of the Phanerozoic history of the earth. Now we must go much further back, to the long early years when everything began, years that must be probed with much less information and much more speculation.

FOR FURTHER READING

Paleoceanography and paleoclimatology have not yet produced much writing useful to the lay observer. For oceanography a nice, but somewhat outdated popular overview is: Turekian, K. K. (1976). *Oceans* (Englewood Cliffs, NJ: Prentice-Hall); and at beginning college level: Ingmanson, D. E. & Wallace, D. J. (1988). *Oceanography: An Introduction* (Belmont, CA: Wadsworth); more advanced are Frakes, L. A. (1979). *Climate through Geologic Time* (Amsterdam: Elsevier); and Kennett, J. P. (1982). *Marine Geology* (Englewood Cliffs, NJ: Prentice-Hall). New, up-to-date, but advanced is: Crowley, T. J. & North, C. R. (1991). *Paleoclimatology* (New York: Oxford University Press).

Not to be missed on sea level and its history: Hallam, A. (1992). *Phanerozoic Sea Level Changes* (New York: Columbia University Press); and on chaos theory: Gleick, J. (1988). *Chaos: Making a New Science* (London: Heinemann).

SPECIAL TOPICS

Arthur, M. A., Dean, W. E., Bottjer, D. & Scholle, P. A. (1984). Rhythmic bedding in Mesozoic–Cenozoic pelagic carbonate sequences, in *Milankovitch and Climate*, A. L. Berger, J. Imbrie, J. D. Hays, G. Kukla & B. Saltzman (eds.), pp. 191–222 (Dordrecht: Reidel).

Arthur, M. A. & Schlanger, S. O. (1979). Cretaceous anoxic events as causal factors in development of reef-reservoir giant oil fields, *American Association of Petroleum Geologists, Bulletin*, **63**, 870–85.

Barron, E. J. & Peterson, W. H. (1989). Model simulation of the Cretaceous Ocean, *Science* **244**, 200–03.

Einsele, G., Ricken, W. & Seilacher, A. (1991). Cycles and events in stratigraphy: Basic concepts and terms, in *Cycles and Events in Stratigraphy*, G. Einsele, W. Ricken & A. Seilacher (eds.), pp. 1–22 (Berlin: Springer).

Fischer, A. G. (1982). The two Phanerozoic supercycles, in *Catastrophes and Earth History*, W. A. Berggren & J. A. van Couvering (eds.), pp. 129–50 (Princeton: Princeton University Press).

Fischer, A. G. & Arthur, M. A. (1977). Secular variation in the pelagic realm, in *Deepwater Carbonate Environments*, H. E. Cook & P. Enos (eds.), pp. 19–50 (Tulsa, OK: Society of Economic Paleontologists and Mineralogists, Special Publication 25).

Hallam, A. (1984). Pre-Quaternary sea-level changes, *Annual Review of Earth and Planetary Sciences*, **12**, 205–43 (a response to Haq *et al.* below).

Hancock, J. M. & Kauffman, E. G. (1979). The great transgressions of the late Cretaceous, *Journal of the Geological Society of London*, **136**, 175–86.

Haq, B. U. (1984). Paleoceanography: A synoptic view of 200 million years of ocean history, in *Marine Geology and Oceanography of the Arabian Sea and Coastal Pakistan*, B. U. Haq & J. D. Milliman (eds.), pp. 201–23 (New York: Van Nostrand Reinhold).

Haq, B. U., Hardenbol, J. & Vail, P. R. (1987). Chronology of fluctuating sea levels since the Triassic, *Science*, **235**, 1156–67.

Hay, W. W. (1988). Paleoceanography: A review for the GSA Centennial, *Bulletin of the Geological Society of America*, **100**, 1934–56.

Heckel, P. N. (1986). Sea-level curve for Pennsylvanian eustatic marine transgressive–regressive depositional cycles, *Geology*, **14**, 330–34.

Kauffman, E. G. (1988). Concepts and methods of high-resolution event stratigraphy, *Annual Review of Earth and Planetary Science*, **16**, 605–54.

Matthews, R. K. (1984). Oxygen isotope record of ice-volume history: 100 million years of glacio-eustatic sea level fluctuation, in *Interregional Unconformities and Hydrocarbon Accumulation*, J. S. Schlee (ed.), pp. 97–107 (Tulsa, OK: American Association of Petroleum Geologists, Memoir 36).

Nance, R. D., Worsley, T. R. & Moody, J. R. (1988). The supercontinent cycle, *Scientific American*, **25**, 44–56.

Pitman III, W. C. (1978). Relationship between eustasy and stratigraphic sequences of passive margins, *Bulletin of the Geological Society of America*, **89**, 1389–1403.

Read, J. & Francis, J. (1992). Responses of some southern hemisphere tree species to a prolonged dark period and their implications for high-latitude Cretaceous and Tertiary floras, *Palaeogeography, Palaeoclimatology, Palaeoecology*, **99**, 271–90.

Ruddiman, W. F. & Kutzbach, J. E. (1991). Plateau uplift and climatic change, *Scientific American*, **264**, 42–50.

Shackleton, N. J. (1987). Carbon isotope record of the Cenozoic: History of organic carbon burial and of oxygen in the ocean and atmosphere, in *Marine Petroleum Source Rocks*, J. Brooks and A. J. Fleet (eds.), pp. 423–34 (Geological Society of London, Special Publication 26).

Sloss, L. L. (1991). The tectonic factor in sea level change: a countervailing view, *Journal of Geophysical Research*, **96**, 6609–17.

Veevers, J. J. (1990). Tectonic–climatic supercycle in the billion-year plate-tectonic eon, *Sedimentary Geology*, **68**, 1–16.

Weedon, G. P. (1993). The recognition and stratigraphic implications of orbital forcing of climate and sedimentary cycles, *Sedimentology Reviews*, **1**, 31–50.

The four-billion-year childhood

The surf is brushing at my steps; I seek
An aged cliff that stands among the sleek
Young chargers of the sea.
Rounds of anemone
And areas held by sea urchins devise
The narrow range in which the tide will rise
And fall; though cliffs themselves
And all the earth's vast shelves
Crumble. And there the mode of permanence
Is framed in the sea-tide's changefull cadence.

Howard Baker, Ode to the Sea

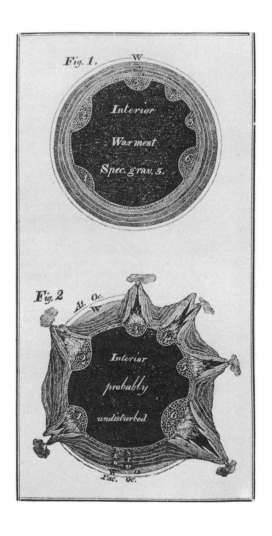

THE YEARS WHEN NEARLY EVERYTHING BEGAN

The early earth is alien, like and yet unlike the other planets we have lately so marvelously explored: a dash of Mars, much of the moon, some of Venus perhaps. From that remote planet to the now familiar world of the Phanerozoic leads a four-billion-year road, nine-tenths of the entire history of the earth.

When history began in the Precambrian, the earth was a ball of cosmic debris: no land nor sea, no mountains nor valleys, no clouds, no winds, no life. When the Precambrian ended, blue seas fringed sandy shores, and myriad tiny green plants drifted in fertile seas, feeding many, diverse, mostly small but already complex animals. The land was bare, but the air was breathable, and the temperature pleasant enough.

How, precisely, was the beginning, those first hundreds of millions of years from which no record remains? No one knows, and we rely on physics, chemistry and our knowledge of the cosmos, seasoned with our imagination, to speculate, to dream. The earth could have evolved in various ways, but proof and disproof have mostly eluded us so far, leaving five hundred million years to a contemplation limited only by a small (but rapidly growing) set of observations.

Another three and a half billion years take us from the oldest rocks to the Phanerozoic when the earth had evolved to nearly its present condition. On the way we encounter life, but the record is silent regarding its origin and not very eloquent about what came afterward. The thread is thin and so faded that once more we are compelled to speculate, testing occasionally against rare facts, rather than, as is the custom of geologists, reasoning from evidence to explanation. Few firm answers have yet been found for even some of the simpler questions, but complex interactions between life, ocean and atmosphere can be dimly discerned. Hypotheses come and go, and the pages that follow are a progress report; they can be no more.

Lest one believe that the Precambrian is important only to scholars who love puzzles without answers, it should be said that economically it is a most important part of the history of the earth. Precambrian rocks contain the lion's share of all known mineral resources and, with the exception of coal and oil, most of what resources remain to be discovered will be of Precambrian age. In practical terms, most questions about the origin of everything are peripheral, but the record of subsequent events is of prime importance.

13

Birth of the solid earth

Astronomy tells us that the solar system is slightly more than 4.5 billion years old, the age of most meteorites. On earth, grains of the mineral zircon have been found that, although they are imbedded in much younger rocks, have an age of 4.3 by. The oldest rocks we know occur in the Slave Lake region of Canada and came into existence 3.96 by ago. Others only slightly younger are at Isua in Greenland, on the Limpopo River in South Africa, and in northeastern India. Staggering numbers that mean little to our own sense of time.

Distorted as they are by time and history, even the oldest rocks confirm the presence of an earth crust, of a sea where sediments were laid down, and of weathering and erosion on the land from which those sediments came. As our knowledge grows, it has become evident that the earth evolved in as little as 500 my from a ball of cosmic dust to a planet with many of its present properties, albeit in different configurations.

Those first 500 million pages are blank, but even the rest is difficult to decipher until we get close to the Phanerozoic. The reasons for this are common to all earth history, but greatly magnified in the Precambrian: our inability to tell time with precision, our frequent failure to read a record of which so much has been lost by erosion, burial, and a metamorphism often so intense that we cannot remove its overprint. Fossils are scarce in the Archean eon and until the end of the following Proterozoic they are stratigraphically useless, while the golden spikes, the isotopic ages, are few and they have uncertainties too long for comfort. For the distant past we must all too often be content when the facts, failing to prove us right, at least do not tell us that we are wrong.

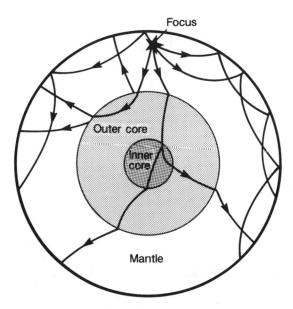

Figure 13.1. Much of what we know about the interior of the earth has come to us by courtesy of earthquakes. Starting at its focus (which need not be at the surface), the earthquake energy travels down and out, but at each internal boundary where the nature of the rock changes, part is reflected and returns to the surface. If we measure the times of arrival of the various waves at many points, we obtain the depths of the boundaries and a good deal of information regarding the nature of the rocks the waves have passed through.

13.1   DOWN BELOW WHERE NO ONE CAN GO

The earth is not a homogeneous body. Its crust is a mere skin covering a 2,900-km-thick mantle divided into a lower and an upper shell, and surrounding a core 7,000 km in diameter (Figure 13.1). The core is hot; precisely how hot we do not know, but already at its outer edge the temperature is about 4,900 °C. The core has two parts, a fluid outer and a probably solid inner core. Both consist of iron and nickel and are very dense, but not dense enough to be pure, so that the metals must be alloyed with something else. Sulfur and oxygen are the best candidates, and if it should be sulfur, the core must have formed at low pressure and therefore very early in the formation of the earth. Oxygen, more likely because it is so abundant in the earth, would imply that the alloying happened at high pressure. That would mean that the core continued to grow throughout the lifetime of the earth.

The outer core is able to flow and is a good conductor of electricity. If it flowed in the presence of an electric current, it would act as a giant dynamo that could very well be the source of the earth's magnetic field and, in a way not yet fully understood, the cause of its polarity reversals.

Why should the outer core flow, what makes the dynamo run? There are two main possible reasons. One is the convection of hot magma rising through the outer core, but that process is not very efficient. More likely is that the inner core expands by freezing (if that is the right word for such a hot place) at its boundary with the outer core. This should leave a residue of hot, light magma capable of rising through the outer core by virtue of its low density. Because magnetized rocks go back at least 3.5 by, the outer core must already have existed then.

About the mantle we know a good deal more from earthquake travel studies, from the analysis of mantle rocks that reached the surface by volcanic or tectonic means, and from laboratory experiments at very high temperatures and pressures. Geochemical considerations based on such exotic elements as neodymium and samarium (also used in dating: Table 2.1), have proved especially interesting. They show that the average composition of mantle and crust combined closely resembles a class of meteorites called carbonaceous chondrites, but with less water and carbon dioxide than those bodies have. When a carbonaceous chondrite melts, a scum, rich in light elements such as sodium, silica, potassium and aluminum, and quite similar in composition to the continental crust, rises to the top. The residue underneath is the upper mantle (Figure 13.2). Below 670 km the lower mantle is not altered by such processes, and pieces brought up by hotspot volcanoes from very deep resemble carbonaceous chondrites that have not lost their light elements.

One puzzle remains. If the upper mantle has lost much of its light material to the crust, it ought to be heavy, heavier than the unaltered lower mantle. Why, then, does it not sink and gather just above the core? Perhaps the shell structure of the earth is unstable, but it is far more likely that some kind of "lightener" has been added to the upper mantle. Water and carbon dioxide subducted into the mantle with an altered oceanic crust and sediments could do the job, but one would rather expect that they would entirely boil away in the andesitic volcanoes of subduction zones.

13.2   BEFORE HISTORY

Earth and the other planets were once thought to have condensed from hot gas, but it is now clear that they were born from a dust cloud, a nebula orbiting the sun. In this nebula, planetesimals, solid bodies ranging in diameter from a few meters to thousands of kilometers, grew by accretion, collided and shattered into smaller fragments, and ultimately combined to form the set of planets we know, leaving the space around

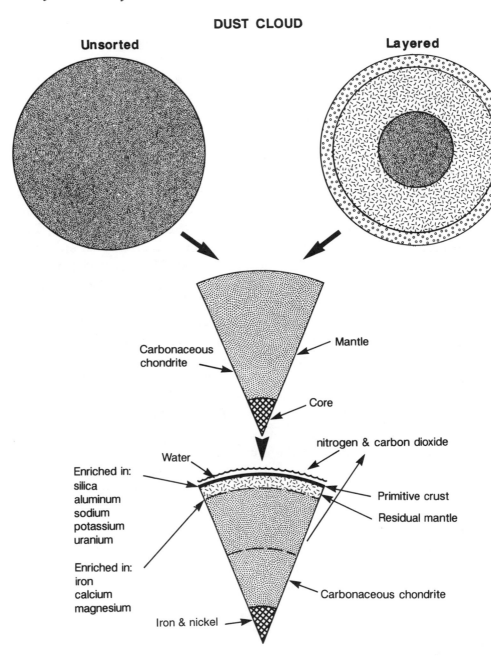

DUST CLOUD

Unsorted   **Layered**

Carbonaceous
chondrite   Mantle

                                   Core

                                   nitrogen & carbon dioxide

                          Water
Enriched in:
silica   Primitive crust
aluminum
sodium   Residual mantle
potassium
uranium

Enriched in:
iron
calcium
magnesium   Carbonaceous chondrite

Iron & nickel

Figure 13.2. The earth formed from a cloud of cosmic dust that either was wholly unsorted or, less likely, already somewhat stratified with the heaviest particles in the center. In either case, melting produced a dense iron and nickel core covered by a shell which chemically resembled a carbonaceous chondrite. The outer part of this shell later melted and separated into a heavy residue, the mantle, and a lighter primitive crust surrounded by an ocean and an atmosphere.

the sun swept remarkably clean. Meteorites are a record of this phase in the formation of the planets, consisting as they seem to do largely of fragments of smashed planetesimals.

In this context, an interesting question is the origin of the moon. Was it once a free-traveling planetesimal captured by the earth? Was it torn from the earth by the attraction of a passing planet, perhaps leaving the Pacific Ocean as a scar as has been claimed? Or might the moon have formed independently alongside the earth by the same process of accretion? For various reasons none of the ideas seems any longer plausible, although they were once strongly held. The current favorite is that the earth, not yet fully formed, collided at a grazing angle with another planet perhaps a bit larger than Mars. The impact blew the mantle of the smaller body away and it became the moon, or perhaps several moons which were later joined. The core itself was swallowed up by the earth. The collision also set the earth to rotate in its present sense, an odd one for the solar system, and gave it its axial tilt, possibly quite a high one. Evidently, if there was no moon, then there would be no Milankovitch theory either. The chemistry of moon and earth rocks is in accord with this scenario, and many scholars regard it as virtually proven, but there are enough problems with it to cause others to reject it out of hand.

To produce the layered earth of today, a period of melting was necessary. Various sources of energy could have brought this about: the gravitational energy released when the dust gathered and condensed into a ball, or the impact energy of space debris as it plunged into the growing planet. In both cases, however, most of the energy would have been radiated away as heat. There is also the friction produced by the tides raised by the sun, the moon and other planets, a tidy amount. Or one might think of more exotic energy sources such as the blanketing effect of a dense green-house atmosphere like the one that keeps Venus so very hot. Together these heat sources might have raised the temperature of the new planet a lot, but not enough to melt it through so that iron and nickel could sink to the center.

Much more energy would have been available from very large impacts, because they penetrate deep enough to trap the heat. In theory, one single, truly huge collision could have been enough to melt the earth catastrophically, causing the dense core to form all at once, and keeping our planet molten for a long time.

Radioactivity is also a large source of energy, ultimately good for several thousand degrees. At the start it came mainly from short-lived isotopes, such as aluminum-26 which decays with a half-life of 0.75 my. This one and others of its kind are long gone, but jointly they made a

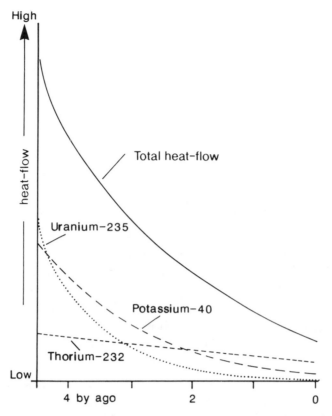

Figure 13.3. In the early years of the earth, the amount of heat generated by the decay of unstable isotopes was much larger than it is now. Many short-lived isotopes have vanished entirely and others, with half-lives of the same order as the age of the earth (thorium-232 for example) or shorter (uranium-235), have greatly decreased in abundance and so produce much less heat now.

major contribution to the heating of the earth, although probably not sufficient by itself to melt the planet through and through.

The long-lived unstable elements, such as potassium, thorium and uranium, were much more abundant then than now (Figure 13.3). Uranium-235, with a half-life of 710 my, for example, has already been through six half-lives, and less than 1/64th of the original amount remains. Still, important as they were for the heat budget of the earth in later years and now, their relative contributions during the first 100 my were not of major importance.

The heat built up continuously during the accretion phase, but when did whole-earth-melting-day arrive? There is plenty of room to argue about that, but it must have happened when the core formed, and there is plenty of chemical and physical evidence that it was very early.

262

On the other hand, even earlier the dust cloud itself might have been sorted into layers under the influence of gravity, with the heaviest elements nearest to the planets-to-be. The core, and also perhaps oceans and atmosphere, might therefore have formed right away, and a subsequent heating phase would be needed only to make oceanic and continental crust out of the primitive mantle. But the layered dust cloud idea, much favored a decade ago, is not so popular now.

Whatever the nature of the dust cloud may have been, we may think of the newly born, primitive earth as a nickel–iron ball wrapped in a "chondritic" mantle of which the upper part at some time melted and formed something resembling the present crust. Precisely when that happened is of paramount importance for the history of the earth surface. New data imply that the first crust appeared after only 50 my, but that is not yet at all certain. For now we may still ponder whether the crust formed early and suddenly or later and perhaps gradually. Or did a wholly different surface exist first?

The notion of an early earth that was covered with a "magma ocean" has inspired many an author of science fiction, but it has not been popular with scientists. Now this image has returned as a metaphor for an extraordinarily active surface, convecting vigorously like the cauldron of Macbeth's witches. Evidence is accumulating that until 4 by ago a primitive crust, floating on a molten mantle, was constantly being bombarded by extraterrestrial projectiles. This crust must have been basaltic because it came from the mantle, but it has completely disappeared on earth. It is preserved on the moon, however, where it shows that it was not yet like the present oceanic crust.

The surfaces of Mercury, Mars and the moon have not been altered by subsequent plate tectonics and display the state that was obliterated long ago on earth. Like the earth, they were once dotted with giant versions of the lava lakes of Hawaii, in which light scum rose continuously to the surface and congealed to form primitive crust, only to be smashed by impacts and dragged down again by vigorous convection. When the surfaces of these magma oceans had finally cooled and solidified, a final bombardment, about 4 by ago, shattered the thin crust one more time. On the moon, the pieces of this crust form the lunar highlands while the maria, the dark blotches visible from earth, are the congealed lava oceans, undisturbed ever since. The moon and Mercury, small and therefore cooling fast, ended their lives at this point, but Mars remained active a while longer, acquiring a crust so thick that it takes the weight of its giant volcanoes without sagging. The much bigger earth, keeping warm, continued to evolve, and four billion years of tectonic upheaval and plate movements have erased the early scenery. For that later history we must turn to the Precambrian rocks themselves.

13.3   THE FIRST CONTINENTS

So some four billion years ago the earth possessed a core and a mantle. It probably also had an atmosphere and ocean of sorts, but the evidence for a continental crust is questionable. And there was no life!

As we have seen (Section 6.2), the continental crust is much thicker than the oceanic and, being rich in granitic rocks as well as sediments and the metamorphic equivalents of both, it has a relatively low density. It also contains about one-third of the heat-producing radioactive elements of the earth. Quartz, very rare in rocks derived from the mantle, is a characteristic and common component of continental igneous and sedimentary rocks. In contrast, the denser but much thinner oceanic crust consists almost entirely of basalt and related rocks that do not yield quartz-bearing sediments upon weathering. The continental crust is of great interest, not only because most of us live on it, but also because of the manner of its birth, a complicated multi-stage process called differentiation that is by no means fully understood.

At first it seemed reasonable to assume that, as the mantle cooled and the scum rose to the surface, a globe-circling shell would form, similar in composition to the continental crust. That view could no longer be sustained when it became clear that the typical "granitic" continental crust needs a much more protracted and complex sequence of processes. To make continental crust, we must recycle oceanic crust many times by melting, cooling, subsidence, burial, and renewed melting. Each time, part of the heavier components remains behind, while lighter ones gather, rise to the surface, and congeal into progressively more granite-like rocks. Once there is land, weathering helps. It retains the most resistant components, the oxides of silica and aluminum as well as minerals such as potash feldspars and above all quartz, and sweeps all solubles into the ocean. By the sustained action of such processes the continents gradually acquired a composition entirely different from that of the oceanic crust and from the primitive crust that was the first cover of the mantle, now seen only on the moon.

When did the formation of a light crust begin, and how long did it continue? How do we go from the lurid image of a surface dotted with giant volcanoes belching ash and noisome gases, rain pouring ceaselessly

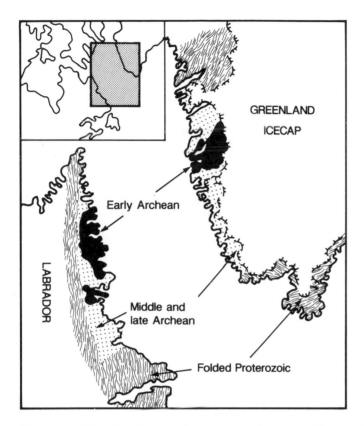

Figure 13.4. The oldest Precambrian crust occupies very small areas. Shown here is the 3.8-by-old Isua Formation (black) in Greenland and Labrador, the source of most of our knowledge about the earliest crust. The rocks may also contain the first traces of life, but that has not yet been proved to everyone's satisfaction.

from dense reddish clouds, lightning glaring on black lavas, to a reasonable facsimile of the present earth?

The earliest Precambrian formations, for example those of the 3.8-by-old craton in Greenland and Labrador (Figure 13.4), contain continental granitic rocks and sediments. The sediments include zircons as old as 4.2 by that must have formed in even older continental rocks. Quartz-rich sandstones (quartzites), prove the presence of land with its own weathering and erosion. These features show that even before 4 by the mantle had begun to segregate a continental-type crust, and also that the geological cycle had operated more than once already.

The Greenland craton is small and might be a freak case, were it not that similar rocks occur in Enderby Land, Antarctica. No more than 100 my later we find other, much larger blocks in Australia and South Africa. The South African Kaapvaal craton seems to have begun as a mid-ocean plateau of oceanic crust that was compressed and uplifted. It then under-

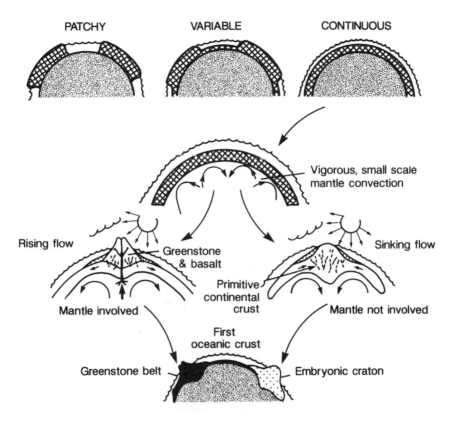

Figure 13.5. We can only guess whether the primitive crust was patchy, of variable thickness, or formed a continuous shell, covered by a continuous ocean. Assuming for simplicity's sake the latter, one may speculate that embryonic continents were formed by vigorous, small-scale mantle convection (center). Above a rising current, melting of the crust gave it a slightly more continental composition, and greenstones formed, but the process of differentiation was delayed by the continuous intrusion of basalt from the mantle. Above a sinking flow the crust would be squeezed and thickened, mantle material could not interfere, and the differentiation leading to true continental crust, aided by weathering on the emerging land, was much faster. The crustal nuclei (cratons), eventually became surrounded by oceanic crust as the mantle was stripped bare.

went weathering and erosion, sediments formed, and the crust thickened until melting took place at its base that produced granites.

To understand what might have happened after the demise of the "magma ocean," let us consider a cartoon, inspired by Robert Hargraves of Princeton University, that begins with a continuous shell of primitive crust under a continuous ocean (Figure 13.5). In those early days, perhaps 4.2 by ago, much more heat was produced than today (Figure 13.3), and it is probable that under the thin crust the mantle was convecting vigorously. Being so hot, the mantle was less viscous than it is today, it flowed faster, and its convection cells were smaller but more numerous. The image of a planet dotted with hotspots like goose bumps on a cold skin is probably more realistic for this time than the linear patterns of modern seafloor-spreading.

Above the rising hot limb of a convection cell (Figure 13.5) the thin crust fissured and volcanoes built from lava separating out of the underlying mantle, while magma coated the base of the crust. Under the weight of the volcanoes, the crust sank until its base reached a level hot enough for it to melt where more light elements distilled off and rose into the crust above. Whenever a volcano grew large, it broke the sea surface; then weathering and erosion caused further differentiation.

In this manner, the primitive crust evolved toward a more continental composition, but slowly, because mantle material was continuously added. Since Iceland is a modern example of this process, we may call it the "Iceland model" of crust generation. The formations representative of this setting would be folded complexes of dark volcanic rocks, associated with shales and sandstones rich in volcanic rock fragments called graywackes. We know them as Precambrian greenstone belts.

Where the cool limbs of two convection cells met and sank, the primitive crust, itself too hot and thus too light to be subducted, was compressed and thickened. This again created a downward bulge which, melting at its base, produced a lighter magma that invaded the crust above. In this case, however, no fresh magma from below was injected, and the crust evolved more rapidly toward a continental composition than above a rising convective flow. And as before, as the buoyancy of the thickening crust increased, land emerged and weathering speeded the process of differentiation. The granitic continental cores, the cratons, may have originated in this way.

Obviously, while the primitive crust was concentrated into ever thicker patches, the mantle was gradually stripped of its cover. In bare spots a true oceanic crust was generated for the first time, in the way it still is on mid-ocean ridges.

Recently, Alfred Kröner has assembled considerable evidence that models of this kind (all kinds of variations are possible) correspond reasonably to what actually did happen, and that thick continental blocks appeared as early as 3.5 by ago. This leads us to expect randomly distributed tiny continents of granitic composition surrounded by greenstone belts of mantle-derived material. The oldest cratons, those of South Africa and Australia, for example, illustrate such early configurations, but they can be explained in other ways as well. Rather than pursuing this exercise further, let us turn to the evidence instead.

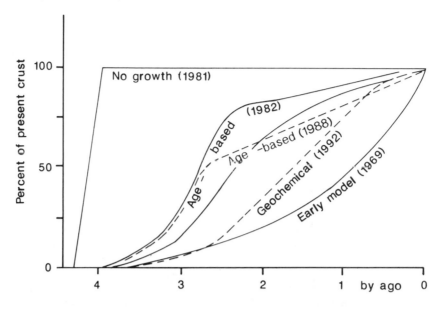

Figure 13.6. The rapid change in our views regarding the growth of the continental volume is displayed here by means of five curves labeled with the source of the evidence and the year in which each was proposed. The unlabeled curve is my own attempt at a compromise. It is clear that we are still a long way from a thorough understanding of this important subject. The 1992 curve based on geochemical considerations is probably a fair representation of the average increase, but the method does not permit us to say whether the growth was speeded up during times of major orogeny.

13.4   A TIME OF GROWTH

Precambrian cratons form the hearts of all continents. Were all continents fully grown shortly after the magma ocean congealed and the celestial bombardment had ceased? Or did the continental crust grow gradually, to reach its present volume only recently (Figure 13.6)? Both possibilities have been staunchly defended, and to settle the issue we must determine the volume of crust as a function of time. The controversy persists because this simple task is actually quite difficult.

Very old rocks, 3.7 by and older, have been found in ever increasing numbers as we have learned to date zircons by means of their enclosed radioactive elements. Zircons were among the first continental materials, and because they are highly resistant to weathering, they survive many geological cycles. More substantial remnants of very old crust are rare, however, and so suggest that the early continental crust was small in volume (Figure 13.4).

Finding old continental crust is not the same as proving the existence of old continents, because we must show that there was land above the sea where weathering and erosion took place. That requires properly

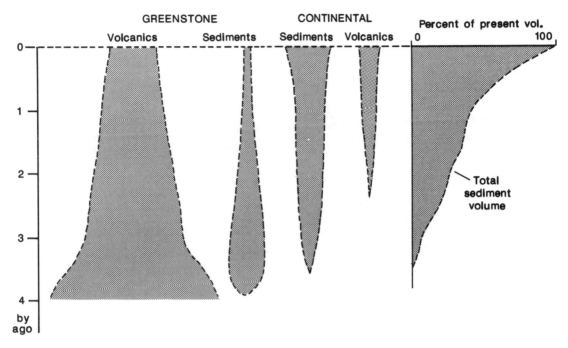

Figure 13.7. The history of Precambrian continental growth is reflected in the type (left and center) and volume of sediment generated on the continents and dumped in the sea. Initially, all sediments were derived from igneous and volcanic rocks of the primitive crust and greenstone belts. Truly continental deposits, identified by a high quartz content, do not appear in abundance until much later. The curve of sediment volume faintly suggests a step-wise increase in the area of land that was being eroded.

dated quartzites, the undeniable products of many cycles of weathering and erosion. Such quartzites do indeed exist, as for example at Isua, proving that at least 3.8 by ago continents, small as they may have been, had emerged from the world ocean.

We can trace continental growth by the changing composition of the sediments they have shed into the sea (Figure 13.7). If most or all continents appeared early above the sea, one would expect little change in the composition of the sediments since that time, even if we allow a generous time for weathering and recycling to convert oceanic rocks into continental ones. But what we find is that, throughout the Archean eon, sediments were derived mainly from oceanic lavas or a more primitive crust. True continental sediments and lavas are scarce until about 2 by ago, and first appear at different times on different continents. It seems, the first continents were small and grew slowly, or mainly during times of major orogeny. The data shown in Figure 13.7, however, do not say whether growth was gradual or intermittent.

Continents also grow by "underplating," the crystallization of mantle-

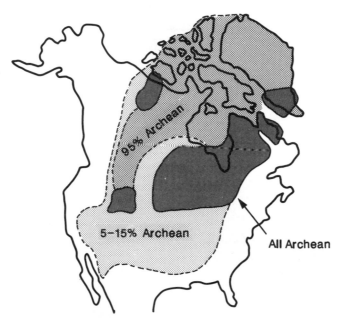

Figure 13.8. The North American continent grew a good deal during the Archean, although quite irregularly. The pattern strongly suggests that it grew mainly through the aggregation of many small continents rather than by simple accretion around continental margins or by underplating of the crust.

derived magma against or within their base. This occurs where hot mantle material is in contact with a cool continental crust above subduction zones, below continental rift valleys or at hotspots. Complicated geochemical considerations provide an estimate of the amount of underplating, but once more the data cannot tell the difference between gradual and step-wise growth.

The total volume of sediments produced over time (Figure 13.7, right) is easier to estimate. Most Precambrian sediments originated on land by weathering and erosion or in shallow seas, and their volume is therefore roughly proportional to the continental area. Subduction and burial distort the result, but the error is small relative to the increase in the total amount of sediment. The estimates confirm that the volume of continental crust has more than doubled in the last two billion years.

Isotopic ages of the Precambrian cratons confirm the long-term growth of the continental crust (Figure 13.6), provided we disregard rocks whose radioactive clocks were reset by later metamorphism. The first, small amounts of crust appeared around 4 by ago, but growth was slow, until a major increase between 2.8 and 2.5 by ago substantially raised the volume. Growth also seems to have been strong between 2.1

and 1.8 by, and again between 950 and 650 my ago. The amount added in the Phanerozoic is usually regarded as trivial, but a recent estimate that one cubic kilometer is added each year suggests that we had better look again.

Thus the first truly large continents had appeared about 2.5 by ago, at the boundary between the Archean and the Proterozoic (Fig. 13.8). It was a momentous event with potentially great, but still obscure biological consequences, as we shall see in Chapter 18.

13.5 TIME AND PLATE TECTONICS

The discussion of continental growth raises the question of plate tectonics; when did it begin, what form did it take, and when did the "classical" plate tectonics of Chapter 6 evolve? Geologists are uniformitarians at heart and when plate tectonics achieved the status of ruling theory, most assumed that, since times immemorial, ridges had rifted, continents drifted, and plates been subducted, in accord with James Hutton's famous phrase that he could see "neither a vestige of the beginning nor the prospect of an end."

That cannot be true, however, because any energy-consuming process that claimed to have been operating without change in mode or pace since the earth began, violates the prohibition against perpetual-motion machines. The second law of thermodynamics demands, and the thermal history of the earth makes clear (Figure 13.3), that the earth's engine has been slowing from the first and must come to a stop some day. Clearly, the dynamic behavior of the earth, and that means plate tectonics, must vary with the speed of its motor. The question is, how and how much?

On an Archean earth with a hot, thin lithosphere, the high heat flow probably favored many hotspots topped by crustal growth of the kind exemplified by Iceland or Hawaii. Mantle convection was in all likelihood vigorous, but the cells may have been small. This suggests many thin, drifting plates, perhaps 100–500 km across, that jostled each other, collided, and at times coalesced (Figure 13.9). Because convection cells and plates were so small, the oceanic crust caught in the collisions would have been young and therefore hot and light. Today oceanic lithosphere usually arrives at a subduction zone cooled by a voyage of thousands of kilometers that lasted more than 100 my, and it is therefore cold and heavy. The Archean lithosphere did not have so far to travel, and may have been too hot and light to be subducted; it could only be compressed and folded.

In time, such "microplate tectonics" would produce plates thickened

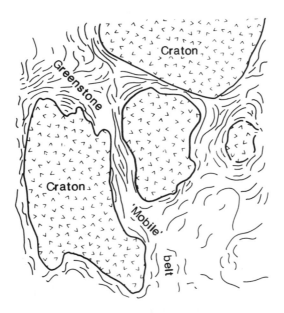

Figure 13.9. The old cores of the continents preserve the imprint of the Archean tectonic style. The terrain, sketched here from a satellite photo, measures 100 km across. It contains four continental nuclei (cratons) wrapped in greenstone belts, so displaying continental "collision" on a scale much smaller than that of the present plates.

by underplating and large enough to straddle a whole convection cell. At the same time the decline in heat production slowed convection and pushed the system toward larger cells. The oceanic plates became larger, had to travel farther, cooled more, and were subducted. Classic plate tectonics had begun.

Preserved from a distant past, about a dozen small Archean cratons, all between 3 and 4 by old, exist within the continents. These ancient nuclei consist of granite, gneiss, and associated sediments and are surrounded by greenstone belts of folded, metamorphosed basaltic rocks and sediments (Figure 13.9). Under a metamorphic overprint, greenstone belts represent deep-water sedimentation, volcanism and the formation of oceanic crust. They give evidence of collisions accompanied by horizontal shortening of the oceanic plates, but paleomagnetists have argued that the cratons themselves had moved very little with respect to each other. If true, this would mean that deformation entirely within plates was the common tectonic style of the time.

Besides, the Archean and early Proterozoic greenstone belts lack an important criterion for subduction. When the cold oceanic crust sinks to great depth, it is metamorphosed at high pressure but at low temperature. This process creates a unique and unusual mineral assemblage called blueschist, and blueschists have never been found in Archean rocks.

Inevitably, however, some of the lithosphere must eventually have become old and heavy enough to make subduction possible. The evidence of this moment suggests that this happened early in the Proterozoic, about two billion years ago. In Scandinavia and Finland, as well as in North America and Australia, we find rock complexes of that age that have a composition typical for island arcs as we know them today. They are associated with subduction and are chemically different from the Archean greenstone belts. Paleomagnetic evidence from 2–1.5-by-old rocks begins to support long-distance continental drift also. Blueschists that are part of later Proterozoic subduction complexes, for example in China, clinch the case for subduction.

Did the earth have plate tectonics before then, more than two billion years ago, or did it not? This is really a semantic problem, and defenders can be found on either side. There is as yet no consensus, but it is increasingly clear that the Archean, probably because of high heat production, had a tectonic style and rock chemistry of its own and different from the Proterozoic and Phanerozoic.

If it seems that in this chapter I have dealt lightly with a very complex problem, I have to admit that this is true, but in its present state of rapid evolution the subject is not kind to attempts at simplification without demanding a lot more detail than the reader can be reasonably asked to absorb.

14

Water for the sea, air for the atmosphere

Let us begin by observing that the origin of ocean water is a problem that is separate from the origin of ocean basins and that it has two parts: the source of the water and the source of the many substances dissolved in it. The key role of the ocean in earth history has been discussed before; its origin and initial composition remain to be considered here.

For the atmosphere, the main issue is its evolution to its present composition. Compared to the rest of the solar system, our atmosphere, with its 78 percent nitrogen, 21 percent oxygen, and minor amounts of carbon dioxide, water vapor and rare gases, is an odd one. It is so odd that the atmospheres of Venus, Mars, Jupiter, Saturn, and of some of their moons offer little help in understanding our own.

14.1   WHENCE THE WATER IN THE SEA?

Whatever model one espouses for the accretion of the earth, the existence at an early time of a magma ocean and the composition of its parent material, the carbonaceous chondrites with their 15–20 percent water, render the early release of large amounts of water and gases inevitable. Total melting is unnecessary; a 50-km-deep magma ocean, itself a conservative guess, would yield every drop of water one needs. Consequently, around 4 by ago the earth had a true ocean, although perhaps not a full one, and an atmosphere, although one quite unlike the present one.

How much free water had the early earth? In 1951, the late W. W. Rubey proposed that the mantle was the source of all water and gases, and that both were vented at the surface by volcanoes and volcanic springs. The total volume of surface water in oceans and lakes, in the ground and the atmosphere, had increased steadily over time, he thought, although initially a bit faster (Figure 14.1). He supported this thesis by

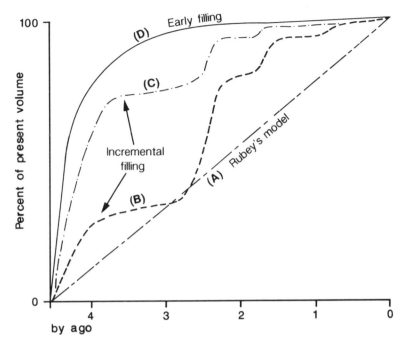

Figure 14.1. The ocean may have filled with water in various ways, but the slow, steady increase suggested by W. W. Rubey (A) is unlikely because of the evidence that the oceans were already quite full in the late Precambrian (Figure 9.1). Most probable is that they filled rapidly (D) when the separation of the primitive crust from the mantle yielded much water. Still, if the continents grew by increments, more water may have been added later, but we cannot say whether the episodic increases were likely to be large (B), small (C), or if they took place at all.

the observation that volcanic fluids contain salts in roughly the right proportions to account for the composition of seawater. His model was consistent with the evidence of the day, but we have since discovered that the record of long-term sea level change (Figure 9.1) conflicts with a steady rise of the sea. Moreover, geochemical studies show that volcanoes exhale mainly recycled ground or ocean water (Figure 14.2); only a tiny fraction of the volcanic steam is "new" water, released from the mantle for the first time.

In fact, we may actually be losing water to the mantle, because the subducted oceanic sediments and crust contain a great deal of water. Subducted water can be stored in many mantle minerals which are able to accommodate considerable quantities. How much of that water later resurfaces by way of volcanic eruptions we do not know, but it is possible that the earth's surface, over the short and the long term, is losing rather than gaining the precious fluid.

Karl Turekian of Yale has noted that, if the mantle does have the composition of the average carbonaceous chondrite as most people be-

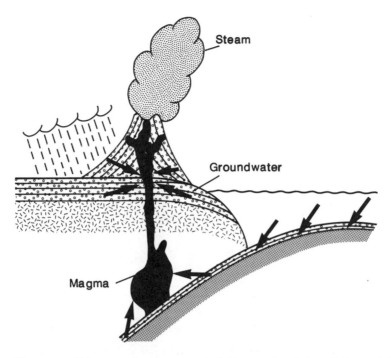

Figure 14.2. Volcanoes release a great deal of steam, but it is mainly recycled groundwater. Especially in volcanoes from the subduction zone, a good part is seawater derived from subducted oceanic crust and sediments. In either case, only a very small fraction is "new" water, derived directly from the mantle.

lieve, it should contain much more water than is now present at the earth's surface. Perhaps our problem is not where the water to fill the oceans came from, but rather what happened to the much larger amount that should have been released during the initial melting, but apparently was not. Bombardment by extraterrestrial objects in the first 500 my must have swept some water away, but the projectiles themselves, a plethora of carbonaceous chondrites and comets, were rich in water too, especially the comets. It is perhaps premature to think that we have a firm grip on even the basic problem of filling the oceans.

Wherever any surplus water went, as far as we can tell today the bulk of the ocean water had arrived by one means or another about 4 by ago (Figure 14.1: C, D). Modest amounts may have been added later, perhaps during intervals of vigorous creation of continental crust. But how did this early ocean acquire its salt?

The oceans contain enough salt to bury North America 10 km deep. The principal ions in seawater are sodium, magnesium and chlorine, with smaller amounts of calcium, silica, potassium and sulfate. Everything else, and I do mean everything, is present in minute quantities, but because the volume of the oceans is so large (1,350,000 km³ of water), the total for each chemical compound is still great, including enough gold to make millions rich, if one could only get it out.

Except for gases such as oxygen, carbon dioxide and sulfur dioxide, the dissolved matter is supplied mainly by rivers, with small amounts from submarine weathering and deep-sea hotsprings. If we take the present annual salt input by rivers and pour it into a world ocean made of distilled water, the present salinity would be reached after about 250 my.

The supply of dissolved substances is roughly proportional to the land area exposed above the sea. In the early Precambrian this area was less than one-tenth of what it is now, and the initial salt influx into the sea was very small. Even taking into account the subsequent growth of the continents, the present salinity could not have been established much before 2 by ago.

At this point we face a different problem. The input of dissolved matter continued, of course, increasing with the growth of the continental mass and, at times of prolonged weathering and erosion, as in the Mesozoic, it may have been higher than today, but the salt content of the oceans has not increased. How is the excess amount being disposed of?

14.2 AN INCOMPLETE HISTORY OF SEAWATER

This question has not been answered satisfactorily, but instead has been joined recently by another, equally important one which had never been seriously contemplated before. Has the oceanic brine, containing a little of just about every element in the universe and a lot of some of them, varied in composition with time?

Obviously, where so much goes in, much must go out too. For some elements we know where they go, but for many, including two of the principal components, the exit route is far from clear. Much salt is contained in ancient evaporites, but even during times of maximum evaporite deposition the effect was small; the 10 percent ocean salt that became rock salt in the Permian and the 6 percent deposited during the Mediterranean salinity crisis are hopelessly inadequate to explain why the present salinity does not reflect two billion years of salt input.

As regards the mix of salts in seawater, most geologists, and many biologists too, believe that it has remained roughly constant since the salinity reached its present value. The reasons for this belief are rather intuitive: because life has inhabited the sea for at least three billion years and direct descendants of the first inhabitants are still found there, one would not expect a drastic change in the chemical environment. Furthermore, the body fluids of most organisms have the same salt concentration as seawater. This makes sense if life evolved in an ocean as salty as it is now, but not if the salinity had varied a great deal over time.

The geochemical evidence is sparse and unsatisfactory. Few minerals are good indicators of ocean chemistry and at the same time stable enough to survive the transformation into sedimentary rocks. What little we know is compatible with, but does not prove a constant salinity over the long term. The average composition of the continental crust has not changed much since the early Proterozoic and that implies an input of constant composition. Unfortunately, that need not be true of whatever process removes dissolved substances from the ocean, and even if the composition was not varied at the input end, it may still have been altered by whatever means saw to the removal of the salts.

These are uncomfortable points, but given the complete lack of evidence, it has always seemed wisest to assume that the salt in the sea had changed little, except for minor excursions during the Permian and late Miocene. At least in this way we avoid the arbitrariness inherent in any other approach.

Here the discussion rested for many years as we reluctantly accepted that an ocean which had never halved nor doubled its salinity, achieved this through a mysterious output process that no one could put their finger on. It is no wonder that no one ever really managed or dared to propose a geochemical drive for the evolution of ocean life.

14.3   DEEP-SEA HOTSPRINGS

A chance discovery has opened new horizons. It came from the mismatch between the heat that ought to be released theoretically where new crust forms on mid-ocean ridges, and the heat flow that one actually observes there. On crust of zero age the shortfall is about 40 percent, relative to the predicted value, and only for crust of more than 5 my do theory and observation agree. The theory is impeccable, the measurement errors are not large, and the conclusion that a large amount of heat (10^{16} kilocalories per year) remains unaccounted for is inescapable. Some process other than conduction through the rocks, the form of heat flow we can measure, is at work here.

There are means to get rid of excess heat other than by conduction. As with a car engine, cold water, circulating through cracks and fissures in the young, hot, oceanic crust should be perfect for the job. Once heated, the water would rise, escaping to the seafloor as submarine hot-

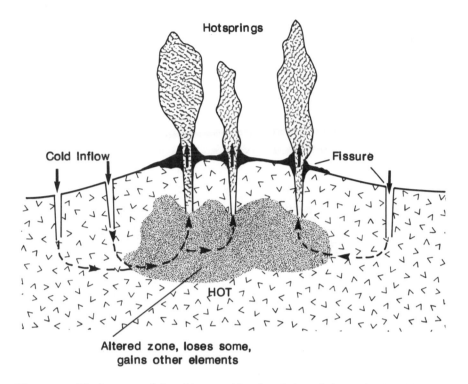

Figure 14.3. The hot crust of the mid-ocean ridges is cooled partly by conduction of heat to the seafloor. Almost as much heat is carried away by cold ocean water that enters fissures and circulates through the hot crust. Once heated, the water returns to the ocean as hotsprings, with exit temperatures that range from a few degrees to 350 °C. While passing through the hot crust the seawater leaves some constituents behind and extracts others, thereby altering the rocks substantially. At the vents, manganese and iron oxides and hydroxides (black) are deposited that may contain valuable amounts of such metals as silver and copper.

springs (hydrothermal springs: Figure 14.3). Predicted on theoretical grounds in the late 1960s, we found the first one east of the Galapagos Islands in the Pacific. It was the 17th of February, 1977, the day of the high point of my career. The springs turned out to be common on mid-ocean ridges, warning us never again to assume that we know most things about the vast realm of the sea. Taken together, the springs can account for the global heat flow deficit. They must, of course, have existed since the first new crust welled up on the first mid-ocean ridge, when seafloor spreading started. Their deposits, not recognized before, have since been found in rocks ranging in age from the Carboniferous to the Eocene, and others are no doubt awaiting us in the Precambrian.

As ocean water circulates through the cooling system, it passes on some of its constituents to the hot rock but, being a corrosive salt solution, it dissolves others from the wall rock. All magnesium is lost to the rock, for example, and possibly some sodium and chlorine. From the rock, the fluid gains silica, iron, manganese, calcium, carbon dioxide, and many other things. The brine, now much different in composition and with a temperature ranging up to about 350 °C, escapes at the surface. Being far out of equilibrium with seawater when it exits, it rapidly deposits manganese and iron oxides and hydroxides that contain many other metals, including copper and silver.

The total flux of water is large; every ten million years, the equivalent of the entire ocean volume circulates through the hot crust. A quantitative estimate of the chemical effect of all springs combined is not easy to make, but they probably furnish nearly all of the manganese and iron in the sea, the latter in its reduced, ferrous state (Section 15.5), up to half the silica, and at least one-sixth of the calcium supplied by rivers.

Therefore, we appear to have found here the long-sought sink for magnesium and also sulfate. Whether that is true for common salt (NaCl) also, the biggest disposal problem of all, is still unclear. We have also located a major source for some important elements, especially ferrous iron which, because it binds oxygen, affects the oxygen content of seawater (Section 15.5).

Clearly, this discovery forces us to rethink the chemical history of seawater. The level of hydrothermal activity depends on the rate of seafloor-spreading. Doubling the spreading rate, not an unreasonable assumption, would double the circulation of cooling water. That would reduce the magnesium concentration by half and greatly increase silica and ferrous iron. The iron, when it entered the sea, would make a strong demand on the available oxygen. In the early Precambrian, the young, hot crust must have had a hotspring activity so vigorous that a different, more variable composition of ocean water, more silica and iron, less oxygen, and a lower salinity, is highly probable.

What we need is hard evidence that hotsprings have indeed produced great variations in the composition of seawater. The evidence is sparse, but may be found in the ancient oceanic crust that occurs on the continents as ophiolites and greenstone belts.

14.4   THE EARLY ATMOSPHERE

If the earth had acquired its present atmosphere from the solar nebula, it would have contained many gases in proportions typical for the sun itself. The noble gases argon, krypton, xenon and neon, which are too large to escape into space and combine rarely or not at all with anything else, show that this was not the case. Neon, for example, should be ten to a hundred times more abundant and the noble gas ratios would be very different.

Even if a primitive atmosphere of hydrogen, methane, ammonia and water once existed, it is almost certain that the collisions of the accretion and following bombardment would have blown it away. A pity, because students of the origin of life like such an atmosphere for the basic building blocks of life that it might have contained (Section 15.1). Alas, it seems that they cannot have it.

So the earth's atmosphere originated somewhat later, but the isotopic ratios of argon and xenon show that it too formed more than four billion years ago. What might it have been like in its pristine state? The noble gases were released by the mantle in the same proportion as today as their isotopic ratios indicate. Nitrogen, the main component, also came from the mantle, because there is no way in which the very large quantities of this rather inert gas could have been added later or from somewhere else. The mantle also exhaled sulfur and hydrogen through volcanoes, probably as the malodorous gas hydrogen sulfide, and water came from the same source. There was no oxygen.

Some of even the earliest Precambrian rocks had weathered on land. Because land plants, the most important weathering agents of the present time, had not yet appeared, only carbon dioxide, dissolved in water as carbonic acid, can have done the job and therefore was present in the atmosphere more than 3.8 by ago. The main source was again the mantle, but the space bombardment that ended *c.* 4 by ago may have added quite a bit. It was probably a good deal more abundant than it is now, because such enormous amounts of it have since then been stored in limestone, coal and oil. If we convert all carbon-containing sediments to carbon dioxide, and distribute the gas between the atmosphere and the ocean in the ratio that prevails today, there would have been 2–3 percent CO_2 in the atmosphere, at least 100 times more than now. That is a high number and it should have caused a terrifying green-house effect (Section 14.7).

Anyway, when the space bombardment ended, the earth had an ocean full or almost full of water and an atmosphere of nitrogen, water vapor, and much more carbon dioxide than it holds now, but the rock record shows that there was probably no oxygen at all.

14.5 OXYGEN

The early atmosphere of the earth thus differed from the present one mainly by its lack of free oxygen, an interesting component because it is absent from all other planetary atmospheres and is the key to all higher animal life. The absence of free oxygen in other atmospheres and on the earth until the early Proterozoic is understandable; it is its presence for the last two billion years that requires an explanation.

Free oxygen is a very reactive element and it is difficult to imagine that any could escape to the surface when the mantle melted, because there was so much hydrogen, carbon, sulfur and iron there, ready to capture each passing oxygen atom. The geological evidence also speaks against the presence of oxygen in the earliest atmosphere and ocean. Many Archean river sediments contain grains of pyrite ("fool's gold") and uraninite, minerals that will not survive long in the presence of oxygen, yet the grains have been rounded, showing that they were transported by running water or wind over a considerable distance. Moreover, Precambrian dolomites (magnesium-rich limestones) contain much ferrous iron that would have been oxidized, if they had formed in seawater containing oxygen. Redbeds, on the other hand, ancient soils and sediments rich in oxidized iron, are very rare until about 2 by ago, telling us that until well into the early Proterozoic weathering took place in an oxygen-free atmosphere.

The absence of oxygen in air and water was fortunate, or otherwise the synthesis of many compounds indispensable for the origin of life would have been impossible. It seems paradoxical that life, now with few exceptions dependent on oxygen for its proper functioning, would not have been able to arise if this element was present, but as far as we know that is the case.

Objections have been raised against this dismissal of free oxygen in the early ocean and atmosphere, and much time has been spent searching for evidence to the contrary. How reluctant we are to think in terms of an earth drastically different from the one we know. The search has mostly failed, but once in a while a piece of evidence points to a possible local occurrence of free oxygen. That does not seriously conflict with the history of oxygen as set forth below, but it is a safe bet that the last word on the subject has not been said.

How do we proceed from a primitive atmosphere free of oxygen to the present one which holds 21 percent of this gas? The process we seek must be efficient, because it generated not only all the oxygen now present in the atmosphere and the oceans, but enough to satisfy all oxygen "sinks" as well, all easily oxidized rocks, elements, and minerals on the earth's surface and in the sea. Only after all accessible iron had been laid to rest as ferric oxide, all carbon transformed into carbon dioxide, all hydrogen oxidized to water, and all sulfur converted into sulfate and sulfide, was it possible for free oxygen to accumulate.

Two processes are serious candidates for this job. Under the influence of the ultraviolet (uv) rays of the sun, water breaks down into oxygen and hydrogen:

$$2H_2O + \text{uv radiation} = 2H_2 + O_2$$

but the end-products will instantly recombine unless one or the other is removed. In the upper atmosphere, where hydrogen, being small and light, escapes into space, an oxygen surplus gradually builds up. At present, this photo-dissociation process generates about two million tons of oxygen per year. It sounds impressive, but had it been the only agent, another 26 by would have to pass before free oxygen could reach its present level. Moreover, the process is self-limiting, because some of the oxygen in the upper atmosphere combines to ozone (O_3), an efficient shield against ultraviolet radiation that would slow the dissociation. That is, of course, a good thing because, as we have all recently become aware, ultraviolet light is very damaging to genetic material and living tissues. Without an ozone shield, the earth's surface would be inhospitable to most life.

The other process is the familiar photosynthesis, used by green plants to split water molecules with the aid of energy from sunlight, and convert carbon dioxide into organic compounds:

$$CO_2 + H_2O + \text{sunlight} = O_2 + CH_2O \text{ (organic matter)}.$$

Provided there are enough green plants, photosynthesis generates copious oxygen, at present about 20 billion tons per year, most of which is used up in weathering and soil formation. The rest, about 20 percent of the amount present in the atmosphere, is consumed for biological purposes or dissolved in the oceans, where it permits life to breathe and decomposes organic matter.

This oxygen-producing (oxygenic) form of photosynthesis is not the only one. Some bacteria and algae use sunlight as an energy source in a different process that does not yield oxygen. It has been argued on biological grounds that this was the first photosynthetic path taken by organisms. If that is true, the sparse evidence for photosynthesis in the early Archean (Section 15.5) might not have entailed biological production of oxygen at all, and we must call on ultraviolet light to explain it. Later Archean and Proterozoic sediments indicate so large a production of oxygen that the only plausible process is oxygenic photosynthesis.

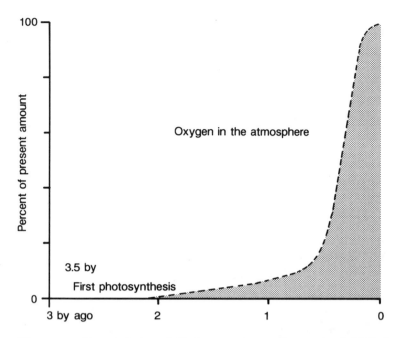

Figure 14.4. Photosynthesis probably began at least 3.5 by ago, but initially it may have been by a process that does not generate oxygen. Alternatively, the first oxygen was used up in the oxidation of earth materials. In any case, free oxygen did not enter the atmosphere until about 2 by ago. More recent estimates indicate that since about 600 my ago the atmospheric O_2 content was equal to or higher than at present (Figure 14.6), but the data base is not yet very solid.

Even so, much time passed before the atmospheric oxygen content reached today's level (Figure 14.4).

14.6   LIFE AND THE ATMOSPHERE

The carbon dioxide content of the atmosphere is but one link in a complex system that also includes the ocean, the rocks at and near the earth's surface, and life itself (Figure 14.5). The system is partly controlled by tectonics. Volcanoes and deep-sea hotsprings supply from the mantle the only new carbon dioxide that enters the system. In this way high rates of seafloor-spreading raise the CO_2 content in the atmosphere and affect the climate. David Rea of the University of Michigan, for example, believes that high tectonic activity in the Pacific caused the unusually warm interval in the Eocene (Section 11.2). Carbon is lost as carbonate when limestones are deposited and subducted, but it is recycled as carbon dioxide by eruptions in the volcanic arc.

Green plants convert carbon dioxide to organic matter and oxygen.

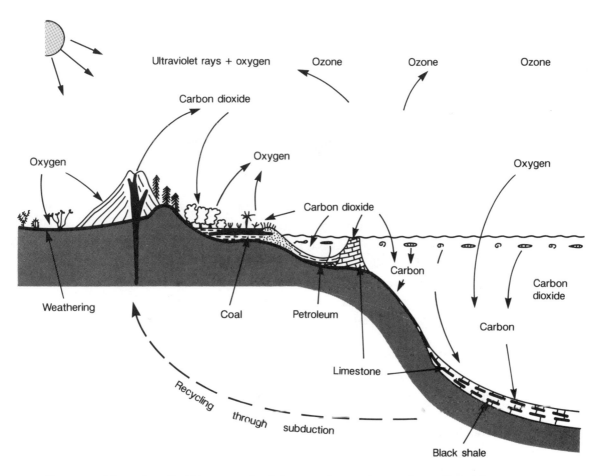

Figure 14.5. The carbon dioxide/oxygen cycle of the atmosphere and oceans is complicated by life. New CO_2 is supplied by volcanoes and submarine hotsprings. Most is consumed by weathering, the rest mainly by photosynthesis and converted into organic matter and oxygen. Some of the organic matter is eliminated from the cycle by burial in sediments. The excess oxygen enters the atmosphere where a small part forms the ozone shield, much is used up in weathering, part consumed in breathing and a great deal dissolved in the ocean. Carbon dioxide also enters the ocean where a large part of it is converted into reef limestones and calcareous oozes, and so removed from circulation.

The oxygen enters the atmosphere and oceans, permitting organisms to breathe, and oxidizes dead organic matter back to carbon dioxide. Some organic matter escapes oxidation by burial in amounts that vary in amount from little to enormous. In doing so it reduces the amount of carbon dioxide in circulation, and increases the oxygen content of air and sea (Section 10.5).

Carbon is thus stored in several reservoirs: as dissolved CO_2 in the ocean, as a gas in the atmosphere, as organic matter in living tissue and in humus, peat, coal, oil and natural gas, as carbonate in limestones, and as alteration products in rock weathering. Weathering consumes a major portion of the new CO_2 in the system, and much of that leaves the cycle permanently, as does much of the carbon stored as limestone. In the long term the trend is towards a gradual decline of free carbon dioxide.

Oxygen also is stored in various reservoirs of which the most important ones are oxidized soils and minerals, and dissolved in water on land and in the sea. Burial of organic matter and sulfur compounds such as pyrite (FeS_2) reduces the consumption of oxygen and raises its level in the atmosphere, whereas erosion followed by oxidation of the buried matter has the opposite effect.

It is obvious that this system, complicated by the role of sulfur, is difficult to study in a quantitative way. Gases in glacial ice (Section 5.5) and stable carbon and oxygen isotopes in ocean sediments (Section 10.5) shed some light on the behavior of the system over the eons, but so far the data fall short of what we need, and one searches for other means.

A budget might be set up, for example, in which the entries are the carbon, oxygen and sulfur reservoirs, and the transfers are tectonic, chemical and biological processes. Even the cursory description of the system given above makes clear that specifying the reservoirs and pathways requires ingenuity, much courage and patience, and endless checking and rethinking of the sparse and insufficient data. To quantify it so that it becomes an account of the history of the system requires far better and much more data than we possess at this time.

The best study of this kind so far has been carried out by Robert Berner of Yale. It yielded an atmospheric oxygen content at the start of the Phanerozoic that is barely lower than it is today (Figure 14.6), and is startlingly high in the Carboniferous when huge amounts of organic matter were buried in coal deposits. The computed CO_2 content is high during the green-houses of the early Paleozoic and the Mesozoic, while low values mark the Carboniferous, Permian and late Cenozoic ice-houses, in accord with our expectations and independent geological evidence (Section 12.4). The high CO_2 of the Silurian and Ordovician, on the other hand, conflicts seriously with the glacial conditions of those times.

Whatever the fluctuations over time of atmospheric CO_2, its continuous removal as weathering products and limestones that are not recycled results in its slow decline in the air and ocean waters. Eventually, there will not be enough left to sustain photosynthesis and life on earth will come to an end. Here is the "prospect of an end" that James Hutton could not see, but that we can now see all too clearly. When will it be? A few years ago, pessimists gave life on earth only 100 my, but their

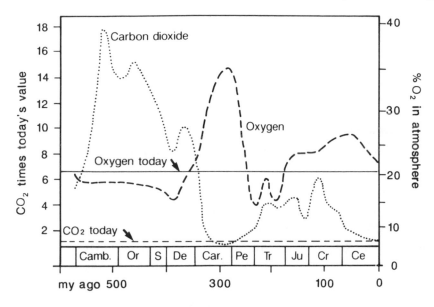

Figure 14.6. Our knowledge of the Phanerozoic history of carbon dioxide and oxygen rests on complex budget calculations that take account of the many reservoirs, fluxes and interactions of oxygen, carbon dioxide and sulfur. The best estimates, made by Robert Berner of Yale University, show startlingly large peaks of carbon dioxide in the early Paleozoic and of oxygen in the Pennsylvanian, the latter due to the burial of a great deal of organic matter as coal. Note that Berner's stratigraphic boundary ages differ slightly from the more recent ones used in this book.

calculations have turned out to be too simple. Taking everything into account, we probably have another billion years, but even that is brief, compared to the five or ten billion we might have had before the sun grows too hot and burns the planet to cinders.

14.7   A WEAK AND PALE SUN

Except during the occasional ice age, the earth has never been very cold, like Mars, nor was it ever overheated like Venus which suffers from a runaway green-house effect. This fine tuning of the earth's surface temperature which keeps it so neatly within the limits that life can tolerate, deserves comment.

There is good evidence that the sun's brightness, the solar constant, has increased by about 25 percent over the lifetime of the earth. Because practitioners of disparate disciplines often communicate less than they should, this fact, long known to solar physicists, escaped the attention of geologists until recently.

The variations of the orbital parameters bring only small changes in the solar heat received on earth, yet they suffice to bring on major

287

climate changes. The impact on the climate of a sun supplying only three-fourths of its present heat ought thus to have been spectacular. Our ignorance of such important factors as the geography of Precambrian lands and the level of their albedos, or of the ocean circulation at that time, renders any estimate of the consequences quite worthless, but it would have been a cold earth indeed. In fact, the earth ought to have been entirely frozen, but the geological record does not bear this out, life persisted, and there is evidence everywhere for running water. Or we should not be here!

What might have compensated for the lack of heat from the sun during the Precambrian? More ocean area and hence a lower albedo and a more even temperature distribution, of course, but those would scarcely have sufficed. A strong green-house effect is what we need, and that is precisely what the high CO_2 content of the Precambrian atmosphere implies. Provided it was negatively coupled to the slowly increasing solar heat, carbon dioxide in the atmosphere may have been the main reason that the earth's surface temperature was kept within a narrow window around 25 °C.

The proposition is a simple one and, although difficult to prove, it has much to recommend it. It also has some disturbing aspects. One is that if we restore to the ocean and atmosphere all carbon contained in coal, oil and limestone, we have too much of a green-house effect. Perhaps that is wrong, because much CO_2 may have been added later by volcanoes and hotsprings, but that smacks unpleasantly of special pleading. Besides, we have not even taken into account other green-house gases, such as water vapor, nor have we dealt with the effect of clouds with a high albedo. Neither effect can be even approximately quantified.

The most troublesome part is that, to keep the temperature within a narrow range, the removal of carbon dioxide from the atmosphere must have been precisely matched to the increase in the solar constant, the luminosity of the sun. The prime removers of CO_2 probably were photosynthesizing algae, and a fine control mechanism by feedback is needed. What that might be is not at all clear.

Can only life achieve this control? Atmospheric CO_2 dissolves in rain, rain falls on land and increases weathering. Weathering produces silica, calcium and carbonate ions that wash into the sea. Organisms incorporate calcium carbonate and silica in their shells and skeletons which after death sink to the bottom. They travel with the plates, are subducted, and CO_2 is recycled to the atmosphere through volcanoes, *ad infinitum* (at least for a while yet).

If the sun increases in strength, the oceans are warmed, evaporation and rainfall increase, and so does weathering. More CO_2 is withdrawn from the atmosphere and the green-house effect is reduced, because the volcanic input depends only on plate motions and remains unchanged. If the surface cools, the inverse happens; with less weathering, less CO_2 is taken out of the atmosphere, the green-house effect increases and the earth warms up.

Two possibilities then, two processes, life and weathering, compete for control of the green-house effect and therefore of the surface temperature of the earth. Weathering is held responsible for three-fourths of the job by some, life for all of it by others. As yet the weathering estimates do not rest on a very solid base, and until better ones come along it is wise to see this process as no more than a very significant contributory factor.

Understandably, the rock record is not very explicit on this point, except for a curious hint that the feedback mechanism, whatever it was, once nearly failed. The terminal Proterozoic ice age left abundant glacial deposits on all continents. As far as we can tell, and that is a necessary caution, it was 200 my long, or there were three closely spaced ice ages, with maxima around 800, 700 and 590 my ago. The geographic distribution of the glacial deposits is surprising. An early reconstruction of the paleogeography (Figure 14.7, top) has most glacial deposits below 60° latitude and a few at or close to the equator. A more recent tectonic fit (bottom) is different; it places a late Proterozoic supercontinent across the South Pole, but one icecap still sits at 30° latitude, far lower than any Paleozoic or Cenozoic ones. Other reconstructions give a much wider distribution of low-latitude icecaps, including one on Australia during the time of the Elatina Formation (Preamble to Chapter 1).

It is early yet for paleogeographic reconstructions of the Precambrian, and a near-global glaciation remains quite likely. Moreover, the Australian G. E. Williams has pointed out that, if the obliquity of the earth at that time had exceeded about 54° (try to visualize that!), the climate zones of the earth would have been reversed; the equatorial zone would carry the icecaps, and the polar regions would be tropical (only in the summer). In the Precambrian, such a high tilt of the earth's axis is not impossible.

However, if there was a low-latitude glaciation, we wonder whether it could have been caused by a large burst of biological activity 900 my or so ago that consumed and buried far more carbon than was wise in view of the still feeble sun. The large amount of limestone of the later

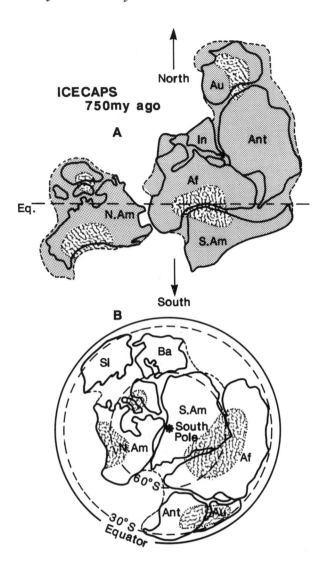

Figure 14.7. The late Proterozoic experienced a major ice age or cluster of closely spaced ice ages that was long-lasting and had icecaps at low latitudes. It may, some think, have been due to excessive consumption of carbon dioxide by photosynthesis at a time when the sun was still weak and the green-house effect badly needed. Diagram A is a 1981 paleogeographic reconstruction that shows many low-latitude icecaps (light wavy shading). A more recent reconstruction (B) puts most ice in more plausible latitudes, but at least one large icecap is situated in the subtropics. In: India; Au: Australia; Ant: Antarctica; Af: Africa; Si: Siberia; Ba: Baltica; N. Am: North America; S. Am: South America. The figure illustrates our need for better paleogeography so that we may tackle these important questions with greater confidence.

Proterozoic supports this suggestion, but better data may demolish the idea in the next few years. If there is one thing that geologists involved in the surface history of the earth want, it is for paleogeographers to get on with their job.

15
The dawn of life

The first steps in the evolution of life need not have involved life itself. Non-biological reactions may have produced organic compounds from which the living cells would later be assembled. This chemical evolution, if it ever happened, has left no trace in the geological record, and neither have the first hesitant steps of life itself. What happened in this critical phase is speculative and controversial, but that is no reason to shun the issue. The arguments depend as much on a definition of life as on the rules of chemistry, and on reasonable although subjective opinions regarding the environment of the early earth, its geochemical condition, and the available sources of energy.

After the chemical evolution there was life, and where one stops and the other begins depends on what is and what is not life. A definition, although of considerable interest, is not needed here, and I shall merely assume that "life" implies an orderly set of biochemical functions, and a reliable way to reproduce and pass on to the offspring the processes that allowed the parents to exist.

15.1   CHEMICAL EVOLUTION

Organisms use a wide range of biochemical compounds that do not require a living factory, although they are usually synthesized there. The right amounts of carbon dioxide, nitrogen, hydrogen, water and a few trace elements will, with application of suitable energy, produce many of the basic blocks in air or water. Most do not form, and few survive in the presence of free oxygen.

Many compounds essential to life have been manufactured in the laboratory under conditions which the experimenters believed to be close to those of the early Precambrian. Amino acids form easily with the help of heat; the kind produced depends on the temperature. The first experiment was carried out many years ago by Stanley Miller at the University of Chicago, who flashed artificial lightning through a flask filled with a mixture of gases. The mix was based on the atmosphere of Jupiter thought by Miller's mentor, Harold Urey, to resemble the primordial earth atmosphere. It did not, but other experiments in more realistic conditions have yielded many biologically important substances.

Synthesizing the more complex and more important organic compounds is a different problem, and one not close to being solved. The most crucial ones are proteins made of long chains of amino acids, enzymes that make the biological factory work, and the key component of life, deoxyribonucleic acid or DNA. DNA contains the genetic information as well as instructions for the biochemical functions of the cell. The instructions are conveyed to the place of execution by the messenger molecule ribonucleic acid or RNA. Both are constructed from long strands of simpler molecules, the nucleotides.

It is a curious property of large biological molecules that they easily combine in the proper sequence. Sidney Fox of the University of Miami has shown that, if we repeatedly evaporate, then wet again a broth of amino acids under the right conditions, they will arrange themselves into simple proteins. What is more, the proteins form microspheres, small, double-walled bodies that look like cells and possess the ability to absorb food from the broth, to grow, to divide and to multiply.

Sometimes large molecules spontaneously form clusters called coacervate droplets, which surround and protect themselves by a tightly bound double layer of water molecules. Like microspheres, the droplets absorb substances from the liquid and can synthesize more complex molecules. But the processes that yield enzymes, nucleotides and then DNA and RNA remain stubbornly out of reach.

If all carbon now tied up in limestone, coal and oil once existed as basic organic compounds, leaving just enough carbon dioxide to keep the earth warm, the whole ocean would have had the strength of a very weak cup of broth. In restricted water bodies such as ponds the concentrations might, of course, have been higher. How did the amino acids, peptides, and innumerable others drifting in these diluted fluids combine to more complex molecules like sugars or proteins? Mere chance encounters would not do, as many a statistically inclined scholar has pointed out, adding that the probability that life could have arisen under such conditions and in the available time was far too small to be taken seriously.

This argument, it seems to me, is specious. If we do not know the processes that operated and their rates, nor the concentrations of the chemicals they operated on, we have little cause to talk about insufficient time. Life did, after all, demonstrably arise. Also, there is no need to call on mere chance. Catalysts favor and enhance certain reactions above all others, and ordered inorganic crystal structures, such as those of the clay minerals, can function as templates on which simpler components line up to form complex ones. Besides, as Louis Lerman of Stanford University has noted, small bubbles are able to perform many chemical tricks, most of them as yet undiscovered. The warm, wind-stirred ocean, its thin surface skin of dust and gas bubbles suffused by ultraviolet light and the warmth of the sun, has much to offer as the birthplace of life's building blocks. There is no reason why some or all of these processes should not be active today, but if they are their output is instantly consumed by a teeming multitude of organisms.

On the primitive earth, energy was available in many forms, lightning, the heat of sun and volcanoes, and ultraviolet light. In excess, however, ultraviolet light is destructive and to be guarded against, once the first life had germinated. Water, carbon dioxide and sulfur existed in the atmosphere in various forms, but probably not the methane, hydrogen and above all ammonia that made the early laboratory experiments a success. This is not a trivial point, because without those the synthesis of simple biological compounds is much more difficult, some think impossible, and certainly much less efficient.

But wait! Some ordinary sands are rich in titanium minerals. The titanium encourages the synthesis of ammonia from atmospheric nitrogen and water with the aid of sunlight, and at times brings a pungent whiff to the desert of Imperial Valley in California. The process seems insufficiently prolific, but its discovery was a warning not to use the word "impossible" too soon.

Fundamentally different ideas are possible where we know so little. A. G. Cairns Smith has compared the chemistry of present life to a highly integrated high-tech system evolved step by step from a low-tech one. We would not expect, he says, to reason back from artillery and jet fighters to the sticks and stones used by Stone Age men as weapons, so why do we try this with biological evolution. To have evolution we need genes and we need them very early, but they need not have been like DNA. Simple proto-genes might, for example, have used clay minerals or other crystal lattices as templates. His ideas have not pleased many, but then, Wegener's insights were not much appreciated right away either.

15.2 THE NEXT STEP

If what we need is warm water in a position protected from such dangers as ultraviolet light, another look at deep-sea hotsprings is in order. This occurred to Sarah Hoffman and John Baross at Oregon State University as soon as news of the discovery of the springs reached them. Deep-sea hotsprings have plenty of thermal energy, and contain carbon dioxide, ammonia and sulfur in various forms including hydrogen sulfide. Also, iron minerals that might serve as catalysts or templates are there in abundance. Hotsprings seemed a fine idea and an unseemly scramble ensued to take credit for it, but it was soon dismissed by the conventional wisdom, mainly on the grounds that the springs would be too hot. That missed the point: some springs are indeed hot, but many are not, and the idea has gradually been taken up again, inspiring more, but so far inconclusive work.

Whatever the processes and wherever the sites, at some time in the remote past isolated pools of water contained enough basic compounds to permit the next step to occur. When would this have been? Some would like it very early when a suitable atmosphere might have existed, but those days hardly seem propitious because of the space bombardment (Section 13.2). The odd term "impact frustration of life" describes the notion that life arose repeatedly, then perished under the hot blasts and poisonous gases of incoming comets and meteorites, until it finally took hold. Deep-sea hotsprings have an advantage here as, sheltered by thousands of meters of water, they would escape the consequences of even an impact large enough to sterilize the earth's surface. Whatever this is worth, we must not expect life to have obtained a permanent foothold until after the bombardment was over. I note parenthetically here that some students of early life perversely believe that meteorites actually brought the first life.

What we have at this point (Table 15.1) are building blocks, but not the crucial step to life itself, the synthesis of DNA and RNA. About a decade ago, the discovery that RNA is capable of making copies of itself raised the exciting possibility of an RNA world where it alone ran the machine and handed down the manual. Later experiments have shown, however, that it is difficult to make RNA perform this trick, and the lovely RNA world went the way of so many other exciting ideas. Having lived through many disappointments of this kind, I sometimes wonder how we retain the optimism with which new ideas are greeted.

Clearly we have a surfeit of times and sites, early, late, the sea surface, volcanic pools, or deep-sea hotsprings, but a dearth of plausible processes

Table 15.1. *Summary of the early history of life*

Time (my)	Events	Consequences
>4,000	Bombardment	Impact frustration of life;
?	Chemical evolution	Organic soup forms, reduces carbon dioxide in atmosphere;
4,000	Competition for organic building blocks	Heterotrophs consume organic soup;
3,800	First life	Ancestor of prokaryotes; genetic apparatus formed;
>3,800	First autotrophs	Carbon dioxide reduced;
3,800?	First organic remains at Isua, Greenland	Earliest Banded Iron Formation (BIF)?
3,500	Photosynthesis brings oxygen; neutralized with ferrous iron	Bacteria and algae lay down stromatolites and BIF;
2,000	First oxygen in atmosphere; enzymes protect cells against oxygen	Ozone shield, redbeds; end of BIF; oxygen-based metabolism;
2,000?	Eukaryotes appear	??
1,000?	Sexual reproduction	Communities more complex and specialized; multicellular organisms;
800	Atmospheric oxygen 10%?	Rapid diversification of higher plants and animals, but still soft-bodied;
600	??	Cambrian diversity explosion; rise of hard skeletons.

and pathways. This should not cause us to think that the search for the origin of life is stuck, far from it. Great progress has been made, but we are still some way away from the next set of hypotheses, while the various ideas in their present form are not well suited to simplification.

15.3 THE FIRST ORGANISMS

Whether the final steps happened on earth by processes not yet conceived of by us or were imported from outer space, one day, four billion years ago, single cells constituting the first true life floated in the organic soup. Heterotrophs like ourselves, they fed on organic matter they had not themselves manufactured, but that was, for the time being at least, reasonably plentiful.

We should not think of these first heterotrophs as simple or invariant.

There may have been much variety from the start, and mutations, abrupt changes in the genetic code due to ultraviolet light, radioactivity or biochemical errors within the DNA, did surely increase it. If a chance genetic shift enabled an organism to subsist on a so far unused substance by means of a new enzyme, it would survive better and longer than others who could not do so. Step by small step, many different organisms, experimenting along many different paths, constructed a single biochemical factory, now used with minor variations by most life forms.

A few questions are in order here. The concept of chemical evolution rests on plausible reactions between reasonable substances, but much is beyond our understanding. When did the organic soup appear? How concentrated was it, and how stable? How long did the various reactions take? With an atmosphere endowed with plenty of carbon dioxide but little or no oxygen, perhaps three million tons of organic substances could have been produced each year, although they need not have been evenly distributed; in the deep-sea it might have been a bit more. Is that enough? How long would the first heterotrophs have been able to subsist on this soup, formed over millions of years, before exhausting it? Simple organisms are adept at rapid adjustment to any available food supply, no matter how copious, and deplete it in no time at all. Was there some mechanism, some balance perhaps between upwelling and a stable stratification of the water that limited the rate of supply of the food? Something of that sort would seem almost imperative, or famine would have soon snuffed out the prospect of lasting life on earth.

Eventually the soup was gone, and to preserve life organisms were needed that could manufacture their own food. This is what autotrophs do, using carbon dioxide, water and various nutrients, with the help of a suitable energy source. A few bacteria obtain energy from chemical reactions; most use photosynthesis and sunlight to provide almost all the food flowing through the web of life from the tiniest green plants to the largest predators.

15.4 PROKARYOTES AND EUKARYOTES

All organisms fall into one of two categories: the simple, small, single-celled prokaryotes, and the more complex single- or multi-celled eukaryotes (Figure 15.1). Eukaryotes hold most of their genetic material in packages called chromosomes inside a nucleus wrapped in a membrane, while prokaryotes carry it loose without a membrane; the process of dividing activates the chromosomes. Only eukaryote cells have small bodies called organelles that each perform a specific biochemical function.

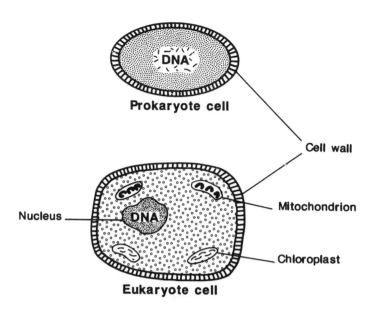

Figure 15.1. Prokaryote cells are small, simple and lack a nucleus. Eukaryotes have a nucleus containing their DNA; they are usually much larger and carry a number of specialized bodies called organelles, such as chloroplasts (plants only) and mitochondria, which contain DNA portions of their own.

The best-known organelle is the chloroplast of green plants which houses the chlorophyll responsible for photosynthesis. Chloroplasts resemble cyanobacteria, once called blue-green algae, and have their own genetic material that is separately passed on to descendant chloroplasts.

According to Lynn Margulis of Boston University chloroplasts began as free prokaryotes living in symbiosis within eukaryote cells, an idea that is widely although not universally accepted. Gradually they became so identified with their hosts that both now reproduce simultaneously and no longer maintain separate existences. Another organelle, the mitochondrion, functions as the main energy plant of the cell; it probably began in the same way, as a symbiotic, respiring bacterium. It is even possible that the tails and hairs that enable many single-cell eukaryotes to swim in search of food may also have been free-swimming bacteria captured through symbiosis (Figure 15.2).

Eukaryotes also differ from prokaryotes in the way they reproduce. Prokaryotes usually multiply by cell division, each part receiving its full share of parental genetic material. This form of reproduction is fairly accurate and any variations come primarily from mutations. Eukaryote cells also divide, but by a complex and elegant process that splits the genetic material in such a way that a second step becomes possible, namely sexual reproduction. In sexual reproduction sperm and egg, the

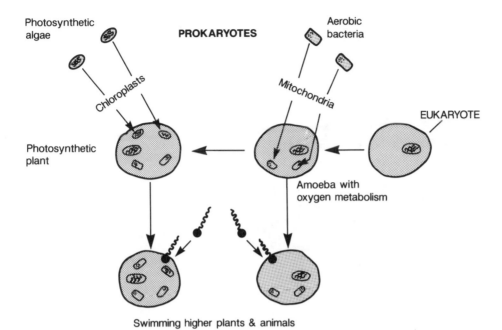

Figure 15.2. Many specialized features of the eukaryote cell may have begun as prokaryotes living symbiotically within the cell. Photosynthesizing algae became chloroplasts, aerobic bacteria took charge of oxygen-based energy housekeeping and became mitochondria, and spirochetes with tails furnished a means of locomotion.

offspring of the divided cell, each acquire half of the parental genetic material. The original amount is restored when fertilization takes place, but the parts come from different parents, giving each cell some qualities of one parent and some of the other. New genetic combinations are so created more gently than by mutation. This important innovation may also have been due to symbiosis with an organism that took charge of the reproductive system.

Should we then conclude that the eukaryotes, which owe so much to symbiotic prokaryote guests, did themselves evolve from prokaryotes? This is indeed the prevalent view, but the relations are not as simple as that. The record is still being deciphered, but it is already clear that the prokaryotes are not a single coherent group and that they and the eukaryotes have followed quite different evolutionary paths (Figure 15.3).

In summary, it seems that life began with a variety of heterotrophs, ever better equipped through random mutations to take advantage of their diverse food supply. Because the air was still without oxygen and there was no ozone shield to protect against ultraviolet radiation damage, an early accomplishment must have been the ability to repair radiation damage. This is something all prokaryotes, and most eukaryotes to some

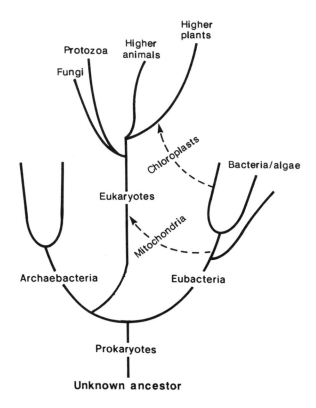

Figure 15.3. This still speculative evolutionary tree of mainly Precambrian life rests on the findings of molecular genetics, a field that has been very successful in elucidating relations between groups of life forms. It suggests that of the two groups of prokaryotes the Archaebacteria are more closely related to the eukaryotes than the Eubacteria and thereby the ancestors of all higher plants and animals. The diagram also incorporates the idea that the eukaryotes benefitted from symbiotic arrangements. The upper part of the tree will be discussed in Chapter 16.

extent, are still able to do, although with diminishing effectiveness the more complex they are. The creation of an ozone radiation shield was thus essential for further evolution.

When the organic soup ran thin, the first autotrophs appeared, either in the chemical or the photosynthetic mode. We still have representatives of each, but the second mode used a much more widely available resource, sunlight, and photosynthesis soon came to dominate, perhaps first in its oxygen-free form (Section 14.4). When it became possible to split water, however, the oxygenic mode rapidly took over, because water is much more abundant than other hydrogen donors such as hydrogen sulfide. The cyanobacteria who acquired this capacity flourished, but the oxygen they produced became a threat to life because so many key components including DNA cannot survive in its presence (Section 14.5). Some organisms must have perished, some developed chemical defenses, but still others went further and adopted an oxygen-based metabolism that gave them a much higher energy efficiency.

And somewhere, sometime, the eukaryotes appeared, rather unobtrusively at first as they built up their collection of symbiotic arrangements.

This is what theory, a few experiments, and much thoughtful speculation suggest as a plausible but by no means proven early history of life. We leave it temporarily here with the means to perfuse air and water with the oxygen that became a major factor in the further evolution of life.

15.5   AN ATMOSPHERE FIT TO BREATHE

What does the Precambrian record have to say to all this? The 3.8-by-old rocks at Isua in Greenland bear witness to the presence of land and sea. Were those seas empty and devoid of life? Carbon found in the Isua rocks is enriched in carbon-12, suggesting but not proving that a process that preferred ^{12}C and so may have been biological was involved. Unfortunately, high temperatures also are able to shift the ratio, and the Isua rocks have been metamorphosed.

Solid evidence appears around 3.5 by ago ("a little later," one often says, but 300 my are not so little) in Swaziland in South Africa and the Pilbara block in Western Australia (Figure 15.4). These well-preserved deposits, laid down in shallow seas on continental crust, contain evidence for life in two forms, as microfossils and as a kind of limestone called stromatolite, rare at first but soon very abundant.

Stromatolites are finely laminated, bulbous masses of calcium carbonate, a few centimeters to meters across. They form when mats of cyanobacteria precipitate thin calcareous layers as a by-product of photosynthesis. The precipitates grow over time into mounds that may coalesce into large reefs. The sometimes exquisite detail of surrounding rocks shows that stromatolites flourished in shallow coastal lagoons and in the intertidal zone. They are rare in the Phanerozoic, but never vanished and still grow in the warm, shallow waters of the Gulf of California and Western Australia.

We cannot prove that the earliest stromatolites were of biological origin, because no fossil cyanobacteria have been found in them. It seems quite likely, however, because in associated sediments the remains are common of what look like small, single-celled organisms, probably also cyanobacteria. The carbon isotopes support a biological origin, reliably this time because the rocks are so well preserved. If life 3.8 by ago was possible, at 3.5 by it is a certainty.

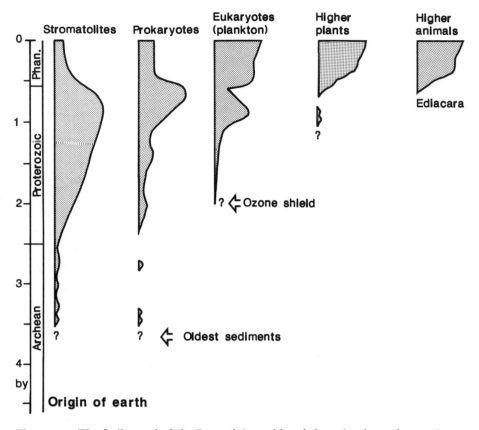

Figure 15.4. The fossil record of the Precambrian, although becoming better known, is not nearly as satisfactory as that of the Phanerozoic. Nonetheless, this diagram, although uncertain about the timing of most events, is much richer in information and much more likely to be approximately correct than we would have thought possible 20 years ago. The Phanerozoic part will be discussed in Chapter 17.

The carbon isotopes imply photosynthesis, but do not say whether it was of the oxygen-producing kind or not, a question which is still being debated on preciously little hard evidence. But if it did yield oxygen, most or even all of it must have been disposed of in sinks. How many oxygen-hungry minerals on land, how much iron dissolved in the ocean had to be oxidized before the first free oxygen appears in the air and the sea? A lot, say some, only a little, say others, both using budget estimates of the kind that are so very difficult to make. Still, a little free oxygen may have appeared here and there, raising the possibility that some heterotrophs in this early ocean might have already adopted an oxygen-based metabolism.

Whatever was the case, the deposition of limestone by life forms soon developed into a success story. Already in the late Archean stromatolites constituted a large part of the mass of sedimentary rocks (Figure 15.4), and when the Proterozoic began 2.5 by ago, stromatolites were everywhere, in Australia, North America, Spitsbergen, and Africa, bearing witness to the practice of photosynthesis by cyanobacteria similar to those that still thrive in our ponds and puddles.

What happened to the oxygen generated by all this activity? Its history is recorded in the Banded Iron Formation (BIF), a set of beautiful rocks unique to the Archean and early Proterozoic. In many colors, blue, brown, red, black, these rocks display innumerable thin laminae of siliceous or more rarely calcareous sediment. The layers alternately consist of silica with some reduced (ferrous) iron, and silica with abundant oxidized (ferric) iron. BIF rocks are voluminous in Australia, Minnesota, Labrador, Russia and Brazil, and, with their one hundred trillion tons of ore, represent 90 percent of the world's iron reserves.

The BIF is a geochemical paradox. Rare element isotopes show that hotsprings, rather than rivers as in the Phanerozoic, dominated the composition of the ocean, furnishing ferrous iron and silica in abundance. Because of the abundance of carbon dioxide the sea was slightly acid, and so the ferrous iron stayed in solution and migrated everywhere. No organisms drew on the dissolved silica for their shells and so its concentration in the Archean oceans was high. Near the continents, themselves sources of dissolved silica too, it may have reached saturation at times and precipitated, taking some of the ubiquitous ferrous iron with it to form the first of the two types of laminae.

To deposit iron in large amounts, however, oxygen is needed. It oxidizes ferrous iron to the ferric state which is insoluble and the iron settles with the silica in quantities adequate to form the iron-rich laminae. The two processes are not exotic, but they are incompatible. How do we explain that ferrous and ferric iron alternately accompanied silica to the bottom, if each demands conditions that exclude the other?

The Banded Iron Formation was deposited in large, shallow continental seas and deep troughs along continental margins. Imagine this seascape, perhaps resembling the Southern California Borderland a little, filled with planktonic algae or bacteria all photosynthesizing and generating oxygen. The oxygen binds the ferrous iron and sends to the seafloor a steady rain of silica-rich, ferric iron mud. At other times, in other seasons perhaps, there are no plankton blooms and so no excess oxygen, and the silica settles mainly by itself, accompanied by just a little ferrous iron. Episodic upwelling in deep basins would fit the fine banding

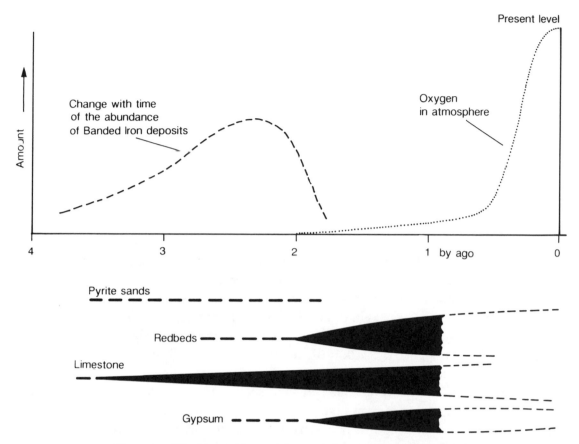

Figure 15.5. The history of oxygen is recorded in many sedimentary rocks. Limestones imply photosynthesis as early as 3.5 by ago, but pyrite sands, not compatible with free oxygen in air or water, do not disappear until 1.5 by later. At that time, the first redbeds bear witness to the presence of oxygen in the atmosphere, and gypsum deposits show that it was dissolved in the sea as well.

of the BIF as well as seasonal plankton blooms in the coastal zone would. Stubborn defenders exist of each model, but the data do not leave much to choose between these two explanations.

In this way the BIF accommodated the oxygen supply for a long time. This may have been fortunate, because it provided time to evolve the biochemical means of coping with this dangerous but, oh, so promising element.

In any case, the course of BIF history is clear. A trace appears 3.8 by ago, before we have proof of life, but oxygen generated from water by ultraviolet light might have done this small-scale job. Less than a billion years later, banded iron rocks became a major deposit on Precambrian continental margins (Figure 15.5). They peaked in abundance between 2.5 and 2 by ago, then vanished suddenly, not long afterwards. A first warning of oxygen in the air comes from sulfur isotopes in 2.2-by-old continental deposits in Canada, and redbeds confirm it 1.8 by ago. At about the same time, the appearance of gypsum (which requires oxygen for its formation) in many marine sediments shows that oxygen had invaded the sea in amounts sufficient to keep all sinks satisfied. There must have been earlier moments when free oxygen was available locally in air or water for a while, but the evidence points to two billion years ago as the time when it became widespread, stayed and steadily began to increase.

For free oxygen to accumulate in air and sea, however, one more condition had to be satisfied. Part of the carbon must be removed permanently or the oxygen released by photosynthesis will be consumed again when, after death, the organic matter decays once more to carbon dioxide. Burial and the transformation of organic matter to black shale, coal and oil were an essential condition for the creation of a breathable atmosphere.

15.6 LIFE IN THE PROTEROZOIC

Oxygen was ringing in a new era in ocean and atmosphere around two billion years ago, although its abundance remained low until late in the Proterozoic. Half a billion years earlier, the great expansion of continents and shallow seas of the late Archean had presented a different challenge to evolution. Did one of these ecological watersheds introduce the eukaryotes to the world? It is an important question that we cannot yet answer.

So much evidence has recently accumulated that the views on Proterozoic life of a decade ago are obsolete. The late Archean and early Proterozoic offer us few fossil-bearing rocks, but two billion years ago, again that momentous date, the record becomes almost continuous. To assume that the previous billion years knew little life would be wrong, however. It is just that so few of the oldest rocks are left and so many are heavily overprinted by metamorphism.

Steeped as we are in the miracles of evolution, it comes as a surprise to find that cyanobacteria, and most other prokaryotes as well, seem to have evolved little during the immensely long last 2.5 billion years. Perhaps they always were perfectly fitted to their environment, or perhaps they are so simple that drastic changes would be difficult to bring about. It is hard to say, but it does not mean that they lack variety. An extraordinary deposit called the Gunflint Chert at Lake Superior in Canada, 1.9 by old and associated with the last BIF, superbly preserves a diversity of prokaryote forms, including odd, complicated ones that foreshadow the future. Although exceptional, the Gunflint is not unique. Other rich, well-preserved fossil assemblages elsewhere suggest new adaptations to the diverse environments of the Precambrian seas.

A thousand million years pass, and the Bitter Springs Chert of Australia and cherts and dolomites in California, some 800 my old, present us once again with a rich prokaryote life in shallow coastal waters. The difference with the Gunflint is not great, but new, complex forms have arisen, some with specialized cells. Here for the first time, enough material allows us to look at the organization of life and its adaptations to several well-defined environments.

Are we close here to the first eukaryotes, to multicellular organisms and thence to higher plants and animals? It is a key question, but how do we tell single-celled eukaryotes from single-celled prokaryotes if both are preserved as microscopic, flattened, roundish blobs in shales? Eukaryotes have a nucleus, but again and again paleontologists have seen their hopes dashed when "fossil nuclei" turned out to be artifacts of fossilization.

In our hunt for the first eukaryote this leaves us only statistical differences based on size. Prokaryotes are normally less than 0.01 mm in diameter, while eukaryotes tend to be much larger, around 0.06 mm or so, although their range is wide and overlaps unhelpfully with the prokaryotes. "Large" cells unlikely to be prokaryotes turn up more than 2 by ago. With due allowance for the feeble statistical data, the largest ones are first seen offshore, their preferred environment for another billion years. A little later, 2.1 by ago, *Grypania*, probably a eukaryote, appears in the BIF of Minnesota, and between 1.9 and 1.7 by ago fossils from China and marks made in sediments by organisms in Australia settle the matter. The eukaryotes are here.

From 1.4 by onward spherical cells of many sizes, some with spines, the acritarchs, appear in Proterozoic shales. Acritarchs are creatures of the open sea, possibly the first phytoplankton. Most are regarded as eukaryotes, but a few prokaryotes are also present. Around 1.1 by ago they diversify, developing specialized cells that some regard as evidence for sexual reproduction. Was it therefore the much touted advantage of sexual reproduction that helped the acritarchs evolve as the cyanobacteria had never done? Again it is too early to say.

A very well preserved fossil assemblage in rocks roughly 700–800 my old from Spitsbergen continues the story. In addition to acritarchs, there are multicellular algae that resemble seaweeds and so might be the first multicellular plants, and new, larger, but still single-celled eukaryotes.

As a group the acritarchs flourished until about 700 my ago when they suffered a mass decline from which the largest, most highly developed forms failed to recover. This crash roughly coincides with the late Proterozoic ice age. I suggested in Section 14.6 that runaway biological productivity, combined with mass burial of carbon, might have brought on the ice age by weakening the green-house protection excessively. Carbon isotope ratios of limestones of this age show an enrichment in ^{13}C that, if the reasoning in Section 10.5 is correct, supports this idea. Burial would not only reduce the atmospheric CO_2 content but also raise the oxygen toward the high early Phanerozoic value (Figure 14.6).

Was it the cold of an almost global ice age that wrote *finis* under most of the acritarchs? Or was it something less obvious, the well-known unhappy impact of too little CO_2 and too much oxygen on the enzymes that drive oxygenic photosynthesis? Again, it is too early to tell.

Between two and one billion years ago then, some major steps in the evolution of life were taken in obscurity. One wonders why they came so late, especially if the eukaryotes indeed descended from the same ancestor that brought forth the much earlier bloom of the prokaryotes (Figure 15.3). What were they waiting for? Oxygen? The right composition of seawater, favorable shallow marine environments, enough prokaryote evolution to produce the correct symbionts? We shall probably not find out easily or soon, because fossil billion-year-old eukaryotes are so much like all other small, shrivelled organic blobs. More, earlier equivalents of the Gunflint would be welcome and may turn up someday. Until then biochemists, biologists and geneticists, rather than geologists, will have the field mostly to themselves.

When we contemplate the Phanerozoic, as we shall do in the final chapters, life seems to dominate the history of the earth. During the Precambrian, the histories of earth and life were more fairly balanced, a minuet, a stately exchange of steps and bows.

PERSPECTIVE

The Precambrian lasted nearly four billion years. Though almost unimaginably long, it does not seem too long for the origin of the earth, the birth of the continents, the formation of oceans and atmosphere, and the first hesitant steps of life. What is astounding is that so much of this, and perhaps life too, had been accomplished before the oldest known rocks formed 3.9 by ago.

The Archean rock record, old, metamorphosed, devoid of usable fossils, is difficult to interpret. The early earth was hot and its lithosphere thin, so it stands to reason that its tectonic processes differed from the present version, but what they might have been is still uncertain. Physics and chemistry play a large part in Precambrian studies, but the discussion in this set of chapters has relied more on what Einstein called thought experiments. Because we have drawn so heavily on the Phanerozoic for parallels, those experiments are also a test of our grasp of the principles of earth history.

Life appeared in the first billion years. Its beginning is pure speculation, but speculation is better than the disconsolate view that it was imported from elsewhere in the universe, a cop-out that merely shifts the problem to a place where we feel no obligation to explain it. Life's early history has an elegant, contrapuntal logic to it. The building blocks of life might not have formed without ultraviolet radiation as an energy source, and would not have survived in the presence of oxygen. Yet as soon as life existed, ultraviolet light became a deadly enemy. Photosynthesis solved that with an ozone shield, but the free oxygen itself became a threat. Once biochemical means had been developed to deal with it, an oxygen-based metabolism developed that has many advantages over one that does not.

Of the many strands that form the history of the early earth, life was the slowest in assuming a major role until, little more than a billion years ago, it blossomed, outpacing every other process on earth in its infinite capacity for change and variety. It is to that world of exploding diversity, and to the time when life took control of the surface of the earth, that we must now turn.

FOR FURTHER READING

The following are accessible to anyone willing to put in a little effort: Briggs, D. E. & Crowther, P.R. (1990). *Palaeobiology: A Synthesis* (Oxford: Blackwell); Cairns-Smith,

A. G. (1990). *Seven Clues to the Origin of Life* (Cambridge: Cambridge University Press); Cloud, P. (1978). *Cosmos, Earth, and Man* (New Haven: Yale University Press); Cowen, R. (1990). *History of Life* (Oxford: Blackwell Scientific); Nisbet, E. G. (1991). *Living Earth: A Short History of Life and its Home* (New York: HarperCollins Academic); Schopf, J. W. (ed.) (1992). *Major Events in the History of Life* (Boston: Jones & Bartlett).

More thorough but more demanding are: Folsome, C. E. (1979). *The Origin of Life: A Warm Little Pond* (San Francisco: W. H. Freeman); Holland, H. D. (1984). *The Chemical Evolution of the Atmosphere and Oceans* (Princeton, NJ: Princeton University Press); Murray, B., Malin, M. C. & Greeley, R. (1981). *Earthlike Planets* (San Francisco: W. H. Freeman).

SPECIAL TOPICS

Baross, J. & Hoffman, S. E. (1985). Submarine hydrothermal vents and associated gradient environments as sites for the origin and evolution of life, *Origins of Life*, **15**, 327–45.

Berner, R. A. (1991). A model for atmospheric CO_2 over Phanerozoic time, *American Journal of Science*, **291**, 339–76.

Berner, R. A. & Canfield, D. E. (1989). A new model for atmospheric oxygen over Phanerozoic time, *American Journal of Science*, **289**, 333–61.

Berner, R. A. & Lasaga, A. C. (1989). Modeling the geochemical carbon cycle, *Scientific American*, **260**, 54–62.

Cairns–Smith, A. G. (1985). The first organisms, *Scientific American*, **252**, 90–100.

Chang, S. (1992). The planetary setting of pre-biotic evolution, *Proceedings, Nobel Symposium on Early Life on Earth*.

Chyba, C. F. (1990). Impact delivery and erosion of planetary oceans in the early solar system, *Nature*, **343**, 129–32.

Clemmey, H. & Badham, N. (1982). Oxygen in the Precambrian atmosphere: An evaluation of the geological evidence, *Geology*, **10**, 141–46.

Conway Morris, S. (1992). The early evolution of life, in *Understanding the Earth*, G. C. Brown, C. J. Hawkesworth & R. C. L. Wilson (eds.), pp. 436–57 (Cambridge: Cambridge University Press).

Davies, D. A. (1992). The emergence of plate tectonics, *Geology*, **20**, 963–66.

Des Marais, D. J., Strauss, H., Summons, R. E. & Hayes, J. M. (1992). Carbon isotope evidence for the stepwise oxidation of the Proterozoic environment, *Nature*, **359**, 605–10.

de Wit, M., Roering, C., Hart, R. J., Armstrong, R. A., de Ronde, C. E. J., Green, R. W. E., Tredoux, M., Peberdy, E. & Hart. R. A. (1992). Formation of an Archean continent, *Nature*, **357**, 553–62.

Edmond, J. M., Measures, C., McDuff, R. E., Chan, L. H., Collier, R., Grant, B. & Gordon, L. I. (1980). Ridge crest hydrothermal activity and the balances of the major and minor elements in the ocean, *Earth and Planetary Science Letters*, **46**, 1–18.

Grotzinger, J. P. & Kasting, J. F. (1993). New constraints on Precambrian ocean composition, *Journal of Geology*, **100**, 235–44.

Holland, H. D. (1990). Atmospheric evolution: Origins of breathable air, *Nature*, **347**, 17.

Horgan, J. (1991). In the beginning . . . , *Scientific American*, **269**, 98–110.

The four-billion-year childhood

Jacob, J. A. (1992). *Deep Interior of the Earth* (London: Chapman & Hall).

Knoll, A. H. (1991). End of the Proterozoic Eon, *Scientific American*, 265, 42–49.

Kröner, A. (1991). Tectonic evolution in the Archaean and Proterozoic, *Tectonophysics*, 187, 393–410.

Kuhn, W. R. & Kasting, J. F. (1983). Effects of increased CO_2 concentrations on surface temperature of early earth, *Nature*, 301, 53–55.

Maynard, J. B., Ritger, S. D. & Sutton, S. J. (1991). Chemistry of sands from the modern Indus River and Archean Witwatersrand Basin: Implications for the composition of the Archean atmosphere, *Geology*, 19, 265–68.

Newsom, H. E. & Sims, K. W. (1991). Core formation during early accretion of the earth, *Science*, 252, 926–34.

Nutman, A. P. & Collerson, K. D. (1991). Very early Archean crustal-accretion complexes in the North Atlantic craton, *Geology*, 19, 791–94.

O'Nions, R. K., Hamilton, P. J. & Evensen, N. M. (1980). The chemical evolution of the earth's mantle, *Scientific American*, 242, 120–33.

Owen, T., Cess, R. D. & Ramanathan, V. (1979). Enhanced greenhouse to compensate for reduced solar luminosity on early earth, *Nature*, 277, 640–41.

Schopf, J. W. (1992). The oldest fossils and what they mean, in *Major Events in the History of Life*, J. W. Schopf (ed.), pp. 29–64 (Boston: Jones & Bartlett).

Stevenson, D. J. (1992). Stalking the magma ocean, *Nature*, 355, 301.

Taylor, S. R. (1990). Not mere scum of the earth, *Nature*, 346, 608.

Taylor, S. R. (1992). The origin of the earth, in *Understanding the Earth*, G. C. Brown, C. J. Hawkesworth & R. C. L. Wilson (eds.), pp. 25–43 (Cambridge: Cambridge University Press).

Walker, J. C. G. (1982). Climatic factors on the Archean earth, *Palaeogeography, Palaeoclimatology, Palaeoecology*, 40, 1–11.

Walker, J. C. G. (1983). Possible limits on the composition of the Archaean ocean, *Nature*, 302, 518-20.

Williams, G. E. (1993). History of the earth's obliquity, *Earth Science Reviews*, 34, 1–46.

Life, time, and change

I will show you time in a handful of life.
                    (with an apology to T. S. Eliot)

Whence and whereto, stranger?
                    Homeric greeting

THE ENDLESS INTERACTION

Two things above all are remarkable about Precambrian life: its enormous duration and its simplicity. The early history of life can be summed up succinctly. Almost four billion years ago came the prokaryotes which were small. Next the eukaryotes appeared; they were not quite so small. Sexual reproduction arrived, followed by the first jellyfish. Six hundred million years remained for the rest of the story.

Late in the Proterozoic, single-celled algae and bacteria became diverse, diverse enough to be useful for stratigraphic purposes. Larger, multicellular organisms arose and evolved in obscurity, until chance finds in beds about 600 my old showed us how rich this multicellular flora and fauna had become. The branches of the tree of life had begun to spread and a mere 100 my later all but a few major categories of life were present. A remarkable progression it was, and a rapid, even explosive one, although defects of the fossil record may account for some of its apparent abruptness.

From then on life evolved steadily toward greater complexity. By the late Paleozoic it had occupied almost every inhabitable corner of the earth, every ecological niche on land and in the sea. Then, spectacularly, it folded, at least in the shallow sea. Expansion followed, until another cluster of extinctions at the end of the Mesozoic. Once more recovery took place, and we find ourselves in the present, on the verge of another catastrophic extinction, this time induced by ourselves.

16

Beyond Darwin

In 1859 Charles Darwin published *The Origin of Species*, a work unique in science for its impact on the spirit of its time and of all times following. Darwin's thesis was startling, although not wholly novel, and it continues to evolve as our understanding of biological mechanisms continues to deepen. It has retained its grip on an overwhelmingly favorable public attention for a century and a half, but the opposition has been as persistent as the concept itself. Ideas of equal importance and sometimes even more directly relevant to the human condition have not always weathered so well.

The theory of evolution is simple, straightforward in its premises, and direct in its consequences. More than a century of biological and paleontological research has added refinement and depth, but has not clouded its basic structure. Virtually all scientists today accept it as true. Lately, some of its aspects have been vigorously debated and new, supplementary ideas have sprouted, but it is a debate of renewal, not rebellion.

Although the concept of evolution touches our lives in many ways, it has been not so much its scientific content that has been troubling as its philosophical consequences. Foremost among the consequences is the realization that, if evolution is really the product of a random selection of variants created by random processes, it is a deeply materialistic concept, leaving no room for loftier views. It is possible to believe that a Creator would create a world ruled by randomness, but for many that is a disconsolate thought. To reject a vital force or a creational design is as difficult emotionally as it is to accept that the evolution from cyanobacteria to humans does not represent progress but merely an increase in complexity. The matter deeply bothered Darwin himself and may well explain in part why he waited twenty years before hc published, and then did so only because A. R. Wallace had arrived independently at the same conclusion.

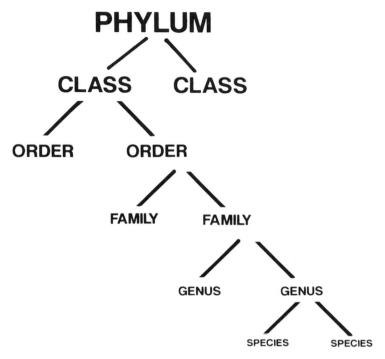

Figure 16.1. The taxonomic system that brings order among the many forms of life is hierarchical, beginning with the most common characteristics for the top categories. Many species are at the bottom of the pyramid and a small number of phyla at the top.

16.1 THE ORGANIZATION OF ORGANISMS

The history of life as we understand it today rests on the labors of many scholars and amateurs since the 16th century and has been known in its essence since the middle of the 19th century. To understand it fully, we need an insight into the classification of organisms, originally developed by the Swede Carolus Linnaeus in the middle of the 18th century.

In this classification (Figure 16.1) the fundamental unit is the species; the members of a species resemble each other closely and can interbreed. Similar species are grouped together in a genus. Each organism bears first the name of its genus, then that of its species: *Homo sapiens*, as contrasted with our extinct ancestor *Homo erectus*. Related genera are combined into families (*Hominidae*); families combine to orders (*Primates*); and orders to classes (*Mammalia*). Living organisms can be classified with the aid of many more characteristics than fossils which are usually known only by their hard parts: a shell, bones, some teeth. The science (or is it an art?) of classifying life forms is called taxonomy, its units are taxa (sing. taxon).

Similarity of form, however, is a treacherous guide, no matter how comprehensive the description may be. Similarity in adaptation, for instance, may produce a close similarity of appearance by convergence, a process that caused five anteaters, who evolved independently on four separate continents, to be very much alike. Within some species there is great variability, while others, possessing diagnostic soft bodies, have indistinguishable shells. Therefore, the formal biological criterion for a species is that its members are able to breed under natural conditions and to produce fertile offspring, but cannot crossbreed with other species. This is straightforward but obviously of little use in paleontology. There are exceptions, and asexually reproducing organisms present a problem. And so, though imperfect, the morphological classification still serves.

Life as we know it began with the prokaryotes, divided into Eubacteria and the Archaebacteria which spawned the eukaryotes. Another divide separates the single-celled from the multi-celled eukaryotes. Among the latter, called for no good reason "higher" plants and animals, we distinguish those that make organic matter by photosynthesis (plants) from the fungi who reduce organic matter to its constituents, and from those that eat plants, fungi or each other (animals). Six kingdoms make up the world of life in all its riches.

In principle, the number of ways in which a living organism might be designed is large, but in practice nature has adopted only a limited selection of the many possible plans. Instead of Kipling's "nine and sixty ways to construct tribal lays," less than half as many designs describe all of life (Figure 16.2). Jellyfish and corals, dissimilar as they seem, are constructed on the same ground plan which is quite different from that of the mollusks (clams and snails) or the echinoderms (starfish and sea-urchins). The basic architectural plans, often best displayed in the embryo, divide the kingdoms of the fungi, higher plants, and animals into phyla (sing. phylum, e.g. Vertebrates).

A useful concept, to be applied frequently later, is that of diversity, the number of species, genera, families, as the case may be, for a given time, place, or condition, for the shallow Ordovician sea, for the whole of the Permian, or for coral reefs.

Morphological similarity can usually, although by no means always, be interpreted as an ancestor–descendant, a phylogenetic relationship. Species of the same genus are thought of as having descended from

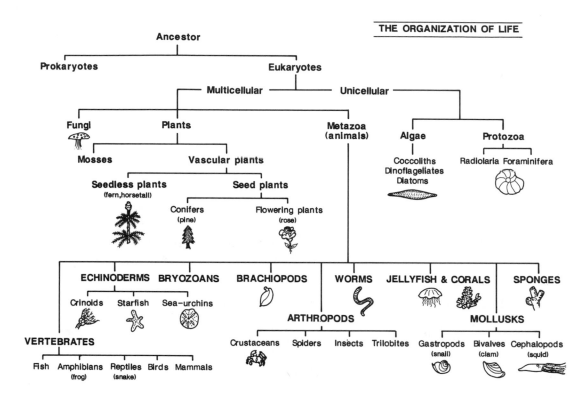

Figure 16.2. The variety of life is far too great to capture it in a single diagram, but this abbreviated one arranges and illustrates at least the most important groups that have a role in this and the following chapters.

the same predecessor, and the taxonomic hierarchy is analogous to a phylogenetic tree connecting progeny to its ancestors. At the level of species and genera, the fossil record and therefore the details of the tree are woefully incomplete. In practice we often display and discuss evolution and the history of life in terms of families and higher taxa.

Making family trees that arrange species, genera, families and orders according to ancestor–descendant relationships or, more precisely, to their degree of similarity, has been a beloved pastime of paleontologists since Darwin. Chapter 17 provides many examples of this art. Recently, this effort has shifted to the new field of molecular genetics which plays an analogous game, comparing taxa by means of similarities and dissimilarities in their biochemistry (Figure 16.3).

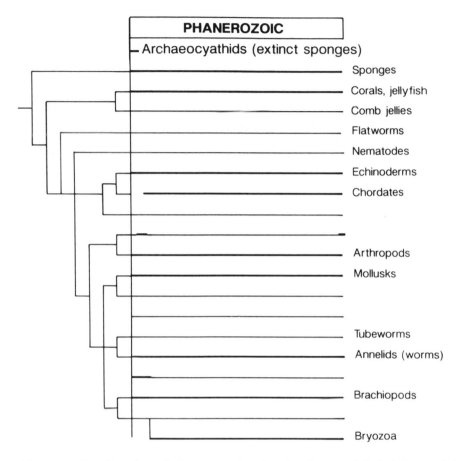

Figure 16.3. Since Darwin, evolutionary trees have been based on morphological characteristics. Molecular genetics plays the same game, but establishes relations between groups of organisms on the basis of similarities and dissimilarities in some of their biochemical compounds. Both kinds of information have been used to compile this tree of the major categories of life forms. The heavy lines indicate an essentially continuous fossil record, thin ones indicate a relationship for which there is no fossil confirmation. Many groups, of which the diagram shows only the archaeocyathids as an example, did not make it beyond the Paleozoic. A few less well known groups have not been labeled.

16.2 PRINCIPLES

Among scientific theories, certainly among those dealing with complex issues, Darwin's theory is a simple one, consisting of two undeniable facts and an inescapable conclusion. The two facts are that organisms vary, and that part of the variation is inherited by their offspring, offspring far too numerous to survive in their entirety. The conclusion is that descendants who vary toward a greater compatibility with their

Life, time, and change

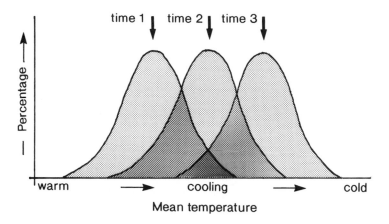

Figure 16.4. The environment selects from a set of individuals of varied characteristics those that best fit the conditions. Many individuals are able to survive at a given mean temperature, although some would prefer it warmer whereas others could tolerate colder conditions. The vertical axis indicates the degree of adaptation, for example expressed by the thickness of fur, ranging from poor to excellent. If the mean temperature drops to a new value (thin arrows), those who have the thicker pelt are better suited to the new conditions and become the dominant variety, whereas others at the warm end of the range, whose coat is too thin, do not survive. A new bell-shaped distribution of adaptations forms around the new best adapted group.

environment will be less likely to die (Figure 16.4). Natural selection causes better-adapted variants to accumulate and so will steer life in that direction.

This says no more than that natural selection weeds out the less fit. To see it as a molding, shaping force as Darwin did, it is also necessary that the variations arise without pattern, just as likely to be of negative as of positive adaptive value. If that were not so, a process internal to the organism would be guiding it to improved adaptation and natural selection would merely discard the failures.

It follows that evolution has no purpose, and no consequence other than to ensure survival of those individuals best suited to their environment or, when the environment is changing, best able to respond appropriately to the change. There is no long-term plan, and any apparent trend beyond the one imposed by natural selection is an illusion.

What lies behind the variability of organisms? Mutations, for one thing: small, sudden changes at specific points in the genetic material. Other variants are due to genetic recombination when half of one parent's genes join half of those of the other. Finally, a small population can lose genes from its pool by accident in a process called genetic drift that is purely random. These are but the most common means by which a menu of variations is put together for nature to choose from.

Most changes so produced are small. They may thicken a pelt when the climate turns cold, or cause brown eyes to become extinct. Is that all there is to it? Must we accept that by infinitely small, random steps the light-sensing cell of some early Metazoa evolved into the human eye?

There are two separate issues here. First, there is the question whether or not changes exist that are not adaptive until they have been fully developed. If the evolution of some complex organ requires many small steps of no immediate use before it can function at all, how can we accept that each of those steps was tested by nature on its adaptive value? Apart from the fact that it has been difficult to point to actual examples of non-adaptive evolution, there are two possible answers to this question.

Usually, it is not a single gene that causes change, but an entire array. A non-adaptive change, or a not-yet-adaptive one, might so ride pig-gyback on an adaptive change: as it gets colder, the pelt thickens, turning lighter in color also. Alternatively, an organ may evolve toward one purpose, then be readapted to a different one. The jaw of the early fishes was fashioned quickly and conveniently from a bone arch designed to support the gills.

The second issue is that of speciation. How does one species evolve into another? Will the selective breeding of dogs some day produce a new species in the genus *Canis*? If that is so, when and how does cross-breeding stop, how does the flow of genes become interrupted to satisfy the rule that only members of the same species may interbreed? The simplest way to accomplish this is to place two populations of the same species in a situation which physically keeps them from mating. The genetic isolation will inexorably cause the two populations to drift ge-netically apart until the differences are such that crossbreeding is no longer possible. The divergence may be the effect of environmental se-lection, of a chain of random mutations, or of genetic drift.

Isolation of populations can be achieved in many ways, but most easily by geography, by migration to an island, by being cut off by a desert or a mountain range or, more spectacularly, by continental breakup and drift. If the isolated population is small, genetic change is rapid and can speedily lead to a large divergence from the parent population. Darwin's finches on the Galapagos Islands, where many species quickly evolved from a few migrants, due to geographic isolation and a diversification of ecological niches, are the classic example of the evolutionary conse-quences of isolation.

Are there means to bring about major changes other than by isolating groups of organisms? The major work of biochemical evolution was completed long ago, and today the possibility is remote that a chance variation might produce a new protein or enzyme of fundamental value.

Now that the days when the basic blocks were created are behind us, evolution at the primary level mainly reshuffles the deck, alters gene patterns, or changes the timing of events in the development of an organism. Herein lies one opportunity for occasional major change. Many genes specify activities, such as the manufacture of an enzyme needed to start the fertilized egg on its way to the mature individual. Others, as regulators, see to it that these actions happen in the right order and at the proper time. A change in a regulatory gene will not introduce a new chemical or a new process, but it may start the activity at a new moment, delaying one, prolonging another. A good part of the evolution from Miocene primates to modern humans had to do with a gradual delay of the maturation process. The large human skull permits a bigger brain, but it passes through the birth canal only with difficulty. Retard the growth of the skull and this problem is solved.

Major evolutionary steps are a different matter. Intuitively one senses that most large, random changes, most postulated "genetic jumps," are unlikely to be beneficial; there is little chance that these "hopeful monsters" will survive, although once in a while one just might.

16.3 ELABORATIONS

Historically, the search for mechanisms that could accomplish change in big leaps has failed, and it might be wise to abandon the chase. The compelling reason why the issue refuses to die is that we are quite able to document minor gradual evolution, but are embarrassingly short of links between higher categories. Thus, at nearly every major branching of the tree we are forced to interpolate (Chapter 17). Darwin was well aware of this lack of paleontological support for his uniformitarian view that evolution was slow and gradual, but he felt inclined to blame the incompleteness of the geological record. We still do so today, but add that paleontologists, bent on defining species, tend to assign intermediate forms to one kind or the other. This reduces the opportunity to recognize transitional types. Still, there is no doubt that the foundation of the theory of evolution lies in biology and genetics, not in paleontology and geology.

Does this mean that there is no role for the science of the earth in the study of evolution? It does not; the labors of paleontologists, whether they work in taxonomy, in ecology, or in stratigraphy, are a cornerstone of evolutionary theory, but less in documenting its mechanisms than in underpinning its consequences. It is paleontology that best illustrates such important phenomena as adaptive radiation where the creation of new opportunities suddenly stimulates the development of many species, each having a different way of taking advantage of them. A case in point is the explosive increase in diversity near the Precambrian/Cambrian boundary (Sections 17.1 and 18.2). The waxing and waning of whole lineages and the processes that are responsible for it are also the domain of the historian of life, the paleontologist. And while we would wish for many more major transitions, the few we have, such as *Archaeopteryx*, undoubtedly an early bird but with strong links to the reptiles, are of enormous importance.

Scientists, like everyone else, learn to live within the limits of their opportunities and except for a few, inside the prisons of their theories. The 20th century has clarified and confirmed the mechanisms of evolution so strikingly and in so many ways that anomalous observations have stayed in the shadows. However, discomfort, even if only intuitive, cannot be repressed forever, and a wave of questioning regarding the validity of a slow, gradual evolution arose two decades ago.

An unbiased look at the fossil record shows that it is quite compatible with a model of evolution that consists of long times of little change, interrupted by divergences so sudden and brief that they might almost be called instantaneous. To a small group of challengers of the establishment, the branches of the phylogenetic tree are horizontal and vertical rather than slanting (Figure 16.5). They believe that this pattern of rapid change and stagnation is real and not due to a defective record, and that long periods of still-stand cannot not be explained by Darwinian evolution. This "punctuated equilibrium" has provoked considerable discussion, partly because of the problem of finding suitable genetic mechanisms, and partly because instantaneous large jumps are so difficult to demonstrate in the fossil record.

The debate, to be sure, was never about natural selection, random variation, or evolution itself at all, and it does not necessarily involve the mechanisms of speciation but only their rates. What might be responsible for the horizontal branching of the evolutionary tree? The most obvious candidate is geographic isolation; I have already alluded to its power. By virtue of the speed of this process and the small size of the area in which it happens, such events might be undetectable in the fossil record, and the resulting changes would seem instantaneous. Another way would be a genetic jump, but we are far from sure that genetic jumps are even possible, let alone exist.

Morphological stagnation, stasis, is real; the horseshoe crab has remained unchanged since the Paleozoic, and so have several species of brachiopods. But as real as stasis is, an understanding of the reasons for

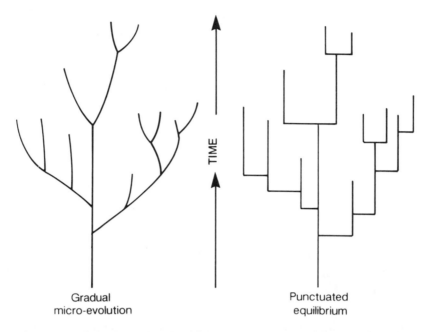

Figure 16.5. The traditional concept of evolution assumes that species gradually diverge from each other, and that change is continuous with the passing of time. The evolutionary tree has sloping branches. Proponents of punctuated equilibrium, on the other hand, regard the formation of a species as a large, instantaneous step followed by stagnation.

stagnation between major steps has proved as elusive as evidence for genetic jumps, and only limited success has been achieved. Gradual small-scale evolution, on the other hand, is common in the record when it is studied in sufficient detail. So some species stagnate and others do not, and we do not know why.

A problem of definition bedevils the discussion. What is instantaneous to the geologist, say 10,000 or even 100,000 years, is infinite time to the geneticist. Speciation can be very rapid. Lake Victoria and other East African rift lakes harbor no fewer than 200 species of small fishes known as cichlids, grouped into many genera, all evolved over the last 200,000 years, and almost all of them endemic. In one lake in Uganda, six new cichlid species have appeared in the last 4,000 years alone. Transfer such rates into the geological record and it is no longer surprising that the difference between the two evolutionary trees of Figure 16.5 is often impossible to demonstrate geologically.

A few proponents of punctuated equilibrium, rationally following through on the idea, have proposed that nature selects for fitness not just the individual but the entire species. If a new species, suddenly evolved, represents a viable adaptation to its environment, it may remain unchanged until the environment itself changes. Then, unable to adapt further, it becomes extinct and its place is taken by another, more adaptable one. As a result, lineages with high rates of speciation would have a clear adaptive advantage over those that stagnate. Any jump, of course, large or small, must be equally likely to be adaptive as counter-adaptive. Otherwise, it would not be nature that selects, but again an internal force steering change in a specific direction.

Lately, the debate has died down, and from the sea full of red herrings and floating strawmen not enough real substance has emerged to change our way of thinking. My personal preference in this matter, and probably that of most geologists, tends toward varying rates of evolution rather than genetic jumps and natural selection by species. But the paleontological record is likely to remain as ambiguous as before.

16.4   NEW APPROACHES

If punctuated equilibrium has turned out to be somewhat of a storm in a teacup in evolutionary terms, are there other new views and approaches worth considering?

Describing the fossil record has always been an intuitive process. It rests on the selection of type specimens for each species, chosen on the basis of criteria devised and judged valid by the investigator. Once type specimens have been selected, all fossils must be assigned to one or another of those. It is clear that, even with greatest care, this approach is biased against intermediate forms. The obvious alternative, which originated in the study of living organisms, is to begin with an unbiased and quantitative description, based on as many criteria as practical, of as many specimens as available. The descriptions can then be sorted by statistical techniques into a hierarchy of units based on diminishing degrees of similarity.

This quantitative form of the Linnaean tradition of defining species, called cladistics, derives its strength from the lack of preconceived notions of what a species should be like, or even what the relevant diagnostic criteria might be. Still, it is not possible to describe anything, no matter what, in its entirety, and unavoidable choices may still introduce bias. The method is also much better suited to the study of living organisms than of fossils, because living organisms afford access to far more and more crucial features. Not surprisingly, cladistic theorists have claimed that the fossil record has no contribution to make to evolutionary history, but this is untrue; it tells its own, often different, but equally valuable story.

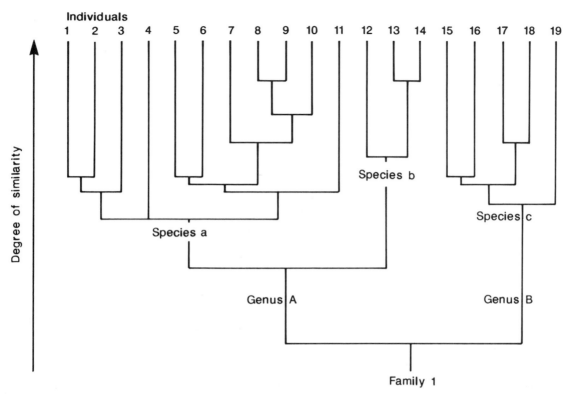

Figure 16.6. If we describe, preferably in numerical form, all (or at least very many) characteristics of a large set of individuals, we can use statistical methods to determine how similar or dissimilar each possible pair is. In the resulting hierarchy of similarity, clusters of very similar individuals may be labeled species, and at the next lower level of similarity, species can be combined to genera. The levels of similarity that separate individuals from species and species from genera are sometimes self-evident and sometimes have to be chosen arbitrarily. Species so defined do not necessarily obey the usual definition that members should be able to breed only within the limits of the species.

The products of the cladistic approach have been more rigorous and hence more easily defended than the family trees of the traditional way of dealing with classification. They have also been much admired, because whatever requires computation seems more solid than what does not. Its insights have not so far fundamentally reordered our views on most evolutionary trees, and it is a matter of judgment whether the similarity–dissimilarity trees so constructed (Figure 16.6) are genetically more valid than those obtained the old-fashioned way.

In the last two decades studies in molecular biology that relate genes to the development of the visible characteristics of the organism have begun to promise new, possibly dramatic changes in the theory of evolution. They have shown that the relations between an organism and its genes are far more complicated than was once thought, but hold out the

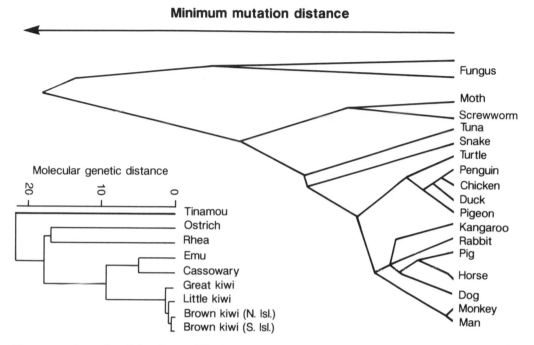

Minimum mutation distance

Molecular genetic distance

20 10 0

Tinamou
Ostrich
Rhea
Emu
Cassowary
Great kiwi
Little kiwi
Brown kiwi (N. Isl.)
Brown kiwi (S. Isl.)

Fungus
Moth
Screwworm
Tuna
Snake
Turtle
Penguin
Chicken
Duck
Pigeon
Kangaroo
Rabbit
Pig
Horse
Dog
Monkey
Man

Figure 16.7. Organisms belonging to different species or genera have different substitutions in the amino acid chains that make up their proteins. From the number of such substitutions the degree of relationship can be determined. Horse and pig are more closely related to each other than to rabbits, and all three together branched from the main stem at about the same time as the ancestors of monkey and man. These similarity relationships can be carried to fine detail as the graph for the flightless birds shows.

prospect of an understanding of the processes that underlie the work of evolution.

We need not pursue these new developments further here, but the same discipline has also contributed a new way to clarify the genetic and evolutionary record. It too is based on quantitative descriptions, but the properties described here are biochemical. The method, briefly alluded to in Figure 15.3 in connection with the ancestry of prokaryotes and eukaryotes, is known as molecular genetics. Over time, small changes accumulate in the amino acid chains of proteins. When species diverge, so do their amino acid chains, the degree of diversion being a measure of their affinity to each other and to the ancestral species.

The sequencing of amino acids is controlled by components of DNA, and the number of DNA substitutions necessary to produce the observed changes can be determined. The more substitutions are implied, the greater the genetic distance (Figure 16.7). Rather astonishingly, very refined analytical techniques have shown that some DNA has survived in fossil material, such as the bones of prehistoric human beings, the leaves of middle Cenozoic plants, or insects trapped in amber. This has opened an entirely new door to the past, because we may thus directly determine evolutionary relationships; what will come of it cannot yet be foreseen.

We can make an estimate of phylogenetic distance based on several proteins; the correspondence is usually good and enhances our confidence in the method. Similar experiments can be done on DNA directly. Living organisms have been the base of most of the information available today, but we can compare the results with the fossil record and so connect them with extinct groups.

It is important to realize that the substitutions on which molecular genetics rest are not related to the morphological changes that form the basis of traditional phylogenetic diagrams. They do not usually have evolutionary consequences, do not constitute a base for selection of the fittest, and their rates are not the same as the rate of evolution of the whole organism.

A vivid example of this independence is offered by the East African cichlid fish we have already met. The 200 species evolved so recently in Lake Victoria are a splendid example of adaptive radiation. They differ little in appearance, but the adaptations to the environment are incredibly diverse. Bottom-grazers, filter- and plankton-feeders, and fish predators, including unpleasant types specializing in scraping scales off their fellow fish, plucking out their eyes, or slurping the young out of mouth-brooding females of their own kind, present as fine an example of the exploitation of all possible ecological niches as one might find anywhere. Yet the DNA contained in their mitochondria varies less than it does within a single species of horseshoe crab which has hardly evolved at all over more than two hundred million years. Clearly, the various kinds of evidence tell us different things about evolution.

Having established phylogenetic trees by this means, what can we say about the time dimension of molecular evolution? We might obviously call once more on the fossil record, but in view of its shortcomings it would be most convenient to have an independent way to date the branches of the tree.

Such a way seems to exist, but it is startling and has met with some skepticism. It rests on the claim that molecular evolution, the accumulation with time of small changes in amino acid sequences, proceeds at a steady rate and can be used as a clock. The clock does not tick at fixed times but, like clocks based on radioactive decay, a certain number of substitutions can be expected to occur within a given time. Counting

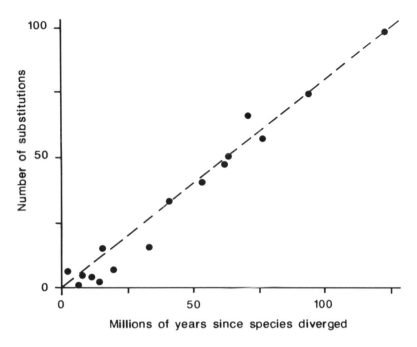

Figure 16.8. Molecular genetics is based on the number of amino acid substitutions in organic compounds such as hemoglobin or DNA. This number, determined on modern organisms, depends on the time elapsed since two species diverged. The divergence time can also be estimated from the fossil record. Notwithstanding the defects of the geological data, a close correlation with the degree of biochemical affinity has emerged. The graph provides encouraging evidence for the usefulness of the molecular clock.

amino acid substitutions is analogous to counting decaying carbon-14 atoms.

It is not self-evident that amino acid substitution must behave in such a controlled manner. There is plenty of evidence that morphological evolution does not proceed at an even rate, and no reason why molecular evolution should do so, or at least not always. The ticking of the molecular clock has not yet been demonstrated to everyone's satisfaction, but a fair array of data now exists that agrees reasonably well with the paleontological evidence. Geologically determined branching points for several taxa (Figure 16.8) correlate decently with their molecular "distances," although we should remember that the chronology of the fossil record is also often beset by uncertainties. There is good reason to believe that the clock runs, but it is possible that it is not highly accurate.

Quantitative classification of fossils and the study of molecular evolution will in another decade or so greatly clarify the structure of the phylogenetic tree. Whether or not this will impose limits on the swirling debate concerning the evolutionary records of specific groups of organisms and on the mechanisms by which evolution proceeds is a different matter.

17

Bones of our ancestors

We are interested here in the place of life in the history of the earth more than in the history of life itself. For this, we need, besides an elementary understanding of the organization of life and of the theory of evolution, a brief account of the course of evolution through time. Armed with that we can then turn to life's conquest of the sea, the land and the air (Chapter 18), and the sudden expansions and large extinctions (Chapter 19) that raise many questions. These questions range from the scientific to the philosophical and metaphysical and have a large impact on our view on the history of our planet.

17.1 EDIACARA!

Late in the Proterozoic, during the Vendian period, an accident of fossilization preserved a rich fauna of soft-bodied organisms in the Ediacara beds of Australia and elsewhere. The Ediacara fauna consists of the first large, multi-celled fossil animals (Metazoa). They were already built on several different ground plans and bear witness to a rapid diversification just before the Phanerozoic began. The Vendian, dated between 610 and 570 my according to the time scale adopted for Figure 2.4, is probably somewhat younger and the beginning of the Cambrian therefore that much later (Figure 17.1).

The imprints of the Ediacaran bodies are not their only remains. Metazoa crawl, walk about, or burrow, and so leave tracks and holes that are often more easily preserved than the animals themselves. In many places these first tracks and burrows appear at about the same time as the Ediacaran fossil (*c.* 575–555 my ago), and in the same order: crawlers first, diggers next. This might mean that the first higher organisms did not come into being much before 600 my ago, long after the eukaryotes appeared perhaps a billion years earlier. Alternatively, the Edi-

Life, time, and change

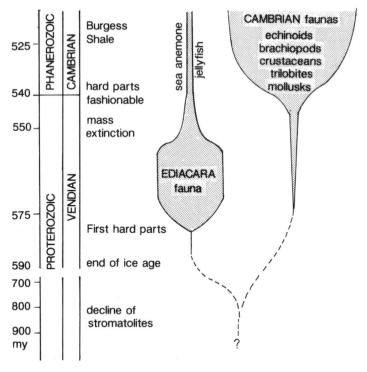

Figure 17.1. *A great increase in the diversity of the higher animals marks the end of the Proterozoic and the beginning of the Phanerozoic, but its details are still controversial. Above is Simon Conway Morris's view (late in 1992) on how this "explosion" might have run its course. It presents a picture wholly different from what was available for the first edition of this book. Note how the chronology, which starts the Cambrian 30 my later, differs from the one adopted in Figure 2.4. Such things are signs of progress and should not disturb us.*

acarans themselves may have descended from as yet undiscovered soft-bodied, multicellular ancestors too small to burrow or leave crawl marks, that fossilized only under special conditions. There is, in fact, evidence for a diversification of eukaryotes and perhaps the first Metazoa around 900 my ago.

The Ediacara faunas inhabited shallow marine environments, a few are found in limestones, and in Newfoundland they are also preserved in volcanic ash layers deposited in deep water. All are soft-bodied, but even so often superbly preserved.

As a whole, the Ediacarans distinguish themselves by their greater size from the Metazoa that followed them. Their strangeness has led Adolf Seilacher of the University of Tübingen in Germany and some others to suggest that they differed fundamentally from any life the earth knows today. "Aliens," he called them, not even to be regarded as animals, an experiment in using a maximum surface area rather than internal transport to deal with the food, gases and fluids needed for subsistence. The body shape of many Ediacarans, flat and with a large surface area, does indeed suggest an attempt to maximize the surface to absorb something from the environment, but what: sunlight?, solutions?, gases? Might some have housed internal colonies of symbiotic algae or bacteria, as the tube-worms (pogonopherans) of deep-sea hotsprings do with symbiotic hydrogen-sulfide oxidizing bacteria? If he is right, the experiment failed.

Now many Ediacarans have been recognized as closely related to the cnidarians which include jellyfish and sea anemones. There are also worms of one kind or another, as well as the possible forerunners of other groups of marine animals. There is even a shelled form, *Cloudina*, with calcareous tubes that foreshadow the hard shells and skeletons of the Phanerozoic. Finally, borings in shells are suspected to have been made by predators.

As a life community, the Ediacarans were very different from what we see today. Many probably grazed on the extensive mats of photo-synthesizing algae and bacteria that formed the bulk of the biomass, and others may have been suspension-feeders. Predators are not in evidence, except possibly for the putative cnidarians, and the food chain was thus short.

In the Cambrian, organisms settled on a limited number of body plans, and to the best of our knowledge no new ones have been added since then. Why nature's ingenuity should have been so vigorously exercised between *c.* 500 and 600 my ago and not ever again, remains a mystery. In any case, there are only three major groups, the radially symmetrical jellyfish and sea anemones (the cnidarians), built from two layers of tissue, the flatworms with three layers and a bilateral symmetry, and the coelo-mates which sport three layers and have a central cavity. The coelomates are by far the largest group and include the diverse body plans of annelids (worms), echinoderms (starfish, sea-urchins), arthropods (insects, crus-taceans), the mollusks, and the latecomers of the lot, the vertebrates (Figures 16.2 and 16.3). Simon Conway Morris of the University of Cambridge has just recently given us one possible view among the many competing ideas of how it all might hang together in evolutionary terms (Figure 17.1).

Whatever the position of the Ediacara fauna may have been, it became extinct, the algal mats of the Ediacara garden, maybe devastated by specialized grazers, diminished, and the now bare mud surfaces were occupied by a new set of bottom-dwellers. In the early years of the Cambrian, the fauna included a suite of tiny shelled animals, some of unknown affiliation and others the first mollusks, as well as archaeocy-athids, reef-building sponges. To our benefit, the new habit of many animals of constructing hard body parts of calcite or iron and phosphate minerals greatly enhanced the geological visibility of what was happening.

Ten million years passed and an immensely diverse life suddenly blossomed. The Cambrian explosion, as it is sometimes called, included in profuse variety the superficially clam-like brachiopods, today minor denizens of the sea, the echinoderms, and above all the now extinct trilobites. The trilobites were early arthropods, a group that also includes the crustaceans, insects and spiders, all eventually very successful animals.

17.2   ONLY IN THE SEA

The sea is life's first environment and its conquest was complete before plants and animals moved onto the land. In the sea, life occupies two distinct, partly separate realms: the upper hundred sun-lit meters of the ocean where floating plankton and pelagic swimmers live, and the bottom, densely inhabited only in shallow depths. Both zones were occupied early; we know of both plankton and benthos in Proterozoic nearshore deposits.

For much longer we remain poorly informed about life in the open ocean, and even more poorly about that of the deep-sea. In the late Proterozoic and throughout the Paleozoic, the planktonic flora of the open sea is mainly represented by the acritarchs, perhaps precursors of the dinoflagellates that are still major and ubiquitous members of the pelagic plankton. There were also single-celled grazers and predators such as the radiolarians, but whether the pelagic realm included anything like the diverse and numerous community that lives there now we do not know. Only in the Mesozoic do we begin to see details, and we see them for the same reason that life in shallow seas abruptly emerged with such clarity in the Cambrian. Quite suddenly, some 175 my ago, many planktonic organisms, plants and animals alike, adopted hard shells and began to proliferate enormously. From then on, a detailed record of pelagic life has been kept in the oozes of the deep-ocean floor which are composed almost entirely of the shells and skeletons of microscopic plankton. We fail to understand the cause of this sudden blossoming and wonder if it means that the ocean had been but sparsely inhabited before. And if so, why?

The fossil record more clearly displays the conquest of the shallow seas that began with the first stromatolites. It took a leap forward 600 my ago with the Ediacaran fauna, followed by the diversity explosion of the Cambrian that brought onto the stage nearly every major category of shallow marine organisms. The exception are the vertebrates, unless a fossil named *Pikaia* is a forerunner of those (see Figure 17.4). Recorded in great detail by their hard skeletons and occasionally by lucky preservation, the new organisms became skilled exploiters of all the opportunities of their environment. They lived on mud, sand and rock, floated in mid-water or at the surface, burrowed, crawled, were mud-eaters, filter-feeders and scavengers. Predators were now present too, as clear traces of attack on some fossils show, but the time of major hunting and killing only came late in the Cambrian, when the earliest cephalopods, cousins of the squid and the octopus, adopted that life style.

Notwithstanding the many adaptations, the scene is dominated by two major groups that are typical for the early Paleozoic: the trilobites and the brachiopods. Both diversified in an amazing number of directions; we find trilobites and brachiopods in just about every role and every environment of the shallow sea, from filter-feeding to burrowing, from rock to sand.

We have encountered before accidents of superb preservation that suddenly show us how limited our view of the richness of life has been up to that point. The Gunflint (Section 15.6) was one of them, the Ediacara another, and the mid-Cambrian Burgess Shale in Canada tops them all. Everything is in the Burgess Shale that one would normally expect for the Cambrian, but detailed study by H. B. Whittington, Simon Conway Morris and D. E. Briggs, all then at the University of Cambridge, revealed in addition an astonishing array of perfectly preserved soft-bodied animals (Figure 17.2), including a variety of worms, arthropods, sponges, brachiopods, and some bizarre animals that seemed unrelated to any known groups. Since then the affinities of many of those to known animal lineages have come to light, but others may have been experiments in diversification that failed. Still others, such as *Pikaia*, if indeed a precursor of the vertebrates and hence our ancestor, had a great (and perhaps regrettable?) future ahead. Well-preserved faunas similar to the Burgess Shale have since come to light from other facies in shallow and also deep water, of which some, from Greenland and China, are as old as the early Cambrian.

The Burgess Shale fauna is not the only Cambrian story of mixed success and failure. Other lineages, although mostly of lower rank, perished before the end of the Cambrian (Figure 17.3), making it a time of experimentation, of trial and error. The Ordovician brought consolidation; the fundamental patterns were set and have served ever since, the failures had vanished and the successes took over the stage. Although a few other major groups did not make it to the present day (Figure

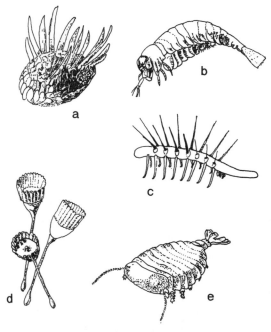

Figure 17.2. In Canada, the middle Cambrian Burgess Shale has, besides the usual trilobites and brachiopods, a great variety of soft-bodied animals preserved in fine detail including sponges, arthropods, mollusks, and bizarre forms not yet assigned to any known group. Here is a selection: (a) Wiwaxia, probably related to the mollusks; (b) Yohoia, looking a bit like a cross between a praying mantis and a shrimp; (c) Hallucigenia, at first a mystery, now regarded as a relative of modern velvet worms; (d) a stalked animal called Dinomischus; and (e) Sidneya, an arthropod almost surely in possession of eyes. Names they all have, but what are they?

17.3), expansion has been chiefly at the level of family, genus and species. It should be remembered, however, that this may in part be an artifact of the way fossils are pigeon-holed.

Among the richest and most interesting of all shallow water environments are reefs, the hard limestone banks and ridges constructed by algae, sponges, or corals growing upward from the seafloor, or by other hard-shelled organisms where they leave mounds of skeletal debris. Reefs have been around since the Precambrian, when calcareous masses produced by the metabolism of cyanobacteria and algae were common, but they truly came into their own in the Cambrian and, with few interruptions, have been a major part of the geological record ever since. Corals existed in the Cambrian, but the Cambrian reefs were constructed by calcareous sponges, the archaeocyathids, and by lime deposited by cyanobacteria.

The experiment failed, and the world had to do without coral reefs until the middle Ordovician. The Ordovician reefs were large and splendid with a fascinating fauna, and their limestones are some of the most

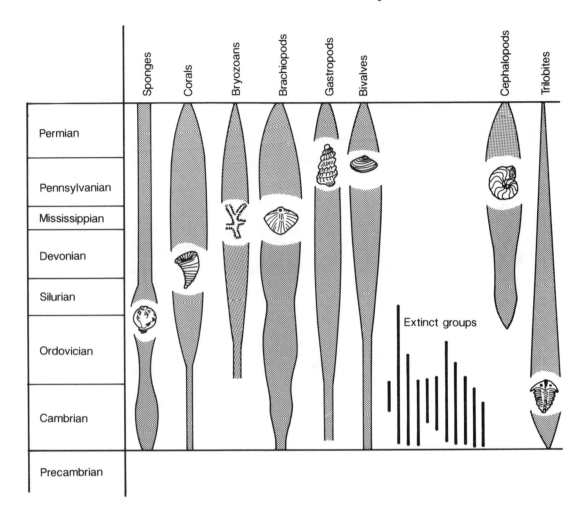

Figure 17.3. Very early in the Paleozoic, all main groups of marine invertebrates had made their appearance. There was a good deal of experimentation, some of it unsuccessful as the many now extinct groups show. The shaded columns in this and following similar diagrams depict the change with time of the abundance or diversity of each group.

durable sedimentary rocks, but they were not meant to continue any more than the reefs of the Cambrian. The Devonian saw the end of them in conjunction with a widespread but otherwise relatively minor crisis of marine life. It took 15 million years for another try, only to have that one fail during the largest marine catastrophe of all time: the great extinction of the late Permian (Section 19.2). Obviously, reefs have merit in this world, however, because new ones, this time mainly built by corals, came to flourish spectacularly in Mesozoic seas, especially in the Tethys between Africa and Laurasia. The story is becoming monotonous:

337

another crisis at the end of the Cretaceous, another recovery, and we come to the reefs of today, once again some of the most beautiful scenery the sea has to offer, and again in danger.

This pattern of expansion and contraction, of success and failure, shows us life's vulnerability, but also its recuperative powers, the stubborn return of a type well suited to its setting, in this case that of clear, warm, coastal waters.

17.3 ACROSS THE SHORE AND INTO THE HILLS

It was a momentous instant in the history of the earth when life crept out of the sea and onto the land. The move was far from easy, because there are many advantages to life in the sea. The sea cradles: there is no need for supporting stems or skeletons, unless one moves about, digs in the bottom, or lives attached to rocks in rough water. The sea protects: against the drying of tissue, sperm, eggs, or embryos, against dangerous radiation, and even against enemies. The sea furnishes water and nutrients in a most convenient way, and sperm and egg can simply be set adrift to find each other. Lakes and rivers are much like the sea, with one important exception: their salt content, so much lower than that of body fluids, makes protection against the loss of salt or the uptake of too much water indispensable for anyone wishing to survive there. An impermeable skin and some kind of kidney to excrete salt are needed. Migration from sea to land by way of freshwater, a useful route in some ways, thus brings its own imperatives.

Plants, in one form or another, must have been the first colonizers of the new space, although perhaps not by very much, because they provide the food. The animals who followed them onto land encountered similar problems, with the need for locomotion, staying upright, and seeing or smelling thrown in for good measure. Many invertebrate groups, although by no means all, successfully transferred from sea to freshwater, but only a few, some worms, the gastropods (snails), and the arthropods, above all the insects, took the next step to dry land. Insects are by far the most successful adaptation to land life ever, accounting for about three-fourths of all animal species existing in the world today, and they will surely be the last group to become seriously threatened by our brutish attempts to rearrange the world to suit ourselves.

The first animals to settle on land were probably small arthropods like the millipedes, already partly prepared for the transition by having a waterproof cover and legs to walk on, but we have no evidence for them until considerably later. Offhand, that seems odd; with the plants available, some freshwater animals, already protected by an impervious skin or some other means against loss of water, ought to have made it sooner. Blame the inadequacy of the record! In fact, although there is no proof, it is likely that a fauna much like the one seen on beaches today survived on animal and plant debris thrown ashore by the waves before there was any growth on land (Section 18.5).

Some animals took advantage of adaptations that had served them well in the water, such as the snail shell which can be closed tight against the low tide and is useful on a dry day on land too, or the impervious chitinous armor of crab or lobster, cousins of the insects. Most innovative were the vertebrates who got an infinity of ingenious uses out of a skeleton. Originally only intended to anchor muscles or protect the body, skeletons have been found equally useful to provide support, to walk and grab with, to chew food, to house a brain, or to play the piano.

When, late in the Cambrian, we encounter the vertebrates, the last phylum to come on stage, they had already separated fully from their still mysterious and vigorously debated ancestors. Although we usually assume that it is the backbone that distinguishes vertebrates from invertebrates, that it not so. The real key is the spinal chord or notochord, the bundle of nerves that lies within the vertebral column. Almost certainly the notochord came first, and even today two groups of animals make do with no more than that, the gelatinous tunicates and the tiny, fish-like lancelet or amphioxus. *Pikaia*, from the Burgess Shale, may have been an early lancelet that swam around freely (Figure 17.4). The physics of swimming dictates that efficiency increases with body length, but that demands something more rigid than a notochord. Perhaps the development of the vertebrate skeleton got under way in this manner.

Once established, the vertebrates, at first only fishes, did well. They began hesitantly in the Ordovician, perhaps even the latest Cambrian, but diversified mightily in the Silurian and Devonian, the true "age of fishes" (Figure 17.4). The early groups, to our eyes, still have an experimental air about them. Small, odd-looking, armored beasts, they show no evidence of an internal skeleton; if they had one, it probably was cartilaginous and has not been preserved. They had no jaws and, notwithstanding their formidable appearance, they must have depended on soft food. On the other hand, some had pectoral fins, a big step forward in stability and maneuvering.

This first evolutionary step soon gave way to other fishes, also armored and finned (Figure 17.4), but equipped with the most useful tool ever

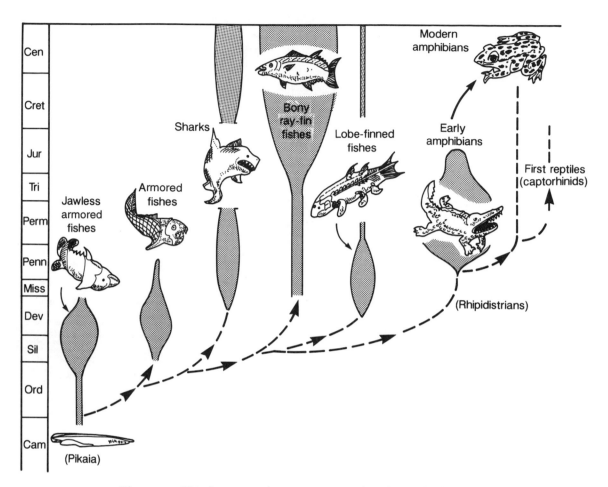

Figure 17.4. Vertebrates were latecomers compared to the marine invertebrates, but the first representatives, the fishes, rapidly became a success. There are many gaps in this record, especially at key branching points, but one major step, from fish to amphibian by way of the lobe-finned fishes, is certain. Paleontology, now more than ever, is in flux regarding the ancestor–descendant relations between many main groups, and what form these relations will eventually take is far from clear.

invented before the hand: a hinged lower jaw. The lower jaw may have developed at least twice, first in the small, spiny fishes of the Silurian and again in the ferocious placoderms of the Devonian.

Eventually the armor disappeared, giving way to light scales which gave the fish more flexibility and better swimming motion. From this stock sprang the sharks who, to this very day, have a cartilaginous skeleton that provides ample muscle support, as long as one does not leave the water. Eventually, however, the ray-finned fishes, the ancestors of all modern ones, developed a bony skeleton, a great advantage for their descendants, the amphibians and reptiles, when they advanced on land and later into the air.

Then came a small, inconspicuous group, the lobe-finned fishes. They survive in the Indian Ocean as that famous living fossil, the coelacanth *Latimeria*, and as lungfishes in tropical lakes and rivers. Lobe-fins have short arrays of bones that support limb-like fins that later smoothly evolved into the stubby legs of the amphibians. A few ray-finned fishes, such as Florida's fearsome walking catfish, manage to get around on their fins too, but they do not walk very efficiently and the tactic has not caught on among the ray-fins.

From the crawling lobe-finned fishes came the amphibians, the first vertebrates to live successfully on land. The earliest, arriving about 370 my ago in Scotland, were still so fish-like that they were long regarded as such, but finds in 1991 show that we are dealing with real amphibians here. These first few do not seem to have been truly happy about their new environment and remained closely tied to the water. One can understand this. They faced drying out, had to obtain oxygen from air rather than water, and kept collapsing under their own weight. They also risked being blown away by the wind, had to walk on paired legs, and needed to hear and see out of the water. Eventually they achieved all of this, something to remember when we look condescendingly at a frog, to become the first true amphibians, the labyrinthodons. In due course those turned into toads and salamanders and the like.

The amphibians postponed the problem of protecting their eggs against dryness by spawning in water as they still do. The reptiles solved that problem with a tough eggshell and an egg that nourishes the embryo until the timid little crocodile is ready to face the world. Exactly when this development took place we cannot say, for the simple reason that fossil eggs and parents are rarely found together. In all likelihood the inventors were small, lizard-like reptiles called captorhinids (Figure 17.5) who lived (and fossilized) in hollow tree-stumps in Carboniferous forests. With that crucial step, the conquest of the land by vertebrates became a foregone conclusion.

The road to complete control of the many and diverse new environments that exist on land went, as usual, by way of a few unsuccessful trials. One of these nevertheless deserves special mention, because it was from the rather lamely named mammal-like reptiles that, as early as in the Triassic, sprang the mammals.

Before they had their turn, however, the world was ruled for more than 200 my by reptiles, and for most of that time by the best known branch, the dinosaurs. Many were perfectly adapted to life on land, others

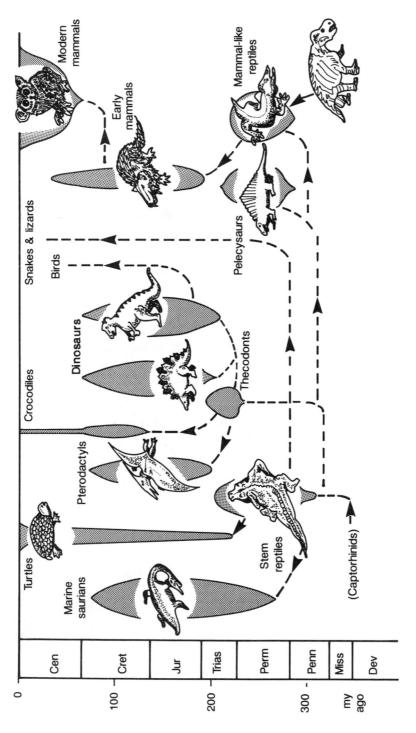

Figure 17.5. The wonderful variety of the reptiles is only inadequately displayed here. Although by no means scarce in the later Paleozoic, the Mesozoic era was clearly theirs to own. Even today reptiles are far from unimportant, demonstrating that a superior design, as exemplified by the mammals, does not by itself guarantee dominance. The mammals started early but, as seems often to have been the case, they had to wait in the wings until the early Cenozoic when they came into their own. Again, the actual ancestor–descendant relationships between the major groups are controversial, and this diagram is only a progress report.

lived in the sea, and some mastered the art of flying. Many achieved at least partial control over their body temperature, enabling them to live in Antarctica, in what passed for a cold climate in the Mesozoic. And some even developed a basic form of internal care for the embryo as mammals do. Bizarre as they appear to the eye of today's beholder, they possessed a perfectly useful series of adaptations to an enormous range of habitats and functions. Some flew, some swam, some ran across the land; some ate meat, others were herbivores, and there were the reptilian equivalents of whales, sharks, tigers and cattle.

Only after 200 my, three times as long as the mammalian domain has so far lasted, did the reptiles begin to decline, and their survivors today, numerous and diverse as they are, are but a shadow of their Mesozoic glory. Still, the descendants of their most famous branch may still be with us. When we feed the birds on our window sill in winter, we are probably feeding the direct descendants of the dinosaurs.

17.4 FROM MOUSE TO MAN

It is not easy to tell from fossil evidence alone just where the amphibians stopped and reptiles began, and neither can we draw a clear line between reptiles and mammals. Some mammals that are still here today, like the duck-billed platypus, are true members of the mammalian class, but they do lay eggs, thus warning us not to be too rigid about what separates us from snakes and lizards. There are, of course, other differences, the placenta, maternal feeding, legs placed under instead of beside the body, hair or specialized teeth, but few traits are unique to mammals alone, and even fewer do fossilize.

The first step in the evolution of the mammals was a quiet one. After a great bloom in the Permian, the mammal-like reptiles left the Mesozoic to the real ones, among them the dinosaurs, but not before, early in the Triassic, they had planted a time bomb in the form of some tiny, inconspicuous, shrew-like creatures.

The little beasts worked out some key problems of life much better than the reptiles did, perhaps in part by being nocturnal. One was an improved system of care for the unborn and the young. Another was a precise control over body temperature that allowed them to put on a sudden burst of energy, to start quickly on a chilly morning, and to live through the winter without lowering their metabolism too much, a dangerous condition. Mesozoic reptiles, living over as wide a geographic range as mammals do now, had the advantage of a warmer climate, while some profited from the heat retention that is possible with a large body.

343

The mammal-like reptiles show by their bone structure that to some extent they could regulate their body heat, and some dinosaurs probably were similarly equipped. But they were not the specialists in heat management that the mammals are, a skill that served the latter well when the climate turned cold in the Cenozoic.

Thus having equipped themselves well during a wait in the wings of more than 150 my, the mammals, small in size and few in number, crossed the border into the Cenozoic to spawn the lions, elephants, pigs and human beings of today.

There they took full advantage of the changing conditions of climate and vegetation, and especially of the rapid evolution of the flowering plants that marks the Cenozoic. The record is very good to us in this period, especially as regards the mammals, and has presented us with splendid examples of adaptive radiation, of which horses are the classic case (Figure 17.6).

Horses are part of a large group called the perissodactyls, of which other members are the rhinos and the tapirs. Curiously, that last group has evolved so slowly that they are essentially living fossils, whereas the development of the horses was one of the most explosive known to us. It is especially well known from North America where, as a result of great changes in temperature and the seasonal distribution of rainfall brought by the Miocene, the early Cenozoic woodlands opened up to form vast savannas and steppes. A new group of plants that had evolved a little earlier, the grasses, was very productive in terms of biomass and was able to develop a high tolerance for grazing. The grasses took advantage of the climate change, and the horses took advantage of the grasslands, a wholly new environment. Their contribution was an adaptation from browsing leaves to grazing that kept the savanna from being overgrown with brush, and another to fast running that kept them in balance with predators.

The earliest horses, for example the small *Hyracotherium*, the size of a medium dog, still subsisted by browsing the leaves on the edges of Eocene woods, but their successors developed ever more specialized teeth and also grew a bit during the Oligocene. When the middle Miocene brought extensive savannas, the first real horses had teeth perfect for grazing grass, legs ready to run fast in open, unprotected terrain, and a body designed for endurance. Savannas are highly varied environments and provide many ecological opportunities in the true savanna (a kind of park land), the grass steppe of the high plains, and the open woodland of the foothills, and the horses took advantage of them all. At their peak some 16 species simultaneously grazed the open prairies of central North

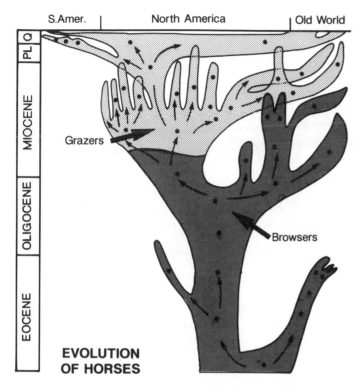

S.Amer. | North America | Old World

PL Q
MIOCENE
OLIGOCENE
EOCENE

Grazers

Browsers

**EVOLUTION
OF HORSES**

Figure 17.6. Horses present us with a spectacular example of explosive radiation as they adapted to the shift from woodland to steppe, prairie and savanna that resulted from the climate changes of the later Cenozoic. The change from dark to light shading indicates the transition from a browsing to a grazing way of life. The small dots and arrows indicate the course of species evolution. The diversity of the Miocene declined sharply with the onset of the northern ice age in the Pliocene (PL) and Quaternary (Q).

America, and it is ironical that the descendants of this famous case of adaptive radiation survive only in the Old World. In America they fell victim to extinction in post-glacial times (Section 19.1), and had to be re-imported by the Spaniards before the Indians could discover their natural aptitude for a horse-based culture.

17.5 THE FINAL STEP?

So where, in this chain of relationships extending back to the Precambrian, do human beings fit? We are members of the hominoids, a subdivision of the primates (monkeys), but are far from alone in that group, because gorillas, chimpanzees, the orangutan and the gibbon, and a number of ancestors of them and ourselves are there too. Between 6.4 and 4.9 my ago, the lines diverge, gorilla and chimpanzee go left (dia-

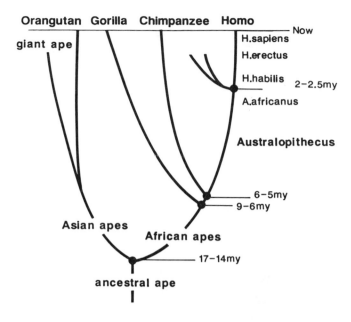

Figure 17.7. The tree of human ancestry. It is worthy of note that the genetic distance between us and the chimpanzee is way off scale as shown here; it is in fact extremely small.

grammatically at least, Figure 17.7) and the hominids turn right. At that moment this human family was down to two genera of which one, the australopithecids, is extinct. The surviving line evolved through several species, and sometimes two at once. Our direct ancestor is *Homo erectus*, the one who stands up straight. More than 100,000 years ago, somewhere in Africa *H. erectus* evolved into *H. sapiens*, the wise or thoughtful one, a premature designation. And here we are, alone in the world, a single species at the peak (or is it the end?) of the line. The rest is archaeology and requires a different book.

The evolution toward modern humans took place against a backdrop of major environmental changes that may have pushed it along or at least facilitated it (Section 11.4). As far as we know, it was played out mainly in eastern and southern Africa, where it began with the rise of the African rift valleys six or seven million years ago. The uplift, together with late Cenozoic climate changes (Section 11.4), brought cooler, drier conditions, the tropical and subtropical forests retreated, and the grassy savanna spread far and wide. This may have driven the earliest hominoids out of their trees and onto the plain where walking on two feet has advantages; it is faster and one's hands are free to carry food or the baby. Was the start of "Man the wise" so simple? Paleontologists do warn us that we must not overdo it when discussing ties between environment and evolution.

There is a great deal more to be said about the evolution of humans thereafter, but this is not the place. The reading list at the end of Chapter 19 makes a few suggestions.

18
Evolution and environment

Because natural selection evaluates currently available variation against current conditions, Darwinian evolution is a process of the moment; it does not see beyond today. Its response to a new demand or new opportunity is to work with what it has, fitting like a good backyard mechanic whatever suitable components are at hand into elegant new adaptations. Because its resources are limited, it sometimes fails and extinction follows, but conditions rarely change dramatically and organisms are flexible; what is here today much resembles what was also here yesterday.

The term environment might be misunderstood as excluding the biological interactions that operate within it, competition for space, food and shelter between individuals for example, or the impact of predators. The biological world is intricately woven together with the non-living environment, and plays as large a role as the motor or the acceptor of change. This is a book about oceans and continents, however, about the face of the earth as it has changed over the eons. When we address here the interaction of the changing environment with the history of life, we are mainly concerned with the impact of the non-living environment on the living one and vice versa.

On a grand scale, the opportunities presented by the earth to its occupants are, or rather were, three. First, there was once, but is no more, the potential of vast, empty territories. Second comes the challenge of environmental change, either from the physical and chemical environment or from the community of life itself. And finally there are the interactions between the organisms themselves. Opportunities such as the colonization of dry land, the drift of continents, changing climates and rising and falling sea levels, impose a pattern that is a proper concern of this chapter. Since the guide is a geological oceanographer, the oceans will get the lion share of the attention.

18.1 CONQUEST OF THE SEA

Environments change on many different scales of space and time. On land, stability is usually brief and homogeneity very local, but offshore changes tend to be more gradual. Consequently, many adaptations to the planktonic and benthonic modes of living at sea have been around since early in the Paleozoic, although the species themselves have not. In the sea, photosynthesis, for example, once performed by prokaryote cyanobacteria, is now mainly the job of eukaryote diatoms and coccoliths, and the top marine predators have successively been cephalopods, sharks, reptiles and mammals. Nevertheless, in many ways life in Paleozoic and Mesozoic oceans was much like life in the present one.

We lack solid information regarding the first colonization of the sea and must think our way through on slender grounds. The early earth had much ocean and little land, but life existed in those boundless waters more than 3.5 by ago. How shall we imagine this ocean to have been? Only here and there broken by land, it should have had a simple current pattern, a sun-lit, warm, upper layer, and gentle surface temperature gradients. The deep water, circulating slowly except during an occasional ice age, contained abundant ferrous iron, silica, and such poisonous substances as copper, arsenic and mercury, all furnished by innumerable submarine hotsprings. Except for the bacteria living in those hotsprings, the abyss should have been devoid of life.

Upwelling, the principal process that returns nutrients to the surface, is largely bound to coasts (Section 10.1). In early oceans with so little land, it must have been of little consequence, and except for its coastal waters the Archean ocean was surely less fertile than the present or even the Mesozoic ocean. The place to be for autotrophic organisms was near the shore, or even in intertidal waters, and the populations must have been small and dispersed to match the local nutrient supply.

By 2.5 by ago the continents had achieved a reasonable size, and shallow seas were widespread. Coasts and shoals impeded the global circulation, and there was upwelling on steep shores of the right orientation. Even though in the absence of a plant cover weathering was not yet very effective, the rivers, now draining large areas, had begun to pour nutrients into coastal waters.

An increasing nutrient supply spurs the diversification and expansion of primary producers to take advantage of the many environments of shallow seas, and this is what the abundant BIF deposits and stromatolite limestones suggest to us. Dense populations invite grazers, and grazers attract predators; together they provide the potential for the evolution of larger, multi-celled organisms. Moreover, upwelling, although guided by coasts, is a deep-water process that encourages offshore planktonic life with its own adaptations. And because upwelling is patchy, geographic isolation might work toward the evolution of new species.

The model gets high marks for logic, but the first evidence for multi-celled grazers does not appear until a billion and a half years later. If larger continents, shallow seas, and a greater nutrient supply were a challenge to the evolution of life in the sea, it makes little sense that signs of diversification and of the arrival of multi-celled animals do not appear until at least half a billion years later.

One possible answer has recently been suggested by Eldredge Moores of the University of California. It rests on the suspicion that the initial oceanic crust was very thick, with the result that the ocean floor floated high, the continents were relatively low, and the world until about one billion years ago was nine-tenths under water. If true, this would push the model for the evolution of higher life through the challenge of shallow seas to a much more recent time, but the evidence is still very thin.

Great importance has sometimes been attached to the arrival of free oxygen in atmosphere and ocean as an evolutionary drive, because it is essential for the development of an oxygen-based metabolism. That condition too was met about two billion years go, long before the acritarchs provided the first evidence for a rising diversity of marine life (Section 15.6).

There are no answers, except to assume that whatever did happen is concealed by flaws of the record, or that it takes a lot of time to make eukaryotes, or some other vague reasoning. It is also proper to note here that we have been casual about time. Connections have been suggested between free oxygen and oxygen-based metabolism, and between the growth of shallow seas and an explosion of algal life, but the fossil and environmental records cannot be pinned down more precisely than within a few hundred million years. It comes therefore as no surprise that ideas of this kind have been around for decades without acquiring substance.

And what about the abyss? No Precambrian deep-sea sediments have been preserved, but those known in the Paleozoic are more often than not black shales, and evidence for benthonic life below a depth of a few hundred meters is lacking until the Mesozoic. In the Mesozoic, however, black shales, though at times abundant, were no longer ubiquitous and since the Oligocene they have been rare. This suggests that throughout the Paleozoic the deep waters were largely anoxic, and that oxygen-carrying deep currents did not develop until afterwards.

In a very broad sense we might therefore take the apparent decrease of black shale over time as confirmation that the oxygen content of ocean and atmosphere gradually climbed to its present level (Figure 14.4), as ever more organic matter was stashed away in ocean sediments. Still, the Carboniferous (Figure 14.6) warns us not to be simple-minded about oceans, upwelling, carbon burial and oxygen, because its coal swamps buried many times more carbon than any marine deposit has ever done. This raised the oxygen in the atmosphere to well above danger level, but there is so far no record of disastrous fires.

The moral of this section is simple, but important. We can theorize a great deal about Precambrian environments, but what we need are quantitative measurements and precise dates. Those should come before the next edition of this book.

18.2 THE CAMBRIAN EXPLOSION

When discussing the sudden blossoming of life in the latest Proterozoic and Cambrian (Section 17.2), we passed over the reasons for this spectacular case of adaptive radiation without stopping. Was it just that the world was still so very empty, a cornucopia of ecological niches patiently waiting for customers?

At the simplest level, these were the main features of the event: (a) an increase of diversity and probably also population density announced by the acritarchs, (b) the appearance of the Ediacara fauna, (c) the sudden exploitation of the advantages offered by hard skeletons and shells made of mineral matter, and (d) the subsequent explosive diversification that brought on stage all but one of the main groups of higher animals.

What was the environmental background against which this was played? A supercontinent had been assembled at a still uncertain late Proterozoic time. Then, some 700 my ago, vast outpourings of plateau basalt began, similar to those that accompanied the early breakup of Pangaea, but the supercontinent seems to have remained essentially intact until its numerous fragments started to drift away some 50 my later. Then many rifts opened (Figure 18.1) and the creation of much young oceanic crust caused a major rise of the sea that peaked about 70 my later in the early Ordovician (Figure 9.1).

The long Proterozoic ice age ended about the time these tectonic events began to happen, and well before the bloom of the Ediacara garden. The marine environments in which the explosion of life took place were thus characterized by wide, shallow seas, long, narrow rift basins generally located in low latitudes, and a newly warmer climate.

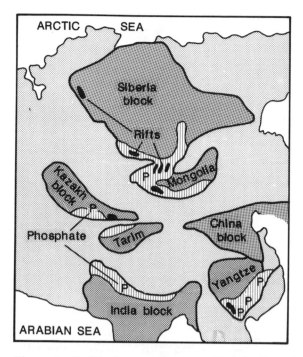

Figure 18.1. In Siberia, many fragments (dark shading) of the late Proterozoic supercontinent are gathered together within younger continental crust. Where their margins are preserved (vertical hatching), they show the initial rifts of the breakup that took place at the end of the Proterozoic. The waters of the new oceans were very fertile as extensive phosphate deposits (P) indicate.

What might we expect from such a world if we are permitted to generalize across a couple of hundred million years? Very high biological productivity for one thing, fed by continental weathering that provided nutrients, and by upwelling on the many continental margins and rift troughs. Today we have just a few such troughs, the Gulf of California, for example, or the Red Sea. They stand out for their exceptional surface productivity, they tend to turn episodically stagnant at depth and so insure the preservation of the organic debris that sinks to the bottom, and they offer a great variety of shallow water environments.

That much organic matter might have been buried is suggested by the carbon isotope ratios (Figure 18.2), but the high $\delta^{13}C$ may also signify a very high biological productivity. The oxygen content of the atmosphere probably rose in consequence of the burial of organic matter, potentially an important factor because of its role in animal metabolism. It is therefore not far-fetched to think that, as long as the long-term oxygen level was still rather low, episodic increases of this gas could have been a significant spur to the diversification of the Metazoa.

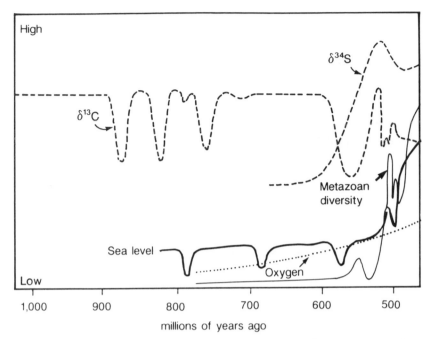

High

$\delta^{34}S$

$\delta^{13}C$

Metazoan
diversity

Sea level

Oxygen

Low

| 1,000 | 900 | 800 | 700 | 600 | 500 |

millions of years ago

Figure 18.2. Major environmental changes between 1,000 and 500 my ago are shown in this graph, where the vertical axis indicates the range from high (top) to low values (bottom). Burial of organic matter seems to have been the rule as the high $\delta^{13}C$ shows, and repeated exhumation and oxidation between 900 and 750 my ago may have been due to the late Proterozoic ice age. The sea level record is not detailed enough to test this. Other alternations between burial and recycling of organic matter mark the Cambrian. Atmospheric oxygen reached its present level by the early Ordovician. A large change in the sulfur isotope ratio ($\delta^{34}S$) suggests (but does not prove) vigorous upwelling and enhanced fertility. The diagram is impressive, but we need to remember that the data points on which it teeteringly rests are few.

Stable isotopes we have not yet encountered appear on stage here in the form of the ratio of sulfur-34 and sulfur-32. In the same way that organisms prefer light oxygen and carbon isotopes over heavy ones, anaerobic bacteria that use sulfate as an energy source prefer ^{32}S over ^{34}S. They precipitate it as pyrite (FeS_2), and thereby enrich their environment, the upper layer of marine sediments and the water just above it, in the heavy isotope ^{34}S. In addition to the high carbon isotope ratio, late Proterozoic and early Phanerozoic sediments show a large $\delta^{34}S$ peak (Figure 18.2). This peak has been interpreted as evidence for a sudden, perhaps even catastrophic upwelling of sulfur-rich deep ocean water. That is possible, but normal upwelling is quite capable of distributing excess ^{34}S, like ^{18}O, throughout the oceans in a less dramatic fashion. Or a high $\delta^{34}S$ might indicate the presence of coastal sediments so rich in organic matter that the oxygen in the pore water was depleted.

Too many choices, for the time being, but reasonable grounds to

353

expect high organic productivity and ample oxygen, the first an induce-
ment for grazers and low-level predators to multiply, diversify and grow,
as the Ediacarans did, the second permitting the diversification of Metazoa in general. Still, it cannot be denied that, as an argument, it is
rather thin.

18.3   SHELLS AND SKELETONS

What about the other, slightly later great event (Figure 17.1), the sudden,
widespread adoption of mineralized body parts? The utility of this move,
for armor, as an anchor for muscles, or as a tool to burrow with, is easy
to see, but the opportunities and pathways are not clear. There was initial
experimentation: the main minerals that were used changed with time,
but the issue is clouded by the habit of phosphate of replacing other
minerals during fossilization. In the earliest Cambrian of Siberia, South
China and Iran phosphatic fossils are common, but then decline, and
calcareous skeletons rise (the archaeocyathid curve in Figure 18.3). Less
than 20 my later calcium carbonate has become the dominant mineral
among invertebrates, as it is today. Phosphate continues in use for bones,
but most of the sixty-odd minerals with which life experimented early
are today of minor importance.

Phosphates are an essential nutrient, and the big phosphate deposits
of late Proterozoic and Cambrian continental margins (Figure 18.1) are
interesting in this regard. Their sources and geochemical pathways are
obscure, but they do coincide with a large $\delta^{13}C$ peak (Figure 18.3), thus
supporting the notion of a time of high productivity. Still, the coincidence in time is, as usual, only approximate and it is wise not to jump
to conclusions regarding evolution.

The interval we are talking about, the transition from Proterozoic to
Phanerozoic, was a time of continental breakup, and swarms of hotsprings must have accompanied the many new mid-ocean ridges (Section
14.3). Such springs add calcium, silica and ferrous iron to seawater, and
so make it easier for organisms to precipitate calcareous and siliceous
minerals, but the geo- and biochemistry of phosphate and iron minerals
(which were also used for hard body parts), is not yet clear.

Hard parts were adopted in many different ways and for many different
purposes. The phylum of the arthropods chose chitin, a hard organic
material, for an external skeleton, reinforced by some with calcium phosphate. Sponges build themselves internal frameworks out of fine needles
of silica and later also calcite. Some brachiopods went for calcium phos-

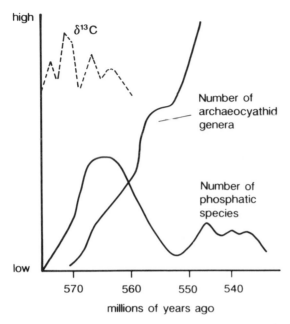

$\delta^{13}C$

Number of
archaeocyathid
genera

Number of
phosphatic
species

low

570   560   550   540

millions of years ago

Figure 18.3. Whether there ever was a shift from a dominant use of phosphate minerals for shells and skeletons to calcium carbonate (exemplified here by the archaeocyathids), is a matter of dispute. The Siberian data of this graph argue in favor of it, and large phosphate deposits (Figure 18.1) show that the element was abundantly available. The phosphate might also be responsible for the high biological productivity that one might infer from the $\delta^{13}C$ record. Much information is now coming out of the former USSR that will alter our perception of the important transition from Precambrian to Paleozoic.

phate, others preferred calcium carbonate as the mollusks do, but used a different carbonate mineral.

In so doing, life slowly began to take biological control over the calcium, carbon dioxide and silica cycles in the ocean, a control that became nearly absolute in the Mesozoic. It is vexing that we do not yet understand the chemistry of this "hard part explosion," nor that of the expansion of calcareous and siliceous plankton in the Mesozoic, an analogous event that fundamentally altered the chemistry of the oceans. The Mesozoic too was a time of young oceans, many hot mid-ocean ridges and presumably myriad hotsprings, but even if this had something to do with it, we are still woefully ignorant of how it actually worked.

18.4   THE LURE OF MUDDY BOTTOMS

We are clearly not doing too well in seeking explanations for the explosion of life at the start of the Phanerozoic. This does not mean that we are chasing mirages, but it does make clear that we have insufficient infor-

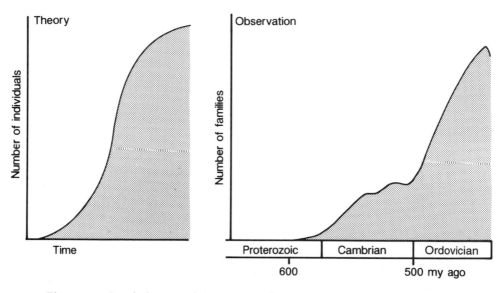

Figure 18.4. Population growth is a matter of continuous doubling of numbers. At first the growth is almost imperceptible, but an explosion follows, as human history so disastrously bears out. The steep rise, after an overshoot, finally flattens to an equilibrium. This is known to statisticians as the logistic curve (left) and it is commonly observed by biologists. John Sepkoski of the University of Chicago, when plotting numbers of families against time for the late Proterozoic and Cambrian, generated a classic logistic curve (right); in fact, one might argue that a second phase of explosive growth took place in the Ordovician.

mation and are thinking in terms that may be too simple. It would not be the first time.

What biological and ecological forces may have been behind the sudden diversity of higher animals and their need to possess hard body parts? Where, or perhaps when, should we look? John Sepkoski of the University of Chicago has argued that the key is not the sudden expansion of life itself, but something that took place a good deal earlier. Diversification proceeds in much the same way as population growth. Once set in motion (Figure 18.4), a population grows at an ever-increasing rate until external conditions put a limit to it. From a vantage point 500 my later, the steep part looks like an explosion.

The standard growth curve applies not only to laboratory cultures of bacteria, but to plankton blooms in the spring, and to the seemingly unstoppable increase of human beings. Sepkoski's point is thus well taken. Still, the increase from Ediacara time to the late Cambrian is so fast that there may be more to the explosion than only the power of exponential growth.

Be that as it may, where should we look for the starting point of the curve? The wide, shallow, and fertile seas of the early Proterozoic are

356

surely too distant for the purpose. A more suitable time might be after one billion years ago (Figure 17.1), but nothing interesting seems to have happened just then to start the curve off.

What else might do? The muddy sea bottom was first occupied extensively by complex communities in Ediacara time or a little earlier. Did the almost complete extinction of the Ediacara fauna start the expansion curve by creating huge vacancies on food-rich, muddy seafloors? Did life diversify in response to the challenge of this rich resource? If indeed the extinction of the Ediacara led to the Cambrian explosion, then the latter event was only the first of the recoveries that followed each major extinction in the Phanerozoic.

Some prefer to call only on biological factors for the big step to the Metazoa, proposing for example the arrival of highly efficient herbivores. In a simple world inhabited only by primary producers, the incentive to adapt is limited, because the main forces of change are geographical and climatic, and those provide large but few environments. The exploitation by grazers, and of grazers by predators, in contrast, raises the pressure to adapt in order to minimize the damage. The mechanism is attractive, but there is not a scrap of evidence that this is what actually happened.

To the physical scientist the need for a purely biological drive seems less compelling, but it is not easy to think of any other carrot or stick that could have brought forth the Metazoa. There may have been a rise in atmospheric oxygen at the right time (but what was the right time?), but no one knows exactly what it could have done. Let us approach the question from a different direction.

Mere clusters of special-purpose cells, as some Metazoa are, float, eat, and even swim, but they are ill-suited to dig in mud, to attach themselves to rocks, or to resist the tearing force of the surf. Most main metazoan groups have solved the problems of leverage and strength by improved designs and stronger materials. Their body cavity, the coelom, provides strength the same way that a cylindrical balloon has flexible strength. They also make good use of a unique material, the strong, flexible fibers of collagen, a protein that needs free oxygen for its manufacture. It is collagen that gives strength to tendons, cartilage, bones and other structural elements in nearly all Metazoa.

If, as their tracks, trails and burrows suggest (Section 17.1), basic metazoan designs suitable for living on or in the sea bottom became first available 600 my ago, the question still is why they should do so. James Valentine, of the University of California, has pointed out that upwelling provides a rich source of food, but that it is a variable source, demanding adaptation to seasonal feast and famine. The seafloor beneath an upwelling zone, on the other hand, stores the organic rain in its deposits, making them an enduring, dependable food reservoir, if one can only get at it. Crawling and burrowing are the answer, but that requires bodies suitably strengthened for this activity.

The advent of crawlers and burrowers brings into play other opportunities. It stirs up organic matter that can be utilized by bottom-dwelling filter-feeders, and the stirring of the mud by innumerable animals crawling and chewing through it alters the structure of the sediment, making it more porous. This provides room for a smaller, more specialized fauna to survive within the pores. The imagined sequence (Figure 18.5) thus begins with small primary producers in surface waters that are taking advantage of upwelling and are harvested by small grazers. The communities are constrained severely by the seasonally fluctuating upwelling, but organic matter accumulates and is preserved in the sediments below. When the eukaryotes appeared, grazers and predators would have evolved, developing eventually into floating, filter-feeding Metazoa like jellyfish. Subsequent development of collagen and a coelom permitted feeding on a dependable year-round supply of food on and in the bottom. There, specialized adaptations to one of the many niches would have been an advantage.

The far greater variety of nearshore seafloor environments compared to surface waters thus might have rendered inevitable a rapid diversification of bottom-dwellers. This scenario assigns a strong evolutionary drive to shallow seafloors, but not until they became accessible due to the development of multi-celled eukaryotes. The complexity of the ensuing ecology then brought biological factors into play as well.

Whether or not this plausible piece of reasoning bears any resemblance to reality I do not know. Subtle changes took place in the early Paleozoic that warn us to keep an open mind. In the Cambrian, bottom-dwelling filter-feeders flourished next to mud-eaters, but soon this peacefully settled community was disturbed by biological bulldozers that plowed the bottom, as many bivalves, sea-cucumbers and worms do today. The disturbance of the surface, as well as the increased turbidity of the water, ever since a fact of benthonic life, caused major reorganization. The filter-feeders moved mainly to the rocks, and the bulldozers monopolized sand and mud.

Predation is another important biological factor. Steven Stanley of Johns Hopkins University has suggested that the advent of predators was at the root of the Cambrian explosion, because experiments suggest that predation enhances diversity by cutting back dominant species and encouraging protective devices such as hard armor, and borings by pre-

Middle Precambrian

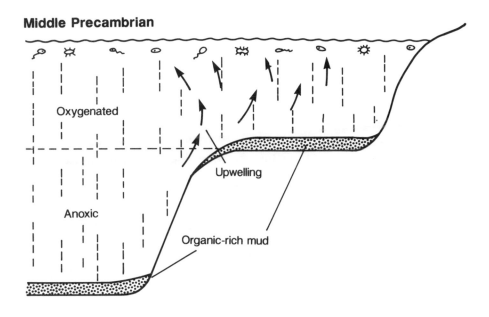

Latest Precambrian

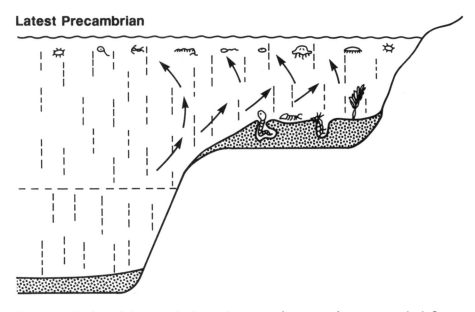

Figure 18.5. In the early Proterozoic, the continents were large enough to exert a major influence on the ocean circulation. This enhanced upwelling and hence the fertility of the surface waters. The result was a rapid expansion of primary producing algae and bacteria that probably led to the deposition of sediments rich in organic matter. The exploitation of this new resource, however, apparently had to wait until the end of the Proterozoic, when multi-celled organisms had developed body structures suited to the arduous job of crawling and burrowing in mud.

dators in *Cloudina* shells from the latest Proterozoic have indeed been reported. Whether this causes diversity or merely maintains it, however, is an unanswered question.

Prokaryote predators may have been around as early as the Archean, and unicellular eukaryote predators seem likely long before the Cambrian, although we have no proof for either. There is only a faint suggestion of predation in the Ediacara fauna, but from the early Cambrian on the existence of predators is demonstrated by damaged trilobite carapaces and general prey debris. Who the predators were is not yet known.

18.5   THE FIRST GREEN SPRING

Imagining land without life is hard, because even our harshest deserts have some of it, but thus was the world for at least four billion years. In the absence of organic compounds to attack minerals, chemical weathering was reduced, and nutrients in soils, lakes and rivers were low. Without a plant cover, rain percolated deep into the soil, while at the surface it evaporated swiftly. Permanent rivers with steady flow were rare, but flash floods causing fast erosion ubiquitous. Think of a very extreme kind of southwestern desert rather than an eastern woodland. The mitigating influence that a plant cover has on temperature and rainfall was absent, and the albedo of the land was everywhere high. This implies an unstable climate with greater extremes of hot and cold, wet and dry, than anywhere today except in a few deserts, and seasonal differences must have been marked. Only the coastal plains, where relief is low and water always near, would have offered a more benign environment.

Nothing demonstrates as dramatically the impact of life on the earth as the colonization of land by plants and animals. Among the first living things on land may well have been bacterial mats and crusts in damp places, followed by lichens, even now the first to occupy newly exposed surfaces, but if they left any traces we have not found them so far. Lichens greatly increase chemical weathering, because they produce acids that dissolve the rock effectively, and so provide a better and more nutritious substrate for other plants such as eukaryotic green algae, liverworts and mosses. The runoff also carries more nutrients, so facilitating the growth of aquatic plants in lakes and coastal lagoons.

The innovations that enabled plants to step from sea to land were not major. Algae do live on land, but are limited to puddles or wet spots. Large water bodies are not required, because plants can make do with the small amounts of water and nutrients usually available within the soil. But for photosynthesis leaves must be raised up to the sun and nutrients extracted from the soil must be carried to the leaves. Liverworts and mosses have stems and roots to do this, but their way of bringing water and nutrients up, and carrying products of photosynthesis down through elongated cells is inefficient and forces them to stay small. For reproduction, they have to place their sperm and eggs in water, even if it is but a mere drop.

Those limitations were decisively overcome by the vascular plants which developed a system of tubular cells for transport of water and nutrients, and spores or seeds for reproduction away from water. Not surprisingly, vascular plants make up by far the largest part of the land vegetation today.

Much of what we know about the earliest plants comes from microscopic spores and later also seeds, preserved in sediments in sufficient variety and abundance to permit the reconstruction of their evolutionary history. The approach is the same as that of the pollen analysis of Quaternary sediments (Section 4.5), the difference being that for most Paleozoic spores we do not know what the whole plant looked like, because leaves, stems, and trunks attached to spores or seeds are rarely preserved.

Until recently the conventional wisdom was that the various adaptations needed for life on land developed all at once in an aquatic environment, so that the higher, vascular plants were among the first colonists. That is no longer held to be true.

The first evidence for land plants are spores from middle Ordovician rocks in Arabia, about 470 my old. They are like the spores of modern liverworts and perhaps tell us that the first adaptation to life on land served to facilitate reproduction. Only a little later, small, moss-like plants named, not very melodiously, *Cooksonia* seem to have managed water transport by means of elongated cells. Mosses and liverworts were the main land plants from the later Ordovician into the Silurian, but spores, continuing as the main source of information, show that late in that period fern-like plants were driving the mosses back to the ecological niches they still occupy. Still, the first remains of stems and leaves do not appear until about 415 my ago.

One need not envisage a landscape covered only with a low green mat, however, because there is plenty of evidence for a group of early trees, the nematophytes, that had stumpy trunks up to one meter thick and several meters long. Having made their entry in the Silurian, they vanished in the Devonian.

No garden is without pests, and land animals go back as far as the middle Silurian. Some 420-my-old remains, possibly of centipedes, have been found, but until the early Devonian the evidence is sparse. At that time a Scottish bog, flooded by nearby silica-rich hotsprings, preserved, in a fine detail that makes the Rhynie Chert a minor Gunflint Chert or Burgess Shale, a fauna of many small arthropods. Early Devonian lagoon sediments in Germany preserved spiders and a fearsomely large millipede.

This is the proper moment to ask why anyone, animal or plant, bothered to move onto the land at all, an unattractive environment in its lifeless mode if there ever was one. It is tempting to look toward specific reasons. Does the land offer opportunities the sea does not? Was there something especially appealing about freshwater? Were there conditions in the sea, overpopulation perhaps, that drove some of the dwellers of the watery realm onto the land?

There has been much and often ingenious speculation, but to me the issue seems quite simple: life goes where the conditions are right and, at the most basic level, this means space, water and food. Space, of course, there was in plenty, but the empty land had neither conveniently usable water nor food. So why did they come?

Perhaps this is not true. Any beachcomber will have noticed how, after a storm, piles of seaweed accumulate on the shore, there to decay slowly, often over many months. Seawrack provides food and saltwater and many marine animals live in the piles, algae and bacteria grow on them, and so do fungi. And most baskers on a sunny beach have noticed the little crustaceans, the springtails that live within the sand. Here, perhaps, we have an easy first step for plants and animals alike, one that offered not only the path landward but also the lure. It suggests that green plants capable of living out of the sea may not have been the first life on land, and that animals pre-adapted in the intertidal zone, such as the crustaceans, were there ahead of them. It would therefore not surprise me if the first life on land predated by a long time the first known land plant fossils, placing its arrival perhaps in the Cambrian.

Enough of this speculation. In the Devonian land plants and their animal users came into their own. Once plants had a solid foothold on land, their expansion was rapid. The Devonian was still a time of large seas, but land plants of many sizes and shapes proliferated on emerging shores, and the first true ferns set the stage for the grand forests (Figure 18.6) that formed the coal beds of the Carboniferous (Pennsylvanian). Giant horsetails, ferns, tree-ferns and many other, stranger species colonized the coastal lowlands where water was plentiful. Equipped with the adaptations to land life that plants still have (except for flowers and fruit which did not come until much later), they occupied a wide range of environments with a diversity of plants to match.

Gradually plants moved out of such convenient settings and during

Figure 18.6. The Carboniferous was a time of extensive swamp forests that left behind thick deposits of organic matter which later turned to coal. The forest shown here is not so lush and not likely to have become a major coal deposit, but it is typical of the exotic nature of the Paleozoic flora. The hills in the background are barren; plants capable of dealing with high, dry ground did not yet exist.

the Mesozoic virtually all climate zones, mountain-tops, hills, semi-arid lands and river plains, acquired a cover of vegetation. It seems that the challenge of even the least suitable open spaces was thus met and that from then on it was mainly competition among plants and between plants and animals that drove further evolution.

The later history of land plants (Figure 18.7) is mainly a history of improving the means of reproduction in a hostile setting. The long road leads from the mosses which must reproduce in water to the horsetails and seedless ferns of the middle Paleozoic, and by way of the extinct seed-ferns to the Mesozoic conifers, ancestors of the pines and firs of our own forests. The ginkgo tree which shades many a suburban street but is extinct in the wild, the cycads decorating green-houses, and the tree-ferns of tropical jungles are survivors from the forests of 200 my ago.

Our present flora, however, consists overwhelmingly of a new class that arose early in the Cretaceous: the flowering plants or angiosperms. One of the many keys to the success of this group is the flower which enlists the help of insects, birds, and many other animals, in addition to the gravity and wind that had served plant fertilization since times immemorial. Another key was seeds that, by refined adaptations to long-

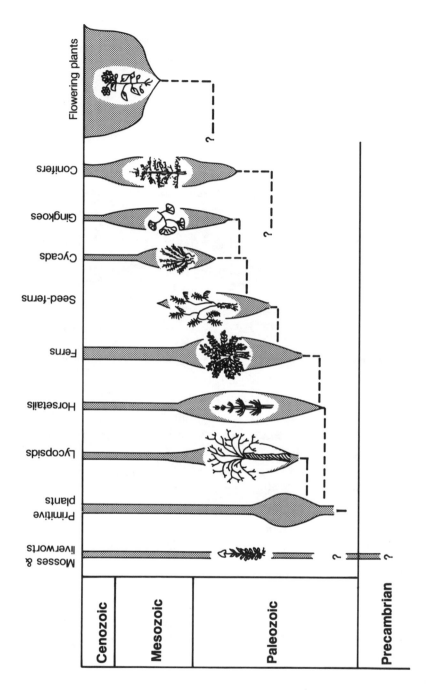

Figure 18.7. The evolution of land plants began with mosses which still needed water for reproduction, and ended with the proliferation of flowering plants that are today the delight of every gardener and nature lover. It is evident from this diagram that more often than not we are in the dark concerning the major branching points of evolution, even when the general trend is clear.

364

distance dispersal and reliable germination, ensured the colonization of distant soils beyond the reach of seeds or spores falling straight down from pine, spruce, or fern. The rise of flowering plants with their need for pollinators and seed carriers spurred many animals to take advantage of the opportunity in sometimes very sophisticated and clever ways. This led to a proliferation of cases of co-evolution of plants and animals. The Cenozoic diversification of insects, of many birds, and of some groups of mammals as well, can be largely attributed to the flowering plants.

The angiosperms also developed a great variety of small herbs and shrublets that suited environments not well filled before. In particular, the advent of grasses in the Cenozoic changed the world. Just imagine for a moment a world without grass! Ferns prefer moist soils and conifers are mainly trees, but only the angiosperms offer greens for all seasons, for all places, for the floors of dry woods, for grassy banks, for ponds, tundras, and alpine meadows. They offer cactuses for the desert, thorny shrubs for the lands of winter rains, and succulents for coastal cliffs. Rocks, saltflats, screes without soil, mountain peaks whipped by vicious winds – they all turned green during the last 100 my.

In the process, plants increased the resistance to erosion of slopes and soils and so stabilized the landscape, while adaptations to saltwater by mangroves and marsh grasses altered the process of silting on muddy shores. Coastal outbuilding accelerated when these ecological communities came into being early in the Cenozoic, and the transport of fine-grained sediments to the deep-sea dropped sharply.

It is clear that the surface of the earth was profoundly altered by the advent of land plants. Chemical weathering was enhanced, soil fertility and the input of nutrients in rivers, lakes and the sea increased, the climate ameliorated, the coastal plains enlarged, the list is endless. In short, except for its control of the CO_2 green-house and the oxygenation of the atmosphere, life has nowhere demonstrated its skills and powers more strikingly as when it took over the land.

18.6 EVOLUTION AND CONTINENTAL DRIFT

Many are the environmental changes that have pushed evolution one way or the other. The climatic deterioration of the late Cenozoic altered, even impoverished the flora and fauna of the northern continents. The coming and going of shallow seas throughout the Mesozoic created and destroyed vast environments and whole faunas and floras. But clear cases of the relation between environment and evolution can also be drawn from continental drift. They illustrate two things especially well: pro-

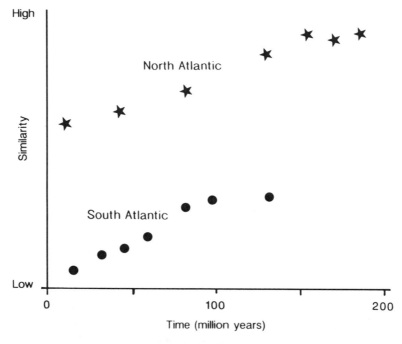

Figure 18.8. During the Mesozoic and Cenozoic land faunas evolved on both sides of the ever-wider Atlantic Ocean. They furnish a convincing example of the impact of continental drift on evolution. Identical early in the Mesozoic, the American and European–African faunas gradually diverged in direct proportion to the width of the Atlantic Ocean. A landbridge in the far north kept the level of similarity in the North Atlantic higher until the bridge vanished in the Miocene.

gressive divergence as faunas became isolated when continents drifted apart, and the consequences of competition with immigrants when lands bearing different faunas and floras collided.

Nature often seems set to frustrate our effort to understand her, but the effect of the increasing distance between continents since the breakup of Pangaea is pleasingly clear. Statistical analysis of the modern faunas of various continents shows that about half of their differences can be attributed simply to the distances between them. If we plot the increase with time of difference between faunas on opposite sides of the widening Atlantic Ocean, we find a straightforward relationship (Figure 18.8). In the North Atlantic, where a connection between North America and Europe persisted until the Miocene, the similarities between faunas on opposite sides remained higher for much longer than in the South Atlantic where separation began much sooner.

When we trace specific animal groups, the consequences of the Mesozoic disruption of the continents become more vivid. Many of the main reptile taxa evolved before the intercontinental distances were too great, and managed to spread far. In the early Mesozoic, however, when the Tethys lay between Laurasia in the north and Gondwana in the south, crocodiles and dinosaurs evolved in Gondwana, while the ancestors of turtles, lizards and snakes arose in Laurasia. Eventually, crossing the Tethys became easier in the west, and in the middle Mesozoic members of all main taxa are found just about everywhere (Figure 18.9). Even so, distinct dinosaur faunas developed on each land mass. On Laurasia, for instance, we find the duck-billed dinosaurs, *Triceratops* and the tyrannosaurs, while on Gondwana other groups occupied equivalent ecological positions. In the Cretaceous, when the continents were even farther apart, the dinosaur assemblages of North America and Asia had not a single species in common.

Mammals entered the world on an undivided Pangaea, and did not really diversify until the continents had fully separated. Their radiations thus started from many isolated centers. Shallow seas divided North America into two parts, Africa into three, and isolated Asia from Europe. North and South America were separated by an ocean, but South America continued to be connected to Australia by way of Antarctica until 50 my ago. This used to present us with a puzzle because only primitive marsupial mammals were known from Australia, although more highly evolved placental mammals could have made it there from South America by way of Antarctica. New finds in Australia have now turned up placental mammals that predate the separation from Antarctica and so solve this puzzle, although it is unclear why they did not survive in the end.

Continents collide as well as diverge, and when they do they may bring together disparate faunas. The classic case is the joining of North and South America, 3.5 my ago. In South America, a fauna rich in large marsupials (Figure 18.10) had evolved during the long isolation of that continent, and achieved a balance between vegetation and herbivores and between herbivores and predators. In North America, the marsupials had been less successful and more advanced placental mammals predominated. When the northerners finally came south across the Isthmus of Panama, this host of apparently more efficient herbivores and aggressive carnivores made short work of their South American competitors. Four entire orders of the local fauna were wiped out in no time at all. Only a few southerners managed the trip north, and even fewer, such as the armadillo, acquired a permanent foothold there. Ironically, South America is today a refuge for several originally northern species that are extinct in the north, such as the llama and the tapir.

The consequences of continental drift are obvious, but we must guard against attributing too much to it. To warn us, nature has provided five

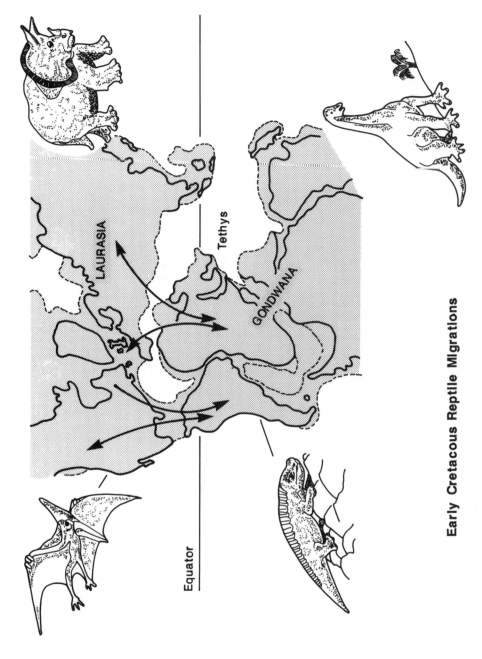

Early Cretacous Reptile Migrations

Figure 18.9. In the Mesozoic, different groups of reptiles arose in the northern and southern parts of Pangaea that had become separated by the Tethys sea. Their even distribution across all continents in the Cretaceous shows that, for many groups, the barrier soon lost its effectiveness.

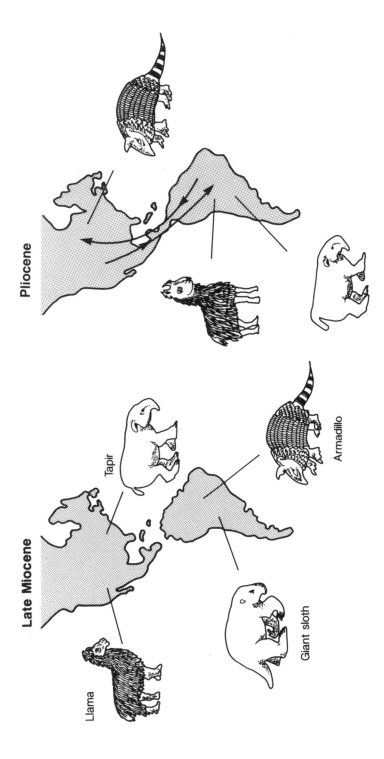

Figure 18.10. Until the emergence of the Isthmus of Panama in the Pliocene, North and South America, separated by a seaway too wide to be crossed by large land animals, had very different faunas. When the bridge formed, the immigration of competitive northern animals made short work of the marsupial fauna in the south. Such typical South American species as the tapir and the llama originated in the north, but only a few southern animals, for example the lowly armadillo, made the reverse trip and succeeded.

369

anteaters on four continents, strikingly alike, but derived from wholly different ancestral stock. They offer a fine example of convergence of form due to a similar mode of living, and a finer one of a common paleontological trap.

19

Crises and catastrophes

The most remarkable events in the history of life are the brief, great contractions, sometimes nearly collapses, that for a while sharply reduce the diversity and sometimes also the abundance of life. Each time in the past, a complete recovery brought a new and different flowering.

It has lately become fashionable to feel guilty about the damage we are inflicting on nature, so guilty that we regard the damage as without counterpart in the history of the earth. But nature, capable of great havoc without our help, is also capable of repair. Time and again tropical rainforests and coastal wetlands have recovered from glacial climatic setbacks and sea level changes so severe that, were they to repeat themselves today, we would despair of the future.

The great dyings stir the imagination and it is easy to see them as catastrophes that demand extraordinary causes. It were well to resist this temptation until we have examined just how catastrophic these events really were and how unusual the causes need to have been. So let us learn from the record what we can.

19.1 DEATH AND RENEWAL

During the early Paleozoic the diversity of shallow marine life continued to increase as specialization and new adaptations gave access to an ever-widening range of environments (Figure 19.1). Stability, interrupted only by two moderate extinctions, lasted more than 200 my, from the middle Ordovician to the Permian, but then, especially in the sea, diversity dropped, slowly at first, then calamitously, until only a fraction of all marine species crossed over into the Triassic. The recovery, interrupted only by two small crises and one major one at the end of the Cretaceous, was steady, and the diversity of marine life has increased ever since the start of the Cenozoic.

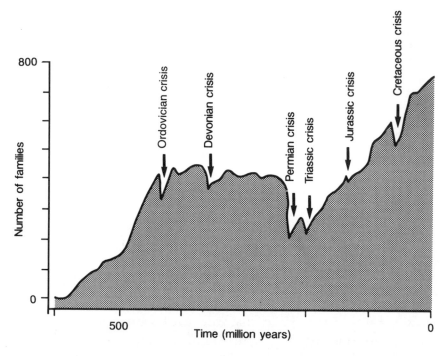

Figure 19.1. The diversity of life, expressed here as the number of vertebrate and invertebrate families known to have been living in the shallow sea, has greatly increased with time. There have been many setbacks, but even the large ones of the later Permian and the Cretaceous were temporary, and were invariably followed by another rise of the diversity of life forms.

An example closer to us is useful. Shortly after the last glacial, many of the famous Pleistocene mammals, the mastodon and the mammoth, the sabertooth tiger and the woolly rhinoceros, were wiped out. In North America, a dozen species vanished, including two kinds of mammoths and the mastodon, two of the three species of bison, all camels and horses, and many others. Similar scenarios, although not quite so comprehensive, were played out at slightly different times in Europe, Australia and South America. Except in Africa and a few other warm areas, relatively few large herbivores and predators did make it across into the Holocene.

This was the time that the climate changed most drastically to its post-glacial optimum, and environment and food resources changed with it. For a few victims this explanation is adequate, but most had survived several glacial–interglacial transitions already. The blame for the extinctions, especially those of the New World, has been laid on the newly arrived human beings who over-hunted a defenseless fauna. The sabertooth tiger and the great bears were not too vulnerable to

372

prehistoric human hunters, but failed to survive the demise of their prey.

The overkill hypothesis fits much of the evidence, but the final verdict is not in. Why, for example, did the North American buffalo survive while two other bison species perished? The issue is important: if human beings were indeed the principal cause of this catastrophe, we have here an early case of what we are doing now. If the human role was a minor one, we have an example of the speed and thoroughness with which a natural disaster can overtake a diverse and seemingly well-established community.

At one time or another, many groups of animals have suffered such setbacks, alone or with others. There were, for example, four crises in the history of the reef corals, but they are still around. The trilobites, on the other hand, having had their day in the Cambrian, declined slowly before they departed in the Permian. Many extinctions are due to the failure of a small group and its replacement by competitors. The nautiloids, shelled cousins of the squid and now represented by just a few species, lost out in the late Paleozoic to their relatives, the ammonites. Both relied on jet propulsion, a suitable way of getting around until swimming fish and other marine vertebrates out-performed them. Thereafter, the ammonites managed a more modest mid-water career until the end of the Cretaceous when a major crisis wiped them out altogether.

Behind some extinctions are changes in the environment, but not even major environmental changes always cause extinctions. The Pleistocene record, for example, is not riddled with mass extinctions, casting doubt on the view that Paleozoic glaciations on Gondwana were responsible for some of the major extinctions during that time. On the other hand, even minor instabilities may bring disaster if the environment is a complex mosaic and its inhabitants are highly specialized.

Extinctions are tediously common throughout the history of the earth as each species rises, flourishes and reaches the end of the road. The evolution of life plays against this backdrop of continuous disappearances and replacements of individual species, above which rise the great extinctions mentioned at the start of this chapter. Affecting entire assemblages of organisms, they have become known as "mass" extinctions, although their magnitude varies. Most have written *finis* under 10 to 25 percent of the taxa under consideration, and are mere dimples on the diversity curve of Figure 19.1. They are surpassed by the true crises of life, the vast destructions that hit the earth once in a while. They are few in number, but dominate the stage, and are the subject of a great deal of discussion and controversy.

Life, time, and change

Most extinctions, however, seem random, their minor, brief causes lost in the shadows of time. Like accidental death, they are the outcome of a brief but fatal mismatch between a species and some quirk of its environment. This led Thomas Schopf of the University of Chicago to suggest that for most extinctions it may be futile to seek a specific cause; species become extinct when their time is up, just as people die when theirs has come. It is not important in the demographic sense whether one's death is due to heart failure, an automobile accident, or "natural causes," but only that people become "extinct" past a certain average age. Such extinctions ought to be distributed randomly over time, and plenty of statistical data suggest that indeed they are.

But what about the major catastrophes, the sudden, large-scale dyings that so much intrigued the 18th and 19th century catastrophists? What lies behind the few great and many smaller but still puzzling global crises, all of them such useful stratigraphic markers for the working geologist?

Mass extinctions, great or small, are not all the same. Some were sudden, some more gradual; some affected only selected taxa, while others devastated all inhabitants of certain environments. Some seem almost arbitrary, wiping out some species while leaving others virtually intact, eliminating the dinosaurs for example, but doing little damage to the mammals. Some coincide with large changes in the environment, but other major shifts of the earth environment seem to have had little impact.

Mass extinctions also have things in common. Many affected life on land and in the sea, although rarely in equal measure. As a rule, tropical faunas appear to be more vulnerable than those of higher latitudes, because they are more specialized. Curiously also, plants have been more resistant to extinction than animals and, as a result, the evolution of plant life has been more gradual than that of animal life. Also, some animal groups, such as trilobites and cephalopods, have been much more prone to crises than others.

When we look for causes, we must ask ourselves, as we did with regard to the Cambrian explosion of life: did extinction follow immediately upon its cause, or was there some earlier, less conspicuous event that made the eventual crisis inevitable?

The common way to diversify for a group of organisms is to rise to a peak then gradually decline: a bell-shaped curve of success and failure. This is not so very surprising, but if we calculate bell curves for many, many taxa (Figure 19.2), we are struck by what seem to be a clustering of disappearances and seek a cause at that time. Should we rather look at clustered starting points, wonder what made that moment so favorable

374

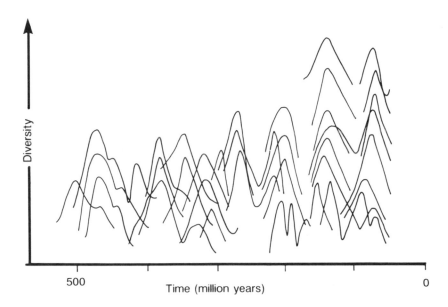

Figure 19.2. The bell-shaped curves of success and failure. This dizzying diagram shows the change with time of the diversity of many groups of organisms: trilobites, various categories of cephalopods (squids) and fishes, many amphibians and reptiles, and so forth. Each group increased gradually, then declined. The collection of curves conveys the impression that certain times were well suited for the simultaneous expansion of many groups, while others were prone to disaster. A crude 30 my spacing seems to be discernible, but see Section 19.6.

to adaptive radiation, and accept that extinction must inevitably follow? The orthodox wisdom does not favor this approach, because it presumes that the interval between rise and fall is approximately the same for many taxa, but with the graphs in front of us I leave the question open.

David Raup of the University of Chicago has wrestled with these problems and has come up with a device grimly called the "kill curve" (Figure 19.3). Species appear on the scene, wait around for a while and become extinct. Given enough data, we can estimate the number of species that become extinct after, say, 10,000, 100,000, 1,000,000 years and so forth. The curve shows that for the vast majority of species the waiting time is quite long; few are given less than a million years to enjoy this earth. Then the curve climbs steeply; 30 percent are gone after ten million years, and almost three-fourths after a hundred million. It seems that most species are secure most of the time, but now and then something happens that wipes them out. Whatever that is, the more severe its impact, the less often it comes.

The 10-my or 100-my mass extinctions of Raup's graph are akin to the 100-year floods we often hear about. Such floods do not return every 100 years; what the term means is that over a long period, perhaps a

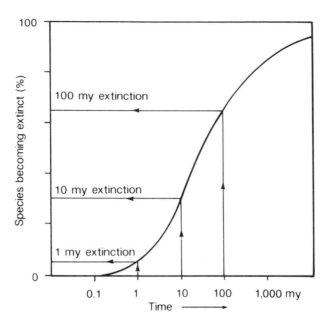

Figure 19.3. David Raup's "kill curve" for species. The longer one waits, the larger the extinction, but it is never quite complete.

millennium, only ten floods will be of this size or larger. They occur without pattern; two may appear in a single year, then none for three centuries. The graph says no more; it predicts no dates, nor does it offer an explanation. If we want causes, we must turn elsewhere.

19.2 THE PERMIAN MARINE COLLAPSE

In terms of the number of animal species, genera and families lost the Permian crisis was probably the largest of them all, but its impact was mainly in the sea. A gradual affair, it began some 255 my ago, but when the Triassic arrived 10 my later, perhaps 95 percent of all marine species had vanished, and only about one-third of the genera and half the families, all much impoverished, survived. The groups that suffered most were the corals, brachiopods and ammonoids, and the greatest damage was done in tropical waters. The record is fragmentary, but only some tens of thousands of marine species reached the Triassic, shipwrecked immigrants straggling to the shores of an uninhabited island. The land vegetation and fauna, on the other hand, suffered little, except for the therapsids, the mammal-like reptiles, who came to the end of their line.

Characteristic for this particular catastrophe was that it was so slow. The extinctions came in waves, followed by adaptive radiations as new species evolved to take the open places, but in the end the world was enormously poorer. This, together with the features listed above, restricts our choice of explanations, but has not limited the inventiveness of baffled paleontologists. Changes in environmental stability, shrunken shallow seas, major sea level changes, a sharp drop in oceanic oxygen, freshwater spills from giant lakes, and impacts of extraterrestrial bodies are only a partial list of what has been proposed.

Let us consider what happens to marine life when conditions alternate between stable and unstable food supplies. Seasonal upwelling is a good example of a rich but unstable food supply: feast when it occurs, famine in the off-season. Big swings from abundance to scarcity and back again favor large populations with short life-spans and a high reproduction rate that allows them a fast recovery from mass mortality. Tolerance for a wide range of conditions and all kind of foods is also an asset. Flexibility, low diversity, and fluctuating numbers are characteristic for this situation.

In a stable environment such as a reef, on the other hand, food is not plentiful, but it is dependably available. This condition calls for specialists, each precisely suited to take advantage of one of the many narrow environmental niches. The population is diverse, but the numbers of individuals are small, they are highly specialized and their adaptability is low. Flip the environment into an unstable condition, and extinctions will be widespread as the specialists find themselves unable to adapt. Convert an unstable environment to a stable one, and the genetically flexible population will explode in a large variety of specialized forms.

James Valentine of the University of California married this concept to continental drift to explain the Permian catastrophe. Supercontinents tend to be dominated by a monsoon climate (Figure 7.7). The strong offshore monsoon winds of the cold season blow the surface water away from the shore, so causing widespread upwelling. Thus the food supply around a supercontinent is likely to be less stable than around the dispersed continents of the earlier Paleozoic or the present. A low diversity of life is to be expected when continents congregate, a higher one when they are scattered. To some degree this is true (Figure 19.4), but not as clearly as the paleogeographic reconstructions available to Valentine 20 years ago suggested.

Moreover, continents move but slowly and it is difficult to believe that they could do so much damage in a mere 10 my. Also, upwelling is common enough even in our present world of dispersed lands (Section 10.1), and an increase of major proportions if all continents were joined together seems unlikely. Still, there are other differences between dispersed and aggregated continents that may help us here.

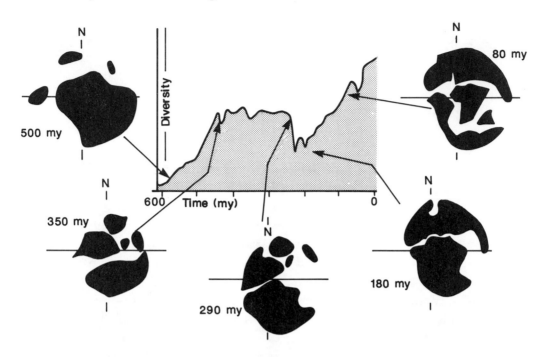

Figure 19.4. James W. Valentine suggested long ago that high diversity is associated with numerous and dispersed continents, while diversity crises coincide with supercontinents. The paleogeography used in his 1972 publication and shown here is now obsolete, and newer versions make it clear that, although there is merit to the hypothesis, the situation is complex and is not to be explained by a single factor.

The late Permian was distinctive not only for its single continent but also for its rapid emergence. About 230 my ago, shallow seas shrank from 43 percent of the total continental area to about 13 percent, about the same as today. The reduction in area paralleled a decline in diversity (Figure 19.5), as one might expect, because less space will accommodate fewer species. The number of bird species nesting in one garden is less than that found in the entire city. However, on close inspection the drop in diversity lagged well behind the reduction in area.

Thomas Schopf, recognizing this explanation as inadequate, proposed another, also related to continental drift. Given a supercontinent surrounded by an ocean, the minimum number of distinct shallow marine biogeographic provinces, each mainly controlled by climatic and oceanographic limits, would be eight (Figure 19.6). In our present geographically complex world, on the other hand, we need at least 18 or 20 provinces to fit the many shallow marine environments, even if we ignore oceanic islands. If we should put all continents together again, a large diversity decline would result because the number of faunal provinces

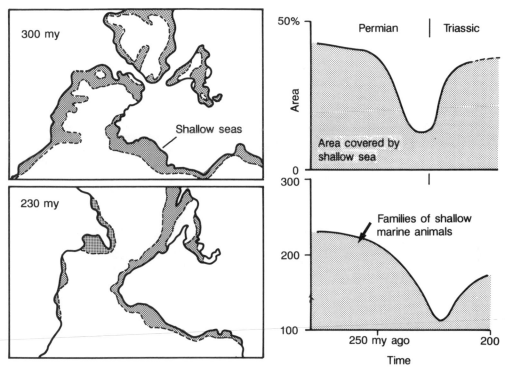

Figure 19.5. Pangaea finally came together in the closing years of the Permian at a time of low sea level. The resulting reduction in the area of shallow seas has been held responsible for the crisis that afflicted marine life at that time. The graphs on the right, however, show that the extinctions came later than the decrease in shelf area.

would be so much smaller. Estimates based on modern faunas show that the effect would be three times larger than that of a reduction in area.

There are difficulties. Oceanic islands might have served as refuges from where shallow seas elsewhere could have been swiftly repopulated. Also, the many alternations of high interglacial and low glacial sea levels of the recent past, endlessly eliminating and re-creating coastal environments, have left no major record of extinctions. We cannot escape the conclusion that these ingenious hypotheses offer at best only very partial explanations.

So what else can one think of, apart from the nowadays fashionable extraterrestrial impact which has been put forward here also but on very questionable evidence (Section 19.4)?

Another key feature of the Permian catastrophe is the prolonged decline which was intermittent as if in response to repeated brief conditions of stress. Characteristic also are the withdrawal of the tropical marine fauna to the Tethys region accompanied by a slow reduction in diversity, and the devastation suffered by reef-building organisms. These suggest

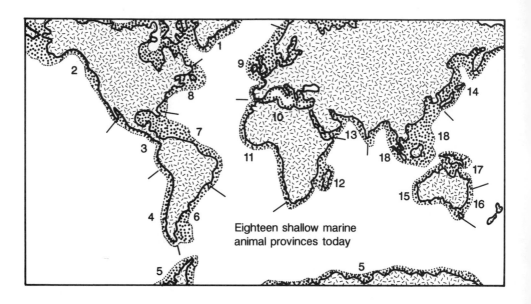

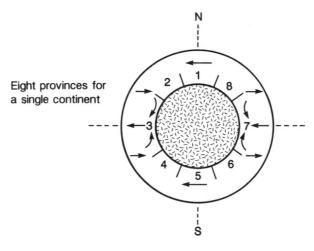

Figure 19.6. Another cause of the Permian marine catastrophe might have been the diminished range of environments that is the result of the simple coastal geography of a supercontinent. A mere eight provinces could account for the environmental range of Pangaea's shallow marine fauna, whereas a minimum of 18 is needed for the modern world (top). The small arrows symbolize ocean currents that disperse the spawn of marine faunas.

that the earth might have cooled much at high latitudes (Figure 19.7). Whether there were polar icecaps is doubtful, but ice-rafted deposits have been found.

The late Ordovician and Devonian mass extinctions (Figure 19.1) share some of these features, especially the strong impact on tropical marine faunas and reefs. Both happened during or at the start of a major ice age (Section 7.3) and, as with the Permian crisis, for a while afterwards

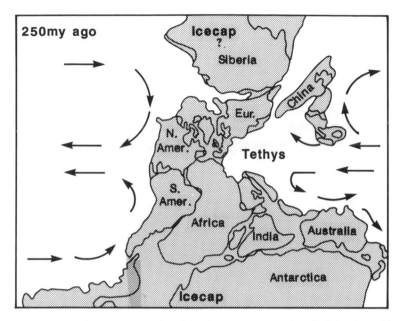

Figure 19.7. The paleogeography of the late Permian and early Triassic world, complete with a sketchy ocean circulation. There was probably an icecap on the southern hemisphere, but the ice-rafted deposits from the northern polar region may imply no more than some coastal glaciers. Still, we are clearly not in the Mesozoic green-house.

the world knew few if any reefs. This persuaded Steven Stanley, of Johns Hopkins University, to attribute all three crises to high-latitude cooling of ocean waters. It is worth remembering, however, that ice age conditions affected Gondwana for much of the middle and later Paleozoic, whereas the extinctions were brief by comparison and far apart in time. Also, the Permian ice age was, at best, rather minor.

A recent suggestion of curious popularity is that a lack of oxygen in the waters of the ocean destroyed the reefs and other marine faunas. Widespread black shales are invoked as evidence for this, as is a sharp drop of the $^{13}C/^{12}C$ ratio in the early Triassic. Coming so late, the latter hardly explains extinctions that had begun 10 my earlier. More serious is that to kill coral reefs the sea should have been oxygen-free practically to the surface. Because the surface layers are well mixed by wind and waves, this is possible only if the atmosphere had also lost nearly all its oxygen. The impact on land animals would have been devastating, and there is no evidence for that. If I may be forgiven a censorious remark, too many ecological proposals that are made with a lack of understanding of how the ocean works, a disregard for careful chronology, and a failure to consider the entire land-sea-atmosphere system are truly irresponsible.

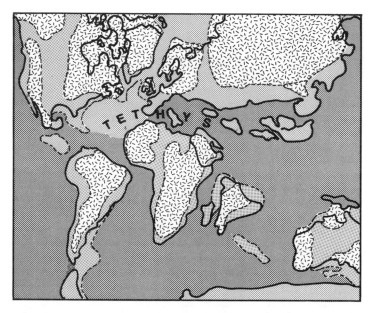

Figure 19.8. The late Cretaceous, 80 my ago, was, as we have seen, a time of vast epicontinental seas (light, dotted shading) and much reduced land areas. The new oceans were still small, and a wide tropical sea, the Tethys, separated the northern from the southern continents. Where the modern continental margins do not coincide with those of the Cretaceous, a dashed line marks the edge of very shallow seas.

19.3   THE GREAT CRETACEOUS DYING

A very different catastrophe occurred close to or at the end of the Cretaceous. Instead of taking its own good time as in the Permian, it was geologically brief, although exactly how brief depends on what group of organisms one is looking at. In the sea it was much less complete than the Permian crisis, although odd, because it affected many different groups in many different ways, and it appears to have affected life on land more strongly.

The marine faunas of the tropical Tethys sea (Figure 19.8) suffered much damage, but most families of shallow marine invertebrates crossed over into the Cenozoic. In the open ocean two subclasses of the cephalopods, the ammonites and belemnites, relatives of the squid equipped with bullet-shaped internal shells, vanished altogether, and so did most large marine vertebrates and large land animals, including the last of the dinosaurs.

The Cretaceous crisis may have been brief, but for many groups it was not instantaneous, nor did they all come to their end at the same time (Figure 19.9). Ammonites and belemnites took four million years

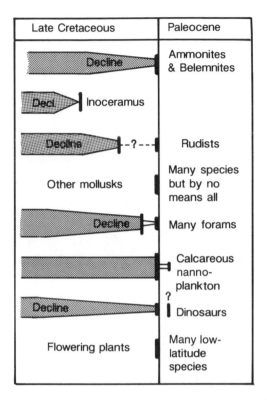

Late Cretaceous	Paleocene
Decline	Ammonites & Belemnites
Decl Inoceramus	
Decline --?---	Rudists
Other mollusks	Many species but by no means all
Decline	Many forams
	Calcareous nanno-plankton
Decline ?	Dinosaurs
Flowering plants	Many low-latitude species

Figure 19.9. The late Cretaceous saw many extinctions, but they occurred at different points throughout an interval of about ten million years. Many were incomplete, and quite a few had been preceded by a prolonged decline. For Forams read foraminifers; the calcareous nanno-plankton consisted mainly of coccoliths.

to decline, and disappeared just before the Cretaceous/Tertiary boundary (K/T boundary). The heavy-shelled bivalves of the genus *Inoceramus* also vanished well before the end of the Cretaceous, as did almost all of the reef-building rudists, mollusks that look deceptively like corals and build reefs. In the open ocean, numerous species of pelagic plankton (foraminifers and the small calcareous nannoplankton) disappeared simultaneously at the K/T boundary.

On land the extinction, except for the elimination of most large animals, was less spectacular than at sea. Earlier, the flowering plants had gone through a major adaptive radiation that had a considerable negative impact on the vegetarian dinosaurs, but that had stabilized well before the end of the Cretaceous.

The late Cretaceous flora of North America and Japan was affected by a cooling that removed many species from the mid-latitudes, but had much less impact farther north. In the Raton basin in New Mexico and some other places, the K/T boundary is marked by a clay layer between freshwater deposits below and a coal bed above. At the boundary, several tree species were temporarily replaced by ferns until the former flora reappeared in the coal. Ferns are good at colonizing new ground, suggesting that a forest of seed-plants and conifers suffered a brief but serious setback. The data is sparse and one should not make too much of it, but if the gap was widespread, a food shortage for large herbivores adapted to a different browse would seem likely.

The most famous of the Cretaceous extinctions is that of the dinosaurs, together with some reptiles of less popularity. Their demise is always depicted as sudden and dramatic, but in Africa, Asia and South America they declined slowly, and survivors into the later Cretaceous are not known. Only in North America do late Cretaceous dinosaur bones occur in stream deposits of the plains left behind by the vanishing Cretaceous sea (Figure 9.5). On the shores of that inland sea, 30 genera of dinosaurs flourished 10 my before the K/T boundary, but declined to about 22 genera 5 my later and 13 just before the K/T boundary. A few may have made it into the early Paleocene, but that is disputed. The problem with large land animals is that, with their low chance of preservation, an argument based on mere absence is always suspect.

As far as the record goes, the K/T boundary crisis merely hastened a long process of dinosaur decline, and the same is true for most marine groups. One might therefore feel that the final demise of less than a score of dinosaur species was a trivial event that does not deserve the name of catastrophe. The oceanic plankton crisis, on the other hand, was sudden and very dramatic (Figure 19.10). Although the Foraminifera had declined for a few million years, all but half a dozen species of the remaining ones vanished, as did the majority of all other plankton. Primary producers, grazers and predators, calcareous forms as well as siliceous ones, suffered great losses in less than 100,000 years, although many species of diatoms and dinoflagellates (another important planktonic primary producer) survived. In contrast, the benthonic fauna was not affected at all.

In many marine deposits the moment of extinction is marked by an unconformity or a thin layer of brown clay (Figure 19.10). Calcareous oozes return above the boundary, but with a different fossil assemblage, and many of the survivors were tolerant of broad salinity and temperature ranges.

It is not easy to say how close in time the extinctions were. Dinosaur bone beds in rivers are not found in juxtaposition with pelagic oozes, and the correlation rests on magnetic reversals. They show that the K/T boundary extinctions occurred within magnetic polarity interval

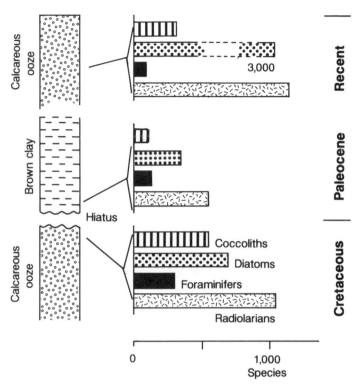

Figure 19.10. The most dramatic of the late Cretaceous extinctions was not the one that put an end to the dinosaurs, but the drastic reduction in the number of species of oceanic plankton, animals (foraminifers and radiolarians) as well as plants (coccoliths and diatoms). The diverse late Cretaceous marine flora and fauna were replaced almost instantaneously with a much impoverished one, and the return to a diversity comparable to that of today was much delayed. One look at the scale at the bottom is enough to understand how truly massive this extinction was. A hiatus usually marks the Cretaceous–Cenozoic boundary, and the first Cenozoic ocean sediments in the oceans are almost always brown clays rather than the calcareous and siliceous oozes of the preceding and following periods.

29R, from 65.5 to 64.8 my ago, an interval 700,000 years long. That is as close as we can get.

The late Cretaceous crisis is a fascinating event and in its great diversity not easily explained. A sober judgment would say that one or more simple, gradual causes, a cooler climate, a sea level change, or an altered vegetation cover, can explain each extinction except the plankton crisis, but the approximate coincidence in time of so many different deaths, even if some seem to have been rather protracted, makes one wonder.

There has been no lack of explanations. A carbon dioxide green-house so hot that it was life-threatening has been proposed, but the evidence suggests a cooling instead. Magnetic reversals have been blamed because,

Life, time, and change

when the magnetic shield briefly fails during a reversal, cosmic rays may do great genetic damage. It makes intuitive sense, but no more extinctions coincide with reversals than would be expected from chance alone.

More extravagant are supernova explosions, poisons from the tails of comets and the fascinating proposition that overcrowding made the dinosaurs so irritable that they refused to copulate. For none of these is there any evidence; in fact, for some, one wonders what evidence there could possibly be.

19.4   THE BOLIDES ARE FALLING!

The death of the dinosaurs has always appealed to the imagination and a more dramatic event than a slow decline of a few straggler species would be delightful. A decade ago, when the currently favored theory, the most dramatic and certainly loudest contender for the cause, appeared on the scene, this wish was fulfilled.

It was put forth by Luis Alvarez, a Berkeley physicist, and his geologist son Walter. In its simplest form it postulates the impact at about 100,000 km/hour of an asteroid or bolide, a large extraterrestrial body some 10 km in diameter. The blast was huge, of course, and made a big hole, but the father and son attributed the damage mainly to a dust cloud thrown so high that it obscured the sun for many months. Photosynthesis came to a halt on land and in the sea, and most plankton, with their short life-spans, became extinct. Actually, it is not clear how marine primary producers, who live only days or weeks, survived at all, unless the duration of the darkness has been grossly overestimated. In any case, there was little to eat for vegetarian dinosaurs, and the return of the sun after several months would have been too late also for those that survived on the carcasses of their plant-eating cousins.

More lurid visionaries speak of acid rain, raised by sulfuric and nitric acids to the strength of battery acid, an environment in which extinction would have been regarded as a blessing.

Over the years, evidence that an extraterrestrial object hit the earth about 65 my ago has accumulated. The brown clay at the Cretaceous/Cenozoic boundary commonly has an anomalously high concentration of iridium, an element rare in terrestrial and marine sediments, and an osmium (another noble metal) isotope ratio, both regarded as indicating an extraterrestrial origin. In addition, the clay often also contains quartz grains that show structural damage of a kind typical for a high-velocity impact.

When the impact hypothesis was announced, the scientific world divided almost instantly into believers and unbelievers. Each side set to work to show that the other was wrong with such diligence that it sometimes seemed that no one was able to spare the time to read what the other party had to say. Innumerable iridium anomalies have been found in marine sediments, while others, taking up the search for the consequences of the impact, looked for and found soot-like carbon particles in the boundary clay and presented those as proof that wildfires swept the world. The number of presumed impact craters has multiplied until almost every even remotely circular feature on the earth's surface has been so labeled. The benefit has been that, after some severe culling, we have realized that more large bolides have struck the earth (and the moon) than we used to believe; however, the K/T boundary crater, spotted many times, has so far failed to pass the test of rigorous close inspection.

The other side, while busily dismantling crater claims, has worked hard to show that the geochemistry of iridium, still poorly understood, is less simple than it was thought. This element is also produced by volcanic eruptions such as those of Hawaii, and was dispersed in large amounts by the plateau basalt eruptions that accompanied rifting, such as the Deccan Traps in India which have the right age for the iridium anomaly at the K/T boundary (Section 7.5). Moreover, iridium also occurs in seawater from which algal limestones scavenge it and produce iridium peaks like the one that coincides with the Ordovician mass extinction (Figure 19.1). Those iridium peaks have nothing to do with bolide impacts.

The search for soot has also been rewarding. It turns out to be quite common in marine clays, telling us that great forest fires burned long before human arsonists appeared on the scene. Even shocked quartz can be due to earthbound processes; it occurs, for example, in the South African volcanic pipes that yield most of our diamonds.

On balance, I would conclude that large bolide impacts have occurred more often than we thought and that it is quite possible that one of them came in at or near the K/T boundary, although not all criteria claimed for the event are unique to impacts. But that is only the beginning. After it has been established that such an unpleasant event took place, we must show beyond reasonable doubt that it coincided with the extinctions and that its demonstrated consequences were capable of causing those. Only then can we responsibly make the claims that continue to be broadcast so liberally for the K/T impact.

On both points there is reason for doubt, as any thoughtful student of Figure 19.9 should realize. Only the plankton crisis is abrupt, occurs at the K/T boundary and is close to an iridium peak, but even here the coincidence in time "within 100,000 years" is not good enough. For all other extinctions, the moment of the event cannot be tied to the boundary clay or an iridium peak, synchroneity is not proven, and many are at the end of a slow decline. As regards the means by which impacts exterminate life, they are almost wholly a matter of conjecture with little solid evidence to go by, and the different forms the extinctions take, as well as the marked incompleteness of some (Figure 19.10), are fair reasons to reserve our judgment.

19.5 GIANT IMPACTS AND GREAT VOLCANOES

While the public has swallowed impacts lock, stock and barrel as a general cause for mass extinctions, many paleontologists still harbor serious doubts. Geologists occupy the middle ground, but to an unhappy degree we have divided into those that believe and those that do not, two camps that tend not to listen to each other. What lessons can we draw from these tales of death and destruction, apart from the conclusion that scientists should communicate more with each other?

The world is being bombarded continuously with space debris ranging in size from microscopic dust to meteorites; more than a hundred meteorites weighing more than a pound hit the earth every year. Bolides much larger than the one that dug Meteor Crater in Arizona strike a few times per million years, and objects of the size proposed for the K/T event might come in once every 50 my or so. The numbers rest, I hope, on good statistics, but whether they also represent solid reality is another matter.

Terrestrial processes able to account for iridium spikes, osmium isotope ratios or shocked quartz have raised the minimum standard of proof, but they do not remove impacts as potential catastrophic events. The argument about the K/T impact is not about its possibility – it is possible – but about its reality, timing, and consequences. The current candidate, a putative impact crater in Yucatan, has increasingly good credentials, but the final verdict is yet to come.

The only reasonable alternatives to bolide impacts so far presented are giant volcanic eruptions of the kind that produced the Deccan Traps in India. Although mainly ignored by those who are committed to the impact hypothesis, the hard work of many geologists has shown that these eruptions, whose existence is not in doubt, are capable of wreaking the same kind of havoc claimed for bolide impacts, and leave similar evidence of their activity. This gives us two processes, both capable of great destruction of life, but which, if any, did do the deed?

That question has two parts. To answer it in the affirmative the time of the impact must be right and its consequences, as we can reasonably infer them, must agree with the pattern of the extinctions. The current state of affairs does not permit us to be complacent on either of these points, and that is why the misgivings of the paleontologists are better justified than the confidence of the public.

Timing is crucial. A bolide impact must **precede** its claimed consequences and, judging by the most plausible effects people have been able to think up, the course of the extinction must be swift, a thousand years or less. That is an interval that we cannot handle with confidence (Section 12.2). Here the giant volcanic eruptions have an edge; although they appear to have been brief, they do allow a few hundred thousand years for their destructive work to run its course.

The decision whether an impact was the sole or main cause of an observed mass extinction also hinges on how one defines that extinction. Is the slow decline of many taxa during the latest Cretaceous (Figure 19.9) to be regarded as part of the mass extinction or not? If we are sure that the asteroid was the cause of the event, the event must exclude all earlier extinctions, but the definition of mass extinction then becomes quite arbitrary.

19.6 SUDDEN DRAMA OR NATURAL INEVITABILITY?

For the moment, this seems a fair summation: (1) bolide impacts have happened, more rarely the larger they were, and some coincided with mass extinctions; (2) impacts are capable of causing mass extinctions, but how and to what extent is not well known; (3) many major extinctions run a course that either rules out an impact or requires multiple causes; and (4) giant volcanic eruptions capable of acting as agents of mass destruction have happened.

Extinctions of species and perhaps genera may, from our distant point of view, be regarded as about as inevitable as the death of the individual. At the other end, the greatest mass extinctions have the quality of great tragedy, unstoppable, some future day perhaps total, and entirely unforeseen.

Or are they? Figure 19.1 carries the suggestion that the mid-size mass extinctions of anywhere from 10 to 30 percent of the species happen roughly, very roughly, every 30 my, and a more complete compilation

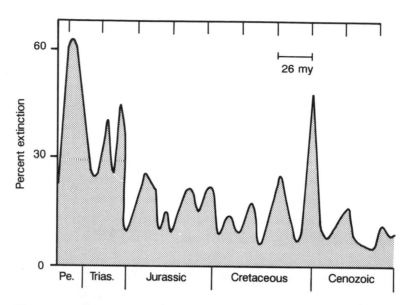

Figure 19.11. Alfred Fischer has suggested that there might be a 30 my episodicity in the state of the oceans. A similar repeat pattern can be faintly discerned in Figure 19.2. This sort of observation inspired David Raup and John Sepkoski to compile the comprehensive analysis of extinctions against time shown here in which they saw evidence of a 26 my periodicity. The top scale is marked in 26 my intervals to make it easier to judge whether this suggestion has merit or not.

of mass extinctions since the late Paleozoic (Figure 19.11) strengthens this suspicion. From this graph John Sepkoski and David Raup concluded that a recurrent catastrophic force struck the earth every 26 my, not as disastrously as the biggest events, but memorably nevertheless, and eliminated many groups of organisms on land and in the sea. The period is close to that of Fischer's cold and warm ocean cycle (Figure 10.15).

The reader, remembering how much geologists love periodic events (Section 12.4), will not be surprised that extinctions 30 my apart (or 32, 28, 26.5 my, your choice) have been welcomed as evidence for the periodic return of an agent of mass extinction. What sort of an agent might this be?

The challenge was taken up expeditiously and there emerged a hitherto unknown body, Nemesis, the death star, a small, dark and therefore unseen binary companion of the sun, capable of disturbing the paths of comets every 28 my. Comets then bomb the earth and other inner planets, fortunately causing only moderate damage most of the time. Nemesis was so catchy a concept that it took a long time to die, even though there is no evidence whatsoever for such a star, nor does it fit current models for the origin and state of solar and planetary systems. Disappointed romantics briefly replaced Nemesis with Planet X orbiting beyond Neptune, but Planet X too has proved to be an inferior brand. And if we return to Figure 19.11 with suspicion in our heart, it no longer seems so periodic, but merely an accident of bad statistics applied to dates with large error margins. But even today, the statistics still as atrocious as before, the concept keeps reappearing.

The extinction of species is the natural background to the history of life, and is nothing to get really excited about. Once in a long while, a major event or combination of them helps this process along in a more spectacular manner. The more dramatic the cause, the less common is the event. There is little here to astonish us. Bolide impacts and giant volcanic eruptions are potential causes of devastation, but it is too soon to say which of the two among the many possible causes for mass extinctions played a dominant role.

What is very clear is that the history of the earth cannot be compared, as we once thought, to a gently flowing stream, but instead resembles an extended set of cataracts and waterfalls. Neo-catastrophism is a better term for it than uniformitarianism.

PERSPECTIVE

The history of life is the history of evolution, and it rests on two uncomfortable truths: the great biological panorama is the fruit of random variety, selected by the environment and its changes, and the road from bacterium to man is not progress, but merely improved adaptation.

The general course of life, from Precambrian prokaryotes to flowering plants, from the thriving fauna of shallow Paleozoic seas to the human condition, is well documented. There is no doubt about the trend, and there are no inexplicable precursors, no Cambrian turtles or Cretaceous elephants. Diversity, except for the occasional setback, has continuously increased, and with it the demands on organisms living in an ever more complicated setting.

The setbacks, occasional though they may be, have been dramatic, and illustrate not only that nature is vulnerable and fallible, but also that science is not always as dispassionate as it ought to be.

Be that as it may, the result of the evolution of life and its interaction with its environment has been a surface of the earth that has been dramatically and forever altered. The world is different today from what it was in the beginning, not only because it is older, cooler, remodeled by plate tectonics, and endlessly refurbished by the geological cycle, but because of life itself. We cannot understand our world without understanding life and its interactions with its environment. It is life that has made a marginally liveable world into a comfortable one and kept it there through the eons. And it is ironic that we, ourselves a product of the co-evolution of life and earth, are well on our way to return our planet to a much less liveable condition.

FOR FURTHER READING

The history of life has spawned an enormous literature of which much is accessible to the non-expert. Richard Cowen's (1990). *History of Life* (Oxford: Blackwell) is easy to read and rich in content. More advanced is Briggs, D. E. & Crowther, P. R. (eds.) (1990). *Palaeobiology: A Synthesis* (Oxford: Blackwell Scientific). Help with the taxonomy and evolution of organisms can be found in Margulis, L. & Schwartz, K. V. (1988). *Five Kingdoms, an Illustrated Guide to the Phyla of Life on Earth* (New York: W. H. Freeman) and Clarkson, E. N. K. (1986). *Invertebrate Palaeontology and Evolution* (Boston: Unwin Hyman). The theory of evolution is cluttered with jargon and not to everyone's taste, but Raup, D. M. & Stanley, S. M. (1979). *Principles of Paleontology* (San Francisco: W. H. Freeman) is manageable.

As regards the evolution of human beings, the literature is large and I cannot begin to present even a selection here. Cowen's book (above: Chapters 18 and 19); and an article by Tobias, P. V. (1992). Major events in the history of mankind, in *Major Events in the History of Life*, J. W. Schopf (ed.), pp. 141–67 (Boston: Jones & Bartlett) can be used for a start.

SPECIAL TOPICS

Albritton, C. C. (1989). *Catastrophic Episodes in Earth History* (London: Chapman & Hall).

Alvarez, W. & Asaro, F. (1990). An extraterrestrial impact, *Scientific American*, **263**, 44–60.

Avise, J. S. (1990). Flocks of African fishes, *Nature*, **347**, 512–13.

Conway Morris, S. (1992). The early evolution of life, in *Understanding the Earth*, G. C. Brown, C. J. Hawkesworth & R. C. L. Wilson (eds.), pp. 436–57 (Cambridge: Cambridge University Press).

Conway Morris, S. (1993). The fossil record and the early evolution of the Metazoa, *Nature*, **361**, 219–25.

Cox, K. G. (1988). Gradual volcanic catastrophes? *Nature*, **333**, 802.

Grieve, R. A. (1990). Impact cratering on earth, *Scientific American*, **262**, 44–52.

Knoll, A. H. (1978). Did emerging continents trigger metazoan evolution? *Nature*, **276**, 701–03.

Knoll, A. H. (1991). End of the Proterozoic Eon, *Scientific American*, **265**, 42–49.

Knoll, A. H. (1992). Early evolution of the eukaryotes: A global perspective, *Science*, **256**, 622–28.

Knoll, M. A. & James, W. C. (1987). Effect of the advent and diversification of vascular land plants through geologic time, *Geology*, **15**, 1099–1102.

MacFadden, B. J. & Hulbert, R. C. (1988). Explosive speciation at the base of the adaptive radiation of Miocene grazing horses, *Nature*, **336**, 466–68.

Norell, M. A. & Novacek, M. J. (1992). The fossil record and evolution: Cladistic and paleontological evidence for vertebrate history, *Science*, **255**, 1690–93.

Ostrom, J. H. (1992). A history of vertebrate successes, in *Major Events in the History of Life*, J. W. Schopf (ed.), pp. 119–41 (Boston: Jones & Bartlett).

Owen, M. R. & Anders, M. H. (1988). Evidence from cathodoluminescence for non-volcanic origin of shocked quartz at the Cretaceous/Tertiary boundary, *Nature*, **332**, 150–52.

Rampino, M. R. & Caldeira, K. C. (1992). Major episodes of geological change: Correlations, time structure and possible causes, *Earth and Planetary Science Letters*, **114**, 101–11.

Raup, D. M. (1991). *Extinction: Bad Genes or Bad Luck?* (New York: W. W. Norton).

Richardson, J. B. (1992). Origin and evolution of the earliest land plants, in *Major Events in the History of Life*, J. W. Schopf (ed.), pp. 95–118 (Boston: Jones & Bartlett).

Runnegar, B. (1992). Evolution of the earliest animals, in *Major Events in the History of Life*, J. W. Schopf (ed.), pp. 65–94 (Boston: Jones & Bartlett).

Shear, W. A. (1991). The early development of terrestrial ecosystems, *Nature*, **351**, 283–89.

Life, time, and change

Stanley, S. M. (1984). Mass extinctions in the ocean, *Scientific American*, **250**, 64–71.

Stanley, S. M. (1987). *Extinctions* (New York: Scientific American Books).

Stebbins, G. L. & Ayala, F. J. (1985). The evolution of Darwinism, *Scientific American*, **253**, 72–82.

Wallace, M. W., Keays, R. R. & Gostin, V. A. (1991). Stromatolitic iron oxides: evidence that sea-level changes cause sedimentary iridium anomalies, *Geology*, **19**, 551–54.

Wilson, A. C. (1985). The molecular basis of evolution, *Scientific American*, **253**, 164–73.

Epilogue

Many and strange are the universes that drift like
bubbles in the foam upon the river of time.
        Arthur C. Clarke, *The Wall of Darkness*

When, in 1830, Charles Lyell set out to slay catastrophism, what he laid in the grave with a stake through its heart was already a ghost. His act of exorcism, however, was so convincing that for many geologists the word catastrophism was forever blasphemy.

Catastrophism, distant cousin of an older, theological view of the history of the earth, had well before Lyell's day evolved into serious science, even by today's standards. Its major practitioners, Louis Agassiz, Georges Cuvier, Roderick Murchison and Adam Sedgwick, were scientists, not living throwbacks to medieval metaphysics. Not for them the supernatural cataclysms that extinguish life wholesale, no repeated acts of creation, nor an earth a mere six thousand years old. They were fine observers whose theories were based on a detailed knowledge of the many crises of life we have discussed. Viewed without prejudice, the evidence was and is, as we have seen, mostly on their side. The gradualists for whom Lyell spoke are compelled to invoke extreme incompleteness of the record if they wish to explain away the evident staccato mode of earth history.

Lyell was well aware of this, but his target was the public rather than the scientific perception of catastrophism, and he deemed a strong and unambiguous statement to be necessary. What he did not foresee was that from this strong statement would spring a seemingly indestructible caricature, one that cast its spell even on Lyell himself who to the end of his life refused to accept the Ice Age because it smacked of catastrophism.

Uniformitarianism has been the subject of much exegesis since then, but catastrophism was permanently relegated to the history of science as an experiment that failed. Lyell's uniformitarianism contains, as Stephen Jay Gould has pointed out, several elements, some still accepted, others no longer regarded as reasonable. There is first a statement of method agreed to by every scientist, that the laws of nature are invariant in space and time. If rocks might have fallen upward in the Precambrian, geology ceases to be possible. Science, any science, can exist only if we assume that nature does not at will violate its own rules. From this statement of method follows one of procedure: the processes we observe in action today are likely to have acted also in the past, and must be the first ones to be invoked in any explanation.

Early 19th-century catastrophists accepted all this with equanimity, as modern geologists do, but Lyell went one step further by claiming that the present was not just the first and best key to the past, but the only one. Few of us would define uniformitarianism quite so narrowly; I only cite the Banded Iron Formation here as an example, even though we all prefer to be parsimonious with deviations from its precepts.

Processes have rates, and here the difference between Lyell and the catastrophists becomes a matter of principle rather than of degree. Lyell unequivocally stated that throughout the history of the earth change had been slow and gradual, never cataclysmic, and drew from this an important but erroneous conclusion: the configuration of the world had remained essentially the same since the beginning. As far as he could see, the course of history consisted of small, gradual variations on a single theme, a view not fully compatible even with the evidence available in his time.

The catastrophists rejected both points. Theirs was a world of major change, effected mainly in brief, dramatic moments, revolutions, explosions of life, great extinctions. They saw the long, patient work of erosion swiftly undone by the rise of new mountains, the slowly matured landscape swept away by a sudden rise of the sea. Their earth history resembled the punctuated equilibrium debated by the paleontologists (Section 16.3) and the record of the rocks is compatible with it.

The disagreement was, as usual, partly a matter of definition, of semantics. Catastrophe, crisis and cataclysm are powerfully emotional words; they convey suddenness, totality, inevitability, blind destruction without hope of salvation. Yet earthquakes or flash floods, both instantaneous events, are neither more nor less catastrophic than ten-year droughts. The impact of a nuclear bomb in England would be no more catastrophic than the Black Plague of the 14th century, which lasted nearly a hundred years. Catastrophes must be uncommon to deserve the name, but totality is not required, and the perception of their instantaneousness must be tuned to the perspective of time. The catastrophic post-glacial extinction of large mammals was surely experienced by contemporary human beings as a slow, almost imperceptible decline.

Even with a tolerant definition, catastrophism would have been anathema to Lyell and later gradualists, but many modern geologists, though affirming the falsehood of catastrophism, invoke similar concepts under other names. Long shorn of its supernatural causes and theological affinities, early 19th-century catastrophism quite resembled the spirit of this book. How close we each stand to a strictly gradualistic or to a more catastrophic view is a matter of degree and personal preference. To draw boundaries serves little purpose, but few would disagree that the time of neo-catastrophism has come.

In preceding chapters we surveyed the historical evidence, noting time and again the cyclic nature of events, and commented on human longing for periodicity. In the 1930s, the great German geologist Hans Stille thought that he saw rhythm in mountain building, and a few years later his Dutch colleague J. H. F. Umbgrove felt the pulse of the earth in the subsidence of basins, sea level changes and climate. They had many followers (Section 12.4), and an episodic return is a feature of many of the earth's processes, but proof that this pulse truly exists and beats regularly has been elusive. Perhaps we desire so strongly to feel the pulse of the earth, because Lyell's gradual and unvarying history is so unbearably monotonous. Put more loftily, is there an underlying harmony among the many sequences of seemingly independent events that make up the history of the earth, or are we listening to the tuning of an orchestra?

It is impossible to answer this last question today. It is statistical in nature and needs a statistical answer, because if we deal with several parallel series of random and mutually independent events, coincidences in time will occur by chance alone. Strange as it may seem, a record 4.5 billion years long is too short for many kinds of events; it contains too few ice ages, supercontinents, and catastrophes of life to determine whether they come and go periodically or merely randomly. It will help if we can show someday that certain events are truly synchronous, but even then, simultaneous does not necessarily mean causally related. Our best chance to find out whether the processes of the earth march to one drummer or to many would be to conceive of a unifying cause, and then test that hypothesis.

There is a much-neglected side to Lyell's uniformitarianism. Lord Kelvin perceived it as a fatal flaw when he pointed out that the earth is not, and cannot be, a perpetual-motion machine. There must be a beginning, there will be an end, and the middle where we live should be different from either, or the second law of thermodynamics will be violated. Whether one prefers a gradualist or a catastrophist view, there must be direction; the engine must run down. Geologists have given much less time to recognizing and describing this inexorable march of time toward the end than to debating possible deviations from its path.

Our ground is not secure when we attempt to discern a direction in the evolution of the earth, direction in a non-teleological sense, of course. The diminishing production of radioactive heat firmly says that there is a direction, a downward one, but apart from the unresolved debate about the nature of early Precambrian plate tectonics and the evidence for a gradual and diminishing growth of the continents, we have little to go by and have thought about it less. The Precambrian past is too dark, the Phanerozoic present too short, to reveal a clear trend in the dynamic behavior of the earth. Perhaps we shall grasp the problem better when we deepen our knowledge of the terrestrial planets, frozen in earlier states of development than the earth.

On the other hand, there can be no doubt about the change of life toward ever-greater specialization and diversity. The theory of evolution says that this course was inevitable once life had been set in motion and that it cannot but continue as long as conditions on earth do not put an end to it.

This cool phrase conceals what is to me the greatest question of all. The earth differs fundamentally from all other planets of the solar system in the presence of liquid water and a gaseous but not poisonous atmosphere, and because conditions at its surface are so well tuned to life's needs. The earth has maintained this narrow environmental window, within which life is possible, for more than 3.5 billion years without failing once. We are not free to argue that this is irrelevant and that, had our planet been differently conditioned, life would have adapted in different ways, because the conditions that permit the necessary biochemical reactions to take place are so narrowly defined and coincide so closely with those the earth maintains. Nevertheless, and though life can be sustained because the earth is the way it is, its origin on the primordial planet is a surprising event, and the persistence of favorable conditions over nearly four billion years is scarcely less so. Yet life came and stayed; to import it from elsewhere merely defers the first problem and does not solve the second.

The issue is so momentous because in 1543 Nicolaus Copernicus taught the world that there is nothing unique about the position of the earth within the universe and that hence man, its observer, does not have a privileged status. This Copernican principle was later generalized to mean that, apart from merely local irregularities such as galaxies, the universe appears the same to its observer no matter from which point it is viewed. The fact that scientific experiments are reproducible even though the earth, between one attempt and the next, moves through time and space, is a powerful illustration of that point.

Of late it has occurred to some that we may have been too zealous in our application of the Copernican principle. No one suggests that earth and man do, after all, occupy the center of the universe, but one can argue with some force that life's existence imposes severe limits on the origin, history, and nature of the universe that contains it. If that is correct, we might be able to explain why the universe is the way it is by reasoning backward from the presence of life and of ourselves, the observers. This is the anthropic principle, first voiced by Brandon Carter, then of Cambridge University. It can be used to advantage to clarify a number of conditions of the universe that otherwise seem arbitrary, and scientists, from long experience, are wary of apparent arbitrariness in nature.

Take, for example, the age of the universe. It should not have been ten times less, or elements like carbon, needed to make physicists, would not yet have been available. Make it an order of magnitude greater, and the suns required to warm planets and sustain life would long since have been extinguished. Reduce gravity ten times, and no planet would hold together long enough for observers to stand on it; increase it ten-fold, and the only suns in the universe would be blue giants, too ephemeral to allow life the time to evolve.

Is it plausible that the universe should be so special, so precisely fit to accommodate life, to house its own observers? There must be an infinity of possible universes in different states that would be equally probable. We would more comfortably accept our special universe if we knew that other, different ones did also exist. Exploring the anthropic principle, physicists have considered this possibility and found that the concept of multiple universes also helps in elucidating other oddities of our own world.

Multiple universes, for instance, illuminate a puzzling aspect of quantum theory. This theory says that the position of an elementary particle, a photon for example, can be described by an equation that gives for each point on its path the probability that we shall find it there if we attempt to detect it. In a light wave, this yields many positions of equal probability; yet we find the photon at only one of those, not at the others. This clash between a probabilistic world and a deterministic one is perturbing. If many positions are equally probable, why should the particle exist in only one of them? Perhaps, one reasons, the equation describes simultaneously an infinity of universes. We, who live in only one of them, find the photon only in one spot.

If an infinity of universes is possible, some with and some without observers, do we assume that they all exist? The answer should, in principle at least, be affirmative, but that would be difficult to confirm. In fact, one might argue that the only possible confirmation of the existence of a universe is that one observes it to exist. It follows then, that only universes that contain observers, that hold "life," need be assumed to be real.

We might carry this kind of reasoning over to the history of the earth. If the presence of life sets limits on the conditions of the universe at its birth, its presence on earth must equally limit the origin and history of our planet and, inevitably, of the solar system. Thus, the issue is not how, in the face of incredible odds, life arose here. Rather, it is the inverse. What can we say about the birth of the solar system and of the earth, starting with the knowledge that life demands certain conditions for its origin and its persistence? What does that tell us about the early and later atmosphere, about the weak and pale sun, and about the many other things that might so easily have gone wrong and wiped out life, but did not do so?

One is easily tempted to see a miracle or a guiding hand in the performance of that infinitely complex system of tectonics, ocean, climate, and life that has maintained a livable condition on the surface of the earth for the past four billion years. It is not easy to accept that the laws of physics and chemistry, the presence of innumerable positive and negative feedbacks in the interactions between the infinite components of that system, and a healthy dose of unpredictable chaos have brought this about, no easier than it is to accept that the evolution of life is a random, directionless process.

The conflict between accepting what science teaches us and what the human heart would like to believe is well illustrated by James Lovelock's Gaia concept that places life in charge of the functioning of our planet. It is a lovely thought, a tempting one too, because it is a form of religion and the human soul requires the comfort of a guided universe; it needs religion. Alas, it is also unnecessary, because the world as it was, has evolved, and now exists, is not inexplicable. It is merely very complex, and life plays a role in it, but not the main one. To understand the world through the science of the earth demands a simultaneous profound understanding of many difficult subjects and a great skill in cross-disciplinary synthesis. Those requirements few if any of us can meet today, and confusion and impatience are the result, but that does not mean that we should turn to mysticism.

Ideas like Gaia or, as an absurd example recently in the news, the smashing of Pangaea by an asteroid, have of late proliferated. Some derive from an unwillingness to accept that understanding the earth is hard, slow work, accomplished in small steps, and not soon finished; they want a shortcut. Some are convinced that ideas are all-important while assembling the evidence is menial work and can be left to others. There is also a deep-rooted (and ancient) need to believe that such a marvelous machine must have a great designer. Behind it all lurks an alienation from the everyday world of science which reacts so slowly, demands proof at every step so unreasonably, discards old ideas only when they are obviously impossible and refuses to reward new genius immediately.

Still, it is everyday science that has created over the centuries elapsed since Pythagoras and Aristotle, Galileo and Newton, Hutton and Darwin, our understanding of the way the world works. The Gaia hypothesis is charming, but it has slight substance, presents no evidence, and offers little more than comfort and love. Gaia-inspired experiments, designed to show that life in the ocean is capable of controlling climate, have failed to find a significant effect. More seriously, taken over the very long term, the influence of photosynthesis, of life, on the oxygen and carbon dioxide budgets of the earth is minor compared to the lifeless process of weathering.

We have drifted across the border into metaphysics and must retreat. That does not mean that we should not admit that, if ever science would remove all of the mystery (and it will not), we should lose much. What it will not do is remove the beauty of the universe and the awe it inspires in its infinite intricacy, as T. S. Eliot said in *Four Quartets*:

> The dance along the artery,
> The circulation of the lymph
> Are figures in the drift of stars.

Glossary

This listing contains the more common terms used throughout the text. For others, reference should be made to the index.

abyssal Used for the environments of the deep-sea floor or the water just above it.

acritarchs Complex, single-celled eukaryotic algae with a preference for a planktonic life offshore.

adaptive radiation Rapid development of new forms to take advantage of new environmental opportunities.

aerosol A suspension of small, colloidal droplets in air.

albedo The proportion of incoming solar radiation which is reflected back by the earth's surface.

ammonites Extinct group of cephalopods (squid, octopus) with a flat, spiraling external shell.

andesite Extrusive volcanic rock intermediate in composition between basalt and granite; major component in subduction zone volcanism.

angiosperms Flowering plants; first arrived in the Cretaceous.

Antarctic Bottom Water Cold, saline water formed at the surface around the Antarctic ice margin; its high density causes it to sink and flow northward just above the bottom.

aphelion Point in the orbit of the earth where it is farthest from the sun.

archaeocyathids Extinct group of reef-building organisms, almost certainly sponges.

arthropods Invertebrate phylum that contains the insects, crustaceans (e.g. crab, lobster), and spiders.

asteroid Body of rock larger than a meteorite, orbiting within the solar system.

asthenosphere Hot, soft layer of the mantle on which the lithosphere floats.

autotroph Organism capable of synthesizing its own organic matter.

Glossary

azimuth   Used here for the direction from a point on earth to the magnetic pole.

back-arc basin   Depression lying behind the volcanic arc above the subducted slab, and apparently spreading; common in ocean–ocean subduction, but not clearly developed in continents.

Banded Iron Formation (BIF)   Precambrian rock consisting of alternating laminae of abundant silica with a little ferrous (reduced) iron and abundant ferric (oxidized) iron with some silica.

basalt   Fine-grained, dark igneous rock rich in magnesium and iron-bearing minerals and poor in silica; characteristic for the oceanic crust.

belemnites   Extinct group of cephalopods (squid, octopus) with a cigar-shaped internal shell.

benthos (adj. benthonic or benthic)   Plant or animal life on the bottom of the sea.

biostratigraphic zone   Geographically extensive fossil assemblage that characterizes rocks of a well-defined age.

black shale   Dark, fine-grained sedimentary rock rich in organic matter; often source bed for oil.

blueschist   A generally fine-grained metamorphic rock formed under high pressure and low temperature; characteristic for subduction.

bolide   Any solid object from outer space; term usually restricted to objects larger than common meteorites.

boreal   Adjective denoting northern regions, northern climates, northern vegetation.

brachiopod   Marine organism with a two-valve shell; resembles the molluskan bivalves (clams), but unrelated to them; prominent especially in the Paleozoic.

catalyst (catalysis)   Chemical that accelerates a chemical reaction without itself participating in the reaction.

cephalopods   Group of mollusks such as octopus and squid; extinct groups are the nautiloids, ammonoids and belemnoids.

chaos theory   Deals with the unpredictable elements in natural systems.

chemical evolution   Term used for the synthesis by inorganic means of basic organic compounds of life, a precursor of the evolution of life itself.

chloroplast   Organelle within a plant cell; the site of photosynthesis.

chondrite, carbonaceous   Type of meteorite rich in water and light elements such as carbon; assumed to resemble the original material of the mantle.

chromosome   Thread-like elements in the eukaryote nucleus that carry the genes.

cladistics   Quantitative method of establishing relationships between life forms that is based on characteristics often taken to imply ancestor–descendant relations.

cnidarians   Group of invertebrates that includes corals, sea pens and jellyfish.

coccoliths Microscopic marine plants (nannoplankton) with calcareous skeletons; an important component of biological productivity in the oceans since the middle Mesozoic.

coelom Cavity within the body of the higher animals; it lends strength, among other functions.

collagen Strong, flexible, fibrous protein that lends strength to tendons, cartilage and bones.

compaction Compression of sediments under the overburden deposited on them; reduces pore space and sediment thickness.

continental shelf Shallow water zone, usually to a depth of a little more than 100 m, separating shore from continental slope.

continental slope Steep edge of the continental block; beyond lies the oceanic crust of the deep-sea floor.

convergence or collision Motion of two plates toward each other. A convergent or collision boundary is either a subduction zone or a mountain range (orogen).

convection current Process of heat transfer by rising hot fluid; once cooled, the fluid moves away laterally and sinks.

core of the earth Innermost unit of the planet, consisting of a solid inner and a probably liquid outer core, both made of nickel and iron.

Coriolis force Drives a moving mass (ocean or air current for example) to the right in the northern hemisphere when viewed in the direction of motion, and left in the southern hemisphere.

correlation Procedure for determining the equivalence of two outcrops of rock, two formations or fossil zones, either in age or by virtue of once having been connected.

craton Stable, Precambrian portion of a continent.

cosmic rays High-energy radiation reaching the earth from space; cosmic rays convert nitrogen to unstable carbon-14, an isotope of great importance for dating archaeological finds.

crust Uppermost solid shell of the earth, with a density lower than the mantle, and separated from it by the Mohorovicic discontinuity.

crust, continental Low-density (2.7–2.8 grams/cubic centimeter) crust that forms the continents; it consists mainly of granite and associated rocks.

crust, oceanic High-density (3.0 gram/cubic centimeter) basaltic crust forming the floors of the oceans.

cyclothem Sequence of shallow marine and coastal sediments, often including coals, that repeats itself vertically.

declination Angle of deviation of a compass needle from true north. In ancient rocks, determination of the paleomagnetic declination gives us the direction to the magnetic pole at the time the rock became magnetized.

Glossary

deformation   The folding and faulting of a rock sequence.

diatoms   Microscopic plants with a siliceous skeleton living in marine and fresh water; major component of the biological productivity in the oceans since the middle Mesozoic.

divergence   Plate-tectonics term describing plates that drift away from each other. A divergent plate boundary is usually a mid-ocean ridge.

diversity   Measure of the variety of life forms in a region, environment, the world, or during a given interval of time; usually expressed as the number of species, genera or families.

DNA   Deoxyribonucleic acid; the carrier of genetic information.

dolomite   Sedimentary rock, usually the product of chemical deposition, consisting of calcium–magnesium carbonate rather than, as does limestone, of calcium carbonate only.

eccentricity   Degree to which the orbit of the earth deviates from a circle.

echo-sounding   Depth-measuring technique that measures the travel time of a sound pulse transmitted to the seafloor.

end moraine   Wall of glacial debris formed at the margin of a glacier or icecap by melting.

epeirogeny, epeirogenic   Gradual uplift or subsidence of large regions without significant deformation by folding or faulting.

epicontinental   Used for seas flooding large parts of a continent.

episodic   Event or process that repeats itself over time without a predictable pattern.

equinoxes   Dates in spring and autumn when night and day are of equal length.

eukaryote   Cell with a distinct nucleus bound by a membrane, and possessing organelles; cell type of higher plants and animals.

Euler's theorem   States that the motion of a point or surface on a sphere can be described as a rotation around an axis through the center of the sphere. On this theorem rests the mathematical analysis of plate movements in plate tectonics.

eustatic   A global change in sea level.

evaporite   Sedimentary rock produced by evaporation of seawater and precipitation of its salts; contains a wide range of components, among which rock salt and gypsum are the most common.

event stratigraphy   Local or regional correlation system based on brief, unusual events such as volcanic eruptions, huge floods or sudden extinctions of life forms.

exotic terrane   Large geological body consisting of rocks of a nature, structure and history so different from those surrounding it that it is assumed to have arrived at its location from a distant point of origin.

facies   Characteristics of sedimentary rocks that reflect the environment of deposition in which it was formed.

fault   Fracture in a rock body along which one side has been displaced relative to the other.

feedback   Occurs when a process is self-reinforcing (positive feedback) or self-limiting (negative feedback).

Foraminifera (foraminifers, or forams for short)   Group of microscopic marine organisms belonging to the Protozoa; they are planktonic or benthonic and are grazers and predators. Their calcareous shells provide a major part of the paleoceanographic and paleoclimatic record.

fore-arc and fore-arc wedge   Zone at the boundary of colliding plates just behind the trench, where sediments and to some extent oceanic crust are being deformed.

formation   Suite of rocks with characteristics distinguishing it from other suites nearby; traceable or mappable over a reasonably large area.

fracture zone   Dislocation at right angles to a mid-ocean ridge; usually marked by a set of transverse ridges and troughs.

gateway or seaway   Connection between two ocean basins.

genetic drift   Random fluctuation in the genetic makeup of a small population that, in time, can cause that population to acquire different characteristics.

geosyncline   Large, elongated depression in the earth's crust in which thick sediments accumulate; deformed into a mountain range.

glacial   Interval of cold climate producing vast continental icecaps.

gneiss   Coarsely crystalline, banded, quartz-rich, high-grade metamorphic rock usually derived from granite or sediments.

granite   Coarsely crystalline, intrusive igneous rock containing mainly quartz and feldspar; typical for continental crust.

graywacke   Poorly sorted sandstone rich in rock fragments (often of volcanic origin) or clay.

green-house earth   Warm state of the earth without polar icecaps and with much smaller temperature differences between polar and equatorial regions than today. The earth has been in a green-house state over much of its existence.

green-house effect   Warming of the climate due to an increase in atmospheric carbon dioxide, which traps heat reflected from the earth's surface.

greenstone belt   Elongated, folded complex of moderately metamorphosed Precambrian sediments and volcanic rocks of oceanic origin.

Gulf Stream   Major warm surface current system issuing from the Caribbean; warms the northeastern North Atlantic, including northwestern Europe as far north as Svalbard.

Glossary

heterotroph Organism incapable of synthesizing its own food; derives needed organic substances from other organisms.

hiatus Gap in the rock record due to erosion or non-deposition.

hotspot Anomalously hot area deep below the lithosphere; source of volcanic activity as a plate passes over it.

hydrothermal Adjective denoting hot waters in the crust, their activity and deposits; here used for springs resulting from ocean waters circulating through hot, new oceanic crust.

ice age Interval, usually lasting many millions of years, during which icecaps cover polar and subpolar lands. The earth has been through at least five ice ages, and is in an ice age now.

ice-house earth State of the earth when icecaps cover the polar regions and equatorial–polar temperature differences are at a maximum.

ice-rafted Denotes sediment particles ranging from silt to boulders in size, transported and dropped by ice into deep-sea sediments.

igneous Rocks formed by cooling from a molten state (magma or lava).

inclination The angle between a suspended magnet and the horizontal plane; it measures latitude.

interglacial Warm interval between two glacial periods.

interstadial Warmer interval within a glacial period, but not warm enough to produce full retreat of icecaps.

isostasy (isostatic compensation) Rise and fall of the earth's crust as a result of changes in its buoyancy due to loading or unloading of, for example, water, ice or sediments.

isotopes Elements with the same number of protons but a different number of neutrons in the nucleus; they have very nearly the same chemical properties, but different atomic weights. Isotopes can be either stable or unstable (radioactive).

jetstream Fast, high-altitude airflow from west to east at mid-latitude, with major influence on seasonal weather as it changes its path from year to year.

lamination (laminated) Sediments displaying very fine bedding on a scale of millimeters.

lava Molten rock, usually so called when it flows from a surface fissure or a volcano; when congealed, it forms an extrusive rock.

little ice age Anomalously cold time during an interglacial interval; no new icecaps form, but mountain glaciers descend far down their valleys. The recent Little Ice Age began in the late Middle Ages and ended in the 19th century.

magma Molten rock existing deep in the earth; when congealed, it forms igneous intrusive and extrusive rocks.

magnetic anomaly   Area of the oceanic or continental crust where the magnetic field deviates from the normal earth's field.

magnetic polarity   Direction of the earth's magnetic field; it reverses from time to time, causing magnetic north to become magnetic south and vice versa.

magnetic polarity reversal time scale   Time scale based on the episodic reversals of the earth's magnetic field. Reversals are dated by radioactive isotopes in volcanic rocks.

mantle   Part of the earth between the crust and the core.

maria   Dark areas of congealed lava on the moon, probably formed by bombardment with space debris, asteroids and meteorites.

margin (active)   Produced at a convergent (subduction) or transform fault plate boundary.

margin (passive)   Formed initially by continental rifting, then widened into a new ocean.

marsupial   Mammal not equipped with a uterus suited for prolonged care of the embryo; the embryo is carried in an external pouch.

metamorphism   Change in composition, mineralogy, or structure of a rock resulting from changes in temperature and pressure; common as a result of deformation, very deep burial, or intrusions of hot magma.

Metazoa   Multicellular animals with differentiated tissues (all higher animals).

micro-continent   Small continental fragment surrounded by oceanic crust.

Milankovitch theory   States that glacial and interglacial climates are caused by periodic variations in the orbit of the earth around the sun.

mobile belt   Former zone of convergence or collision.

Mohorovicic discontinuity (Moho)   Boundary between crust and mantle marked by a sharp downward increase in the velocity of earthquake waves.

monsoon   Climate of alternating wet and dry seasons, caused by seasonal differences in airflow between a large continent and the adjacent ocean.

mutation   Genetic change that results in offspring with new characteristics; increases variety within a species.

normal fault   Fault along which one block moves vertically with respect to the other.

notochord   Rod of cartilage that supports the body; evolved to the spine of the vertebrates.

nutrients   Trace elements (e.g. nitrogen, phosphorus, potassium) essential for the biological production of organic matter.

obliquity   Angle between the rotational axis of the earth and the plane of its orbit.

Glossary

ophiolite   Assemblage of oceanic basalts and associated deep-sea sediments considered to have formed at a mid-ocean ridge..

orbital theory   See Milankovitch theory.

organelle   Distinct body within the eukaryote cell with a special function, e.g. photosynthesis or energy housekeeping.

orogen   Mountain range formed by deformation.

orogeny   Mountain building, involving deformation, metamorphism, uplift and the intrusion of igneous rocks; takes place during plate collisions, especially those involving two continents.

oxygen isotope stage   Zone in the rock record defined by changes in the oxygen isotope ratio.

ozone shield   Upper stratospheric layer rich in ozone (O_3) that protects the earth and life on it from the deleterious effects of ultraviolet radiation.

paleomagnetism   Record of the ancient magnetic field of the earth in rocks.

palynology   Study of pollen and spores, often used as a means to understand the vegetation and climate of the past.

Pangaea   Supercontinent formed and completed in the Paleozoic.

Panthalassa   World ocean surrounding Pangaea.

pelagic   Of the open ocean, used for sediments, organisms.

perihelion   Point on the orbit of the earth or any other planet where it is closest to the sun.

periodicity (adj. periodic)   Designates events that predictably repeat themselves.

per mil (parts)   Parts per thousand.

photosynthesis   Synthesis of organic compounds by organisms, using sunlight as the energy source.

phytoplankton   Microscopic floating oceanic plants.

planetesimal   Small planetary body.

plankton (adj. planktonic)   Organisms, mainly microscopic plants and animals, that drift in the surface layers of the oceans.

plateau basalt   Thick, extensive sheets of mantle-derived basalts, also known as flood basalts, usually associated with the onset of continental rifting.

pluvial   Climatic interval characterized by increased rainfall and/or decreased evaporation. Formerly thought to have affected non-glaciated regions during glacial periods.

polar wandering   Assumption that the magnetic poles of the earth have wandered far and wide in the past as a result of plate movements.

pollen analysis See palynology.

precession Shift with time of the points on the earth's orbit where the equinoxes occur.

predation (predators) Preying of animals upon one another.

primary producers Photosynthesizing autotrophs forming the base of the food chain.

primary (biological) productivity Production of organic substances by organisms using sunlight (mainly) or chemical reactions (rarely) as a source on which the entire pyramid of life depends for its food.

primates Order of mammals that contains human beings as well as chimpanzees, gorillas, monkeys, lemurs, etc.

prokaryote Cell lacking a membrane-bound nucleus and organelles; bacteria and certain algae.

punctuated equilibrium Model for evolution combining sudden, essentially instantaneous changes with long periods of no change.

quantum theory An advanced field of physics that rests on the assumption that energy comes in discrete packages.

quartz Silicate mineral (SiO_2) especially common in granite and gneiss and in sandstones derived from granite; highly resistant to weathering and characteristic for continental crust.

quartzite Sandstone with more than 90 percent quartz; it implies a source in continental crust.

radiolarians Microscopic marine Protozoa with siliceous shells.

reflector A boundary between strata that reflects energy pulses transmitted by an earthquake or by an artificial sound source (in seismic reflection profiling).

regression Seaward shift of the shore due to a fall in sea level, a rise of the land, or sedimentation.

rift (rift valley) Trough between two zones of normal faulting, with a down-dropped central block. It may signal continental breakup, and also occurs on mid-ocean ridge crests.

RNA Ribonucleic acid; conveys instructions from DNA to the cells or organs where they are to be carried out.

salinity Measure of the salt content of seawater.

savanna or steppe Tropical or subtropical grasslands with few, if any, trees.

seismic tomography Use of earthquake data to produce three-dimensional images of the structure of the interior of the earth.

semantics Study of meaning in language.

Glossary

sequence analysis and sequence stratigraphy   Use of transgressive and regressive deposits displayed on seismic reflection records for stratigraphy, and especially for sea level studies.

shale   Fine-grained sedimentary rock consisting of silt and clay; tends to separate in thin sheets along bedding planes.

slate   Hard, fine-grained metamorphic rock formed from clay; its fine lamination is due to pressure, not to depositional bedding.

solar constant   The amount of solar radiation arriving at the earth as measured just outside the atmosphere; presumed to have increased significantly since the earth was formed.

solar cycles   See sunspots.

solstice   Date of the longest or shortest day, the days on which the sun is farthest from the equator.

speciation   Process by which new species are produced.

stadial   Colder phase within a glacial interval.

stasis   Used in evolutionary theory for a state of no change or equilibrium.

steppe   See savanna.

strata   Rock layers or beds.

stratigraphy   The study of bedded (stratified) rocks and their arrangement in space and time.

stratosphere   Zone of the atmosphere extending from about 9 to 24 km (30,000 to 80,000 ft).

strike-slip fault   Fault along which the displacement of rock bodies is mainly horizontal.

stromatolite   Finely layered mound of limestone produced by an algal mat; particularly common in the Precambrian, and one of the earliest signs of the presence of life on earth.

strontium isotope dating   Dating rocks by means of the ratio of the two strontium isotopes ^{87}Sr and ^{85}Sr.

subduction   Process by which a plate, always an oceanic one, is dragged under an adjoining plate; subduction compensates for the continuous creation of new oceanic crust on mid-ocean ridges.

sunspots   Dark blotches on the sun indicating unusual solar activity; they vary periodically in number and intensity (solar cycles) and may influence climate, but the mechanism of the interaction is not known.

supercontinent   Continent composed of all continental crust fragments extant on the surface of the earth.

superocean   Single ocean surrounding a supercontinent.

suture   Join or weld left from a former collision of two continents.

taxonomy   Classification of organisms into categories (taxa, sing. taxon).

tectonics   Study of the movement and deformation of the earth's crust and mantle.

Tethys   Large, low-latitude embayment or gulf on the east side of Pangaea. Later a vast gulf separating the northern and southern fragments of Pangaea during the Mesozoic.

thermocline   Boundary between surface water and deep water in the ocean, usually at a depth of 100 to 200 m, and marked by a large, abrupt downward drop in temperature.

thermodynamics   Subdiscipline of physics dealing with the relations between various forms of energy.

thermohaline circulation   Deep-water circulation of the oceans driven by density contrasts due to high salinity and low temperature.

time series analysis   Quantitative study of variables that change with time, for example climate.

trade-wind   Northeasterly, easterly, and southeasterly winds blowing between $c.$ 15° and 30° N and S latitudes.

transform fault   Fault with horizontal movement connecting two segments of a mid-ocean ridge; more broadly, a plate boundary along which plates move horizontally past one another without either convergence or divergence.

transgression   Landward move of the shore produced by a rising sea or a sinking land. Erosion of the shore is self-limiting and does not normally cause a significant transgression.

tree-ring dating   Most trees annually add a layer of tissue to their trunks, so producing a set of annual tree-rings. Counting tree-rings and dating individual rings by means of ^{14}C decay is used to calibrate the radiocarbon time scale in calendar years. Tree-rings also record climate changes.

trench   Deep trough lying just seaward of the line of contact of two converging plates.

trilobites   Fossil arthropods characterized by a body that is laterally and longitudinally divided into three lobes.

tundra   Treeless plain at high latitudes characterized by low vegetation and having a marshy surface in summer because it is underlain by permanently frozen ground.

turbidity current   Fast-flowing water mass with a high density due to much suspended sediment; its deposits are called turbidites.

unconformity   Surface separating two beds of sedimentary rock and representing a gap in time during which sedimentation ceased or erosion removed the rock record.

Glossary

underthrusting   Process pushing one rock body underneath another along an inclined fault plane. It is usually accompanied by intense deformation and some metamorphism and is common at the collision edge of two plates.

uniformity or uniformitarianism   Theory that earth processes have throughout been due to the same processes operating in the same combinations and at the same rates as today.

upwelling   Water rising to the surface from intermediate depth, usually as a result of surface water being driven away by wind or currents; the main agent in recycling nutrients to primary production in the surface waters of the sea.

Vail curve   Popular name for the curve describing Paleozoic, Mesozoic and Cenozoic sea level changes constructed by staff of the Exxon Corporation.

varves   Thin, annual layers of fine-grained sediment; formed for instance in glacial lakes as a result of the annual melting and freezing of glacier fronts.

vascular plants   Plants equipped with a system of vessels to transport water, nutrients and organic compounds.

weathering   Decay of rocks at the surface of the earth under the influence of wind, water, plants, temperature changes, etc.

Wilson cycle   Plate-tectonic cycle beginning with the formation of new oceanic crust on mid-ocean ridges and ending with the subduction of the now old crust into the mantle.

wobble   See precession.

Younger Dryas   Interval when glacial conditions returned briefly, long after the icecaps of the last glacial had begun to retreat.

zircon   Zirconium silicate; a very resistant mineral, and the only remnant of the earth's most ancient rocks.

Sources of illustrations

TITLE PAGES

Itinerary
W. F. A. Zimmerman, *De Wonderen der Voorwereld*, p. 1, Rotterdam: D. Bolle, 1892.

Foundations
Charles Lyell, *Lectures on Geology Delivered at the Broadway Tabernacle*, frontispiece, New York: Greeley & McElrath, 1833.

Climate past and present: the Ice Age
M. Grouner, *Histoire naturelle des glaciers de Suisse*, Tableau X, Paris: Panzkouche, 1777.

Drifting continents, rising mountains
Thomas Sutcliff, *The Earthquake of Juan Fernandez as it Occurred in the Year 1835*, plate 2, Manchester.

Changing oceans, changing climates
Louis Figuier, *La terre avant le déluge*, p. 430, Paris: Hachette, 1872.

The four-billion-year childhood
Amos Eaton, *Geological Textbook for Aiding the Study of North American Geology*, 2nd ed., p. 55, Albany: Websters & Skinners, 1832.

Life, time and change
G. F. Richardson, *An Introduction to Geology*, revised by Thomas Wright, frontispiece, London: H. G. Bohn 1855.

Epilogue
B. Faujas Saint Fond, *Natuurlijke Historie van den St. Pietersberg bij Maastricht*, colophon, Amsterdam: Johannes Allart, 1802.

TEXT FIGURES

Most illustrations are based on concepts and data in the common domain, but in the construction of the following I have made use of specific data or ideas for which I am grateful to the original authors and their illustrators.

Sources of illustrations

Figure 2.6. Standard curve from Rundberg, Y. & Smalley, P. C. (1989), *American Association of Petroleum Geologists, Bulletin*, **73**, 298–308.

Figure 3.1. Data from Lamb, H. H. (1969), in *World Survey of Climatology*, H. Flohn (ed.), pp. 173–249 (New York: Elsevier). The lower graph is based on data from Jacoby, G. C. & d'Arrigo, R. (1989), *Climatic Change*, **14**, 39–60.

Figure 3.2. Adapted from Parry, M. L. (1975), *Transactions of the Institute of British Geographers*, **64**, 1–13.

Figure 3.3. From Imbrie, J. & Imbrie, K. P. (1979), *Ice Ages: Solving the Mystery*, Fig. 43 (Garden City, NJ: Enslow Publishers).

Figure 3.4. Based on an anonymous article in *Mosaic*, July–August 1977, 35–41.

Figure 3.10. Modified from Fig. 4.10 of Kana, T. W. *et al.* (1984), in *Greenhouse Effect and Sea Level Rise*, M. C. Barth & J. G. Titus (eds.), pp. 105–50 (New York: Van Nostrand Reinhold).

Figure 4.2. Adapted from Fig. 14 of Williams, D. F. *et al.* (1988), *Palaeogeography, Palaeoclimatology, Palaeoecology*, **64**, 221–40.

Figure 4.3. Data from CLIMAP (1976), *Science*, **196**, 1131–37.

Figure 4.5. Modified from Fig. 10 of McIntyre, A. & Kipp, N. G. (1976), *Geological Society of America, Memoir*, **145**, 43–76; with new data added.

Figure 4.6. Data from CLIMAP (1976), *Science*, **196**, 1131–37; from Peterson, G. M. *et al.* (1979), *Quaternary Research* **12**, 47–82; and COHMAP (1988), *Science* **241**, 1043–52.

Figure 4.7. Data from Smith, G. I. & Street-Perrott, F. A. (1983), in *Late Quaternary Environments of the United States 1*, S. C. Porter (ed.), pp. 190–214 (New York: Longmans); and from Benson, L. *et al.* (1992), *Palaeogeography, Palaeoclimate, Palaeoecology*, **95**, 19–32.

Figure 4.8. Data from Woillard, G. M. & Mook, W. G. (1982), *Science*, **212**, 159; and from Fig. 48 of Imbrie, J. & Imbrie, K. P. (1979), *Ice Ages: Solving the Mystery* (Garden City, NJ: Enslow Publishers).

Figure 4.10. Adapted from Fig. 19 of Curray, J. R. (1960), in *Recent Sediments of the Northwestern Gulf of Mexico*, F. P. Shepard, F. B. Phleger & T. H. van Andel (eds.), pp. 221–66 (Tulsa, OK: American Association of Petroleum Geologists); Fig. 1 of van Andel, T. H. *et al.* (1967), *American Journal of Science*, **265**, 737–58; and various other sources.

Figure 4.11. Main curve from Fig. 2 of Fairbanks, R. G. (1989), *Nature*, **343**, 637–42.

Figure 4.12. Data from Imbrie, J. *et al.* (1984), in *Milankovitch and Climate*, A. L. Berger *et al.* (eds.), pp. 269–306 (Dordrecht: Reidel); and Fig. 1 of Chappell, J. & Shackleton, N. J. (1986), *Nature*, **324**, 137–40.

Figure 4.13. Based on maps by Andrews, J. T. (1987), in *North America and Adjacent Oceans during the Last Deglaciation*, W. F. Ruddiman & H. E. Wright (eds.), pp. 13–

38 (Boulder, CO: Geological Society of America); and Porter, S. C. (1989), *Quaternary Research*, **32**, 245–61.

Figure 5.3. Adapted from an idea of W. F. Ruddiman & A. McIntyre.

Figure 5.4. From Figs. 10 (left column) and 11 (left column) of Berger, A. & Loutre, M. F. (1990), *Quaternary Science Reviews*, **10**, 297–319.

Figure 5.5. From Fig. 5 of Kutzbach, J. E. & Otto-Bliesner, B. L. (1982), *Journal of Atmospheric Science*, **39**, 1177–88; and data from Street, F. A. & Grove, A. T. (1979), *Quaternary Research*, **12**, 83–118.

Figure 5.6. Adapted from Fig. 2 of Broecker, W. S. (1987), *Nature*, **328**, 123–26; and data from Street-Perrott, F. A. & Perrott, R. A. (1990), *Nature*, **343**, 607–16.

Figure 5.7. Modified from Fig. 1 of Lorius, C. *et al.* (1990), *Nature*, **347**, 139–45.

Figure 6.1. Based on data of McElhinny, W. H. (1973), *Palaeomagnetism and Plate Tectonics* (Cambridge: Cambridge University Press).

Figure 6.4. Based on Fig. 8 of Gahagan, L. M. *et al.* (1988), *Tectonophysics*, **155**, 1–26.

Figure 6.8. Adapted from the map *The Age of the Ocean Basins*, by W. C. Pitman III *et al.* (Boulder, CO: Geological Society of America, 1974).

Figure 6.9. Top: data from Barazangi, M. & Dorman, J. (1969), *Bulletin of the Seismological Society of America*, **59**, 369–80; bottom: data from Minster, J. B. (1990), *Nature*, **346**, 218–19, and from Gordon, R. G. & Stein, S. (1992), *Science*, **256**, 33–42.

Figure 6.12. From Fig. 9.16 of van Andel, T. H. (1992), in *Understanding the Earth*, G. C. Brown, C. J. Hawkesworth & R. C. L. Wilson (eds.) (Cambridge: Cambridge University Press).

Figure 7.1. Modified from Smith, A. G. & Briden, J. C. (1977), *Mesozoic and Cenozoic Paleocontinental Maps*, Map 13 (Cambridge: Cambridge University Press).

Figure 7.3. Adapted from Fig. 12 of Caputo, M. V. & Crowell, J. C. (1985), *Bulletin of the Geological Society of America*, **96**, 1020–36.

Figure 7.4. Compiled from many publications by John C. Crowell.

Figure 7.5. Modified from Figs. 14 and 17 of Smith, A. G. & Livermore, R. A. (1991), *Tectonophysics*, **187**, 135–79.

Figure 7.6. Based on Fig. 3 of Boucot, A. J. & Gray, J. (1983), *Science*, **222**, 571–81.

Figure 7.7. Base maps of Scotese, C. *et al.* (1979), *Journal of Geology*, **87**, 217–78; and of Ziegler, A. M. *et al.* (1979), *Annual Reviews of Earth and Planetary Sciences*, **7**, 473–502.

Figure 7.8. Data from White, R. S. & McKenzie, D. P. (1989), *Scientific American*, **261**, 44–55; and other sources. Base map from Smith, A. G. & Briden, J. C. (1977),

Sources of illustrations

Mesozoic and Cenozoic Palaeocontinental Maps, Map 12 (Cambridge: Cambridge University Press).

Figure 7.10. Based on various publications by Kevin Burke *et al.*; plateau basalts and plumes after White, R. S. & McKenzie, D. P. (1989), *Scientific American*, **261**, 44–55.

Figure 7.12. Modified from Fig. 1 of Duncan, R. A. (1991), *GSA Today* **1** (10).

Figure 7.13. Modified from Figs. 19 (top) and 2 (bottom) of White, R. S. & McKenzie, D. P. (1989), *Journal of Geophysical Research*, **94**, 7685–729.

Figure 8.1. Adapted from Burke, K. *et al.* (1977), *Tectonophysics*, **40**, 69–99.

Figure 8.3. Downgoing slab structure partly based on Fig. 15 of Bradley, C. & Kidd, W. F. S. (1991), *Bulletin of the Geological Society of America*, **104**, 1416–38.

Figure 8.4. Adapted from Sleep, N. H. (1990), *Nature*, **347**, 518–19.

Figure 8.6. Based on Scotese, C. *et al.* (1979), *Journal of Geology*, **87**, 217–78.

Figure 8.7. Adapted from an anonymous article in *Mosaic*, March–April, 1981, 33.

Figure 8.8. Adapted from Molnar P. & Tapponnier, P. (1977), *Scientific American*, **236**, 30–41.

Figure 8.9. From an idea of Peter Bird (1978), *Journal of Geophysical Research*, **83**, 4975–88.

Figure 9.2. Base maps from Scotese, C. *et al.* (1979), *Journal of Geology*, **87**, 217–78.

Figure 9.3. Adapted from Fig. 5 of Hallam, A. (1984), *Annual Review of Earth and Planetary Sciences*, **12**, 205–43.

Figure 9.5. Base map from Smith, A. G. & Briden, J. C. (1979), *Mesozoic and Cenozoic Palaeocontinental Maps*, Map 7 (Cambridge: Cambridge University Press).

Figure 9.6. The seismic reflection record is from the shelf off Walvis Bay, southwestern Africa, R. V. *Argo* Circe Expedition, 1968 (Scripps Institution of Oceanography, La Jolla, CA).

Figure 9.8. Simplified and adapted from various publications of P. R. Vail, B. U. Haq and others, including Haq, B. U., Hardenbol, J. & Vail, P. R. (1987), *Science*, **235**, 1156–67.

Figure 9.9. Adapted from Pitman III, W. C. (1978), *Bulletin of the Geological Society of America*, **89**, 1389–1403.

Figure 9.10. Diagram on the left adapted from Fig. 2 of Gurnis, M. (1992), *Science* **255**, 1556–58; graph on right from Fig. 6 of Kelley, P. *et al.* (1992), *GSA Today*, **2** (5).

Figure 10.7. After Fig. 12 of van Andel, T. H. (1979), in *Historical Biogeography, Plate Tectonics, and the Changing Environment*, J. Gray and A. J. Boucot (eds.), pp. 9–25 (Corvallis, OR: Oregon State University Press).

Figure 10.8. Compiled from various papers in *Cretaceous Palaeogeography*, E. J. Barron (ed.), *Palaeogeography, Palaeoclimatology, Palaeoecology*, **59**, 1–214, 1987; and miscellaneous sources.

Figure 10.9. From data in Dietrich, G. *et al.* (1975), *Allgemeine Meereskunde*, 469–74, 3rd ed. (Stuttgart: Borntraeger).

Figure 10.10. Adapted from Fig. 5 of Fischer, A. G. & Arthur, M. A. (1977), in *Society of Economic Paleontologists and Mineralogists, Special Publication*, **25**, 19–50.

Figure 10.12. Simplified from Fig. 1 of Shackleton, N. J. (1987), in *Marine Petroleum Source Rocks*, J. Brooks & A. J. Fleet (eds.), pp. 423–34, Geological Society of London, Special Publication 26.

Figure 10.13. Adapted from Thompson, S. L. & Barron, E. J. (1981), *Journal of Geology*, **89**, 143–68.

Figure 10.15. Adapted from Fig. 1 of Fischer, A. G. & Arthur, M. A. (1977), *Society of Economic Paleontologists and Mineralogists, Special Publication*, **25**, 19–50.

Figure 11.1. Composed of modified Figs. 6, 8, 10, and 11 of Haq, B. U. (1984), in *Marine Geology and Oceanography of the Arabian Sea and Coastal Pakistan*, B. U. Haq & J. D. Milliman (eds.), pp. 201–31 (New York: Van Nostrand Reinhold).

Figure 11.2. Land temperature data from Wolfe, J. A. & Upchurch, G. R. (1987), *Palaeogeography, Palaeoclimatology, Palaeoecology*, **61**, 33–77; marine data from various sources.

Figure 11.3. Data courtesy of N. J. Shackleton, University of Cambridge.

Figure 11.4. Adapted from data of Haq, B. U. *et al.* (1987), *Science*, **235**, 1156–67.

Figure 11.5. Adapted from Figs. 1 and 7a of Ruddiman, W. F. *et al.* (1989), *Journal of Geophysical Research*, **94**, 18379–91.

Figure 11.6. Adapted from an unnumbered figure of Ruddiman, W. F. & Kutzbach, J. E. (1991), *Scientific American*, **264**, 42–50.

Figure 12.1. Modified from Fig. 1 of Kauffman, E. G. (1988), *Annual Review of Earth and Planetary Sciences*, **16**, 605–54.

Figure 12.3. Simplified from Fig. 3 of Einsele, G. *et al.* (1991), in *Cycles and Events in Stratigraphy*, G. Einsele, W. Ricken & A. Seilacher (eds.), pp. 1–13 (Berlin: Springer).

Figure 12.4. Adapted from Figs 3.3 and 3.4 of Weedon, G. P. (1992), *Sedimentology Review*, **1**, 31–50.

Figure 12.5. Modified from Fig. 7 of Shackleton, N. J. *et al.* (1990), in *Transactions of the Royal Society of Edinburgh, Earth Sciences*, **81**, 251–61.

Figure 12.6. Adapted from Fig. 7 of Ruddiman, W. F. *et al.* (1986), *Geological Society of London Special Publication*, **21**, 155–73; and Fig. 8 of Moore Jr., T. C. *et al.* (1982), *Marine Geology*, **46**, 217–34.

Figure 12.7. Data from Berger, A. *et al.* (1992), *Science*, **255**, 560–66.

Sources of illustrations

Figure 12.8. Adapted from Umbgrove, J. H. F. (1947), *The Pulse of the Earth*, Table II (The Hague: Nijhoff).

Figure 12.9. Adapted from Fig. 7.1 of Fischer, A. G. (1982), in *Catastrophes and Earth History*, W. A. Berggren & J. A. van Couvering (eds.), pp. 129–50 (Princeton: Princeton University Press).

Figure 12.10. Modified after Fig. 6 of Veevers, J. J. (1990), *Sedimentary Geology*, **68**, 1–16.

Figure 13.4. Simplified from Fig. 1 of Nutman, A. P. & Collerson, K. D. (1991), *Geology*, **19**, 791–94.

Figure 13.5. Based on a hypothesis of Hargraves, R. B. (1976), *Science*, **193**, 363–71.

Figure 13.6. Based on Fig. 1 from Stevenson, R. K. & Patchett, P. S. (1990), *Geochimica et Cosmochimica Acta*, **54**, 1683–97; and Fig. 8.7 of O'Nions, R. K. (1992), in *Understanding the Earth*, G. C. Brown, C. J. Hawkesworth & R. C. L. Wilson (eds.), pp. 145–64 (Cambridge: Cambridge University Press).

Figure 13.7. Data from Ronov, A. B. (1964), *Geochemistry* **8**, 715–43; and Fig. 10.2 in Garrels, R. M. & Mackenzie, F. T. (1971), *Evolution of Sedimentary Rocks* (New York: Norton).

Figure 13.8. Adapted from Fig. 1.11 in Howell, D. G. (1990), *Tectonics of Suspect Terranes* (London: Chapman & Hall).

Figure 13.9. Drawn from a LANDSAT photograph of Groves, D. I. *et al.* (1981), *Scientific American*, **245**, 66.

Figure 14.4. Adapted from Fig. 2 of Cloud, P. (1976), *Paleobiology*, **2**, 351–89.

Figure 14.6. Adapted from Fig. 20 of Berner, R. A. (1991), *American Journal of Science*, **291**, 339–76; oxygen curve from Fig. 3 of Berner, R. A. & Canfield, D. E. (1989), *American Journal of Science*, **289**, 333–61.

Figure 14.7. Top based on a map of Morel, P. & Irving, E. (1992), *Journal of Geology*, **86**, 535–62; bottom on Fig. 3 of Dalziel, I. W. D. (1992), *GSA Today*, 2 (11).

Figure 15.3. Adapted from diagrams by Woese, R. (1981), *Scientific American*, **244**, 163–92; and Fig. 5.7 of Nisbet, E. G. (1991), *Living Earth* (London: HarperCollins Academic, London).

Figure 15.4. Data from an anonymous article in *Mosaic*, **9**, March–April 1979, 5; from Vidal, G. (1984), *Scientific American*, **250**, 48–58; from Schopf, J. W. (1978), *Scientific American*, **239**, 84–92; and from Fig. 2.7 of Schopf, J. W. (1992), in *Major Events in the History of Life*, J. W. Schopf (ed.) (Boston: Jones & Bartlett).

Figure 15.5. Data for the Banded Iron Formation from Cole, M. J. *et al.* (1981), *Journal of Geology*, **89**, 169–184; oxygen curve from Cloud, P. (1976), *Paleobiology*, **2**, 351–389; pyrite from Maynard, J. B. *et al.* (1991), *Geology*, **19**, 265–68.

Figure 16.3 Modified from Fig. 3.6 of Runnegar, B. (1992), in *Major Events in the History of Life*, J. W. Schopf (ed.), pp. 65–94 (Boston: Jones & Bartlett).

Figure 16.7. Adapted from a drawing in *Mosaic*, March–April 1979, 21; and from Fig. 4.1 of Sibley, C. G. & Ahlquist, E. (1987), in *Molecules and Morphological Evolution: Conflict or Compromise?*, C. Patterson (ed.), pp. 95–121 (Cambridge: Cambridge University Press).

Figure 16.8. Adapted from a drawing in *Mosaic*, March–April, 1979, 21.

Figure 17.1. Adapted from Fig. 1 of Conway Morris, S. (1993), *Nature*, **361**, 219–25; and miscellaneous sources.

Figure 17.2. Drawings adapted from Figs. 3.18, 3.31, 3.46, and 3.58 of Stephen Jay Gould (1990), *Wonderful Life – The Burgess Shale and the Nature of History* (Johannesburg: Hutchinson Radius); except c. which is courtesy of Simon Conway Morris.

Figure 17.6. Adapted from Fig. 1 of MacFadden, B. J. & Hulbert Jr., R. C. (1988), *Nature*, **336**, 466–68.

Figure 17.7. Adapted from Fig. 6.1 of Tobias, P. V. (1992), in *Major Events in the History of Life*, J. W. Schopf (ed.), pp. 141–67 (Boston: Jones & Bartlett).

Figure 18.1. Simplified after Fig. 1 of Ilyin, A. V. (1990), *Geology*, **18**, 1231–39.

Figure 18.2. Modified from Fig. 22.10 of Conway Morris, S. (1992), in *Understanding the Earth*, G. C. Brown, C. J. Hawkesworth & R. C. L. Wilson (eds.), pp. 436–57 (Cambridge: Cambridge University Press).

Figure 18.3. Adapted from a figure in a letter by Brasier, M. D. (1990), *Nature*, **347**, 521–22.

Figure 18.4. Right side adapted from Fig. 4.26 of Cowen, R. (1990), *History of Life* (Oxford: Blackwell Scientific).

Figure 18.8. Based on data of Fallaw, W. C. *et al.* (1980), *Geology*, **7**, 398–400; and (1980), *Journal of Geology*, **88**, 723–29.

Figure 18.9. Based on a drawing by Kurtén, B. (1969), *Scientific American*, **220**, 54; with base map from Smith, A. G. & Briden, J. C. (1977), *Mesozoic and Cenozoic Palaeocontinental Maps*, Map 8 (Cambridge: Cambridge University Press).

Figure 18.10. Based on a figure of Hallam, A. (1982), *Scientific American*, **227**, 56–66, using a more modern base map.

Figure 19.1. Adapted from Fig. 2 of Raup, D. M. & Sepkoski, J. J. (1982), *Science*, **215**, 1502.

Figure 19.2. Compiled from graphs of Thompson, K. S. (1976), *Nature*, **261**, 578–80.

Figure 19.3. Modified from Fig. 4.5 in Raup, D. M. (1991), *Extinction: Bad Genes or Bad Luck?* (New York: Norton).

Figure 19.4. Modification of a concept of Valentine, J. W. & Moores, E. M. (1972), *Journal of Geology* 80, 167–84, 1972, on paleogeographic maps of Scotese, C. *et al.* (1979), *Journal of Geology*, **87**, 217–78.

Sources of illustrations

Figure 19.5. Base maps of Scotese, C. *et al.* (1979), *Journal of Geology*, **87**, 217–78; lower right from Fig. 8 of Schopf, T. J. M. (1979), in *Historical Biogeography, Plate Tectonics and the Changing Environment*, J. Gray & A. J. Boucot (eds.) (Corvallis, OR: Oregon State University Press).

Figure 19.6. Based on Figs. 1 and 2 of Schopf, T. J. M. (1979), in *Historical Biogeography, Plate Tectonics and the Changing Environment*, J. Gray & A. J. Boucot (eds.) (Corvallis, OR: Oregon State University Press).

Figure 19.7. Based on maps of C. Scotese *et al.* (1979), *Journal of Geology*, **87**, 217–78.

Figure 19.8. After Fig. 9 of C. Scotese *et al.* (1988), *Tectonophysics* 155, 27–48; and a diagram of Stanley, S. M. (1987), *Extinctions*, p. 101 (New York: Scientific American Books).

Figure 19.9. Modified from a graph of Stanley, S. M. (1987), *Extinctions*, p. 151 (New York: Scientific American Books).

Figure 19.10. Data from Tappan, H. & Loeblich, A. R. (1973), *Earth Science Reviews*, **9**, 207–40.

Figure 19.11. Adapted from Fig. 15.9 in Cowen, R. (1990), *History of Life* (Oxford: Blackwell Scientific).

Index

Bold page numbers indicate a definition or description of the term.

abyssal circulation: from Antarctica, 200; Cenozoic, 225; Cretaceous, 207; initiated, 225

Acadian–Caledonian orogeny, 159

acritarchs, **306**; mass decline, 307; Paleozoic, 334; radiation, 357

adaptive radiation, **322–323**; horses, 344–345; Proterozoic-Cambrian, 335

aerosols, **59**, 89; climatic role, 59; in ice cores, 101

Africa: lake levels, 75, 96; rift, 138; savanna and human evolution, 226, 346; vegetation changes, 226, 229

Agassiz, Louis, 66, 397

age of the earth, 27, 257

albedo, **58**; Cenozoic change, 90, 226, 228, 229; and climate, 38, 95, 214, 228; and continental drift, 214, 228; glacial–interglacial, 90, 95; and mountain building, 229; of various surfaces, 58, 64, 214

algae, blue-green, *see* blue-green algae

Alps: mobile belt, 151, 164; origin of ice age theory, 66

Alvarez, Luis and Walter, 386

Amazon delta, 181

Americas joined, impact: on

climate, 226; on faunas, 367; on ocean circulation, 221, 226

amino acids, **292**; form microspheres and coacervates, 293; substitution of, 328

ammonia: in atmosphere, 294; and chemical evolution, 294; in hotsprings, 295

ammonites, 373; Cretaceous extinction, 282

amphibians, evolution, 341

Anatolian transform fault, 121, 124

Andes, Andean range, 114, 151, 164; and climate, 229; continent–ocean collision, 151

andesite: **114**, 118; explosive volcanism, 154; in mobile belts, 151; source of magma, 116; and subduction, 114

animals, higher, 317

annelids, 333

anomaly, magnetic, *see* magnetic anomalies

Antarctica: Archean craton, 265; enters ice age, 66, 225, 232; first glaciers, 222, 223; Oligocene icecap, 223, 228; polar position, 221; temperate woodland, 205, 221; west icecap, 81

Antarctic Ocean: Antarctic Bottom Water forms, 223, 225; circulation, 200; Circum-Antarctic Current starts, 221, 223, 228; fertility, 201–202; moisture supplied by North Atlantic Deep Water, 225

anthropic principle, 401

Appalachians: collision with Avalonia, 159–160; geosyncline, 157; history, 157–159; mobile belt, 151; problems, 160

arc: double, 156; fore-arc, **152**; sedimentary, 152; subduction of, 155; volcanic, 114, 151, 154–155

archaeocyathids, 333, 336; radiation, 336

Archaeopteryx, 23, 323

Archean, **257**; age, 271; anoxic atmosphere, 281; convection, 271; cratons, 265, 270; early ocean, 349; first life, 283, 301, 305; first photosynthesis?, 301; greenstone belts, 267, 279, 273; lithosphere hot, 271; microplate tectonics, 271–272; oceanic crust, 271; oldest crust, 257, 265; tectonic style, 272

Arctic, Canadian, *see* Canadian Arctic

Index

Index

phylogeny, phylogenetic, **317**; distance, 327–328; tree, 318, 323, 328

phylum, **316**

phytoplankton and acritarchs, 220, 334

Pikaia, 335; ancestor of vertebrates?, 339

Pinatubo volcano and climate, 89

planetary circulation, **56**; always present, 56; deflected by high mountains, 229

planetesimals, **259**; form earth, 261

Planet X, 391

plankton, planktonic, **58**, 349; acritarchs, 306, 351; adaptations, 349; diversity, 351; mass extinctions 384; Mesozoic, 334, 355, 384; Proterozoic, 306, 334; and upwelling, 349

plants: adaptations for land living, 360–361; environmental impact, 365; evolution, 362–365; fertilization by insects, birds, 365; first colonizers on land, 338, 361, 362; first multicellular, 306; flowering, 344, 363–365; grasses, 229, 344; vascular, 361

plateau basalt, **138**, 144; breakup of Pangaea, 138; Proterozoic, 351; *see also* hotspot, mantle plume

plates, tectonic, *see* tectonic plates

plate tectonics, **108**, 121; on early earth, 271–272; emphasis on ocean basins, 116, 196; geometry, 125; of planets, 263; reconstructs paleogeography, 130, 175, 289

Pliocene: Isthmus of Panama closed, 220, 226; onset of northern icecap, 226

pluvial, 75

polar: easterlies, **54**; path,

magnetic, 109–110; wandering, 109–110

polarity reversal, magnetic, *see* magnetic polarity reversals

pollen analysis, *see* palynology

population ("standard") growth curve, **356**

post-glacial, 51; carbon dioxide, 99; cold spells, 50; mammal extinctions, 372; sea level rise, 81–83; Younger Dryas, 86, 89

prairies: Little Ice Age, 49; Miocene origin, 226, 229; radiation of horses, 344

Precambrian, **5**, 255; continental growth, 268–271; economic role, 255; and history of life, 305–307; ice ages, 88, 289; low latitude icecap, 289; magma ocean, 263, 266, 274; oldest rocks, 257, 265; tracks and burrows, 331

precession, **94**; change over time, 94; climatic role, 92; cycle, 91, 94, 242

precipitation: future regional patterns, 62; influenced by high mountains, 230; late Cenozoic changes, 226, 229; reduction during glacial period, 76

predators: Cambrian, 345, 358–360; cephalopods, 345; as evolutionary stimulus, 348, 357, 360; first, 333, 345, 358–360; planktonic, 334, 349, 354

present key to the past, **4**, 9, 397

primates, **316**; origin of humans, 345–346

primitive crust, **263**; evolution, 264, 266–267; Iceland model, 267

principles of stratigraphy, 17

processes, geological, rates, 23–25

productivity of oceans, **198–199**; low in Cretaceous, 210

prokaryotes **297**; evolution, 300, 305, 317; predators?, 360; reproduction, 298

proteins, **293**

Proterozoic, **257**; first free oxygen, 284–285; first plate tectonics, 273; fossil-based stratigraphy, 313; greenstone belts, 267, 272, 280; ice age, 80, 289, 307, 351; island arcs, 273; multicellular organisms, 331–332, oceans, 349, 351; phosphates, 354; subduction, 273; supercontinent, 289, 351

pulse of the earth, 244, 399

punctuated equilibrium, *see* evolution

quantum theory, 401

quartz: in continental crust, 264; impact origin?, 386; shocked earth origin, 387, 388

quartzite, **264**; and continental emergence, 264, 269

Quaternary, **53**; extinctions, 372; outdated views, 66; present status, 68–69; regional subdivisions, 84

radiation, adaptive, **322–323**, 328, 344–345, 351

radioactive dating, *see* dating

radioactivity: for dating, 29; energy for early melting, 261–262

radiocarbon dating, **30**; and calendar years, 52; calibration, 52; and cosmic ray flux, 30, 52

radiolarians, 334

rain-forest: Cenozoic, 371; destruction of, 60; in glacial times, 77

rates of geological processes, **23**; erosion, 25; glacial sea level changes, 25, 81; importance, 26; plate motions, 24; seafloor-spreading, 118; sea level